Short-Period Binary Stars: Observations, Analyses, and Results

Astrophysics and Space Science Library

Recently Published in the ASSL series

Volume 352: *Short-Period Binary Stars: Observations, Analyses, and Results*, edited by Eugene F. Milone, Denis A. Leahy, David W. Hobill. Hardbound ISBN: 978-1-4020-6543-9, September 2007

Volume 351: *High Time Resolution Astrophysics*, edited by Don Phelan, Oliver Ryan, Andrew Shearer. Hardbound ISBN: 978-1-4020-6517-0, September 2007

Volume 350: *Hipparcos, the New Reduction of the Raw Data*, by Floor van Leeuwen. Hardbound ISBN: 978-1-4020-6341-1, August 2007

Volume 349: *Lasers, Clocks and Drag-Free Control: Exploration of Relativistic Gravity in Space*, edited by Hansjörg Dittus, Claus Lämmerzahl, Salva Turyshev. Hardbound ISBN: 978-3-540-34376-9, September 2007

Volume 348: *The Paraboloidal Reflector Antenna in Radio Astronomy and Communication – Theory and Practice,* by Jacob W.M. Baars. Hardbound 978-0-387-69733-8, July 2007

Volume 347: *The Sun and Space Weather,* by Arnold Hanslmeier. Hardbound 978-1-4020-5603-1, June 2007

Volume 346: *Exploring the Secrets of the Aurora,* by Syun-Ichi Akasofu. Hardbound 978-0-387-45094-0, July 2007

Volume 345: *Canonical Perturbation Theories – Degenerate Systems and Resonance*, by Sylvio Ferraz-Mello. Hardbound 978-0-387-38900-4, January 2007

Volume 344: *Space Weather: Research Toward Applications in Europe*, edited by Jean Lilensten. Hardbound 1-4020-5445-9, January 2007

Volume 343: *Organizations and Strategies in Astronomy: Volume 7*, edited by A. Heck. Hardbound 1-4020-5300-2, December 2006

Volume 342: *The Astrophysics of Emission Line Stars*, by Tomokazu Kogure, Kam-Ching Leung. Hardbound ISBN: 0-387-34500-0, June 2007

Volume 341: *Plasma Astrophysics, Part II: Reconnection and Flares*, by Boris V. Somov. Hardbound ISBN: 0-387-34948-0, November 2006

Volume 340: *Plasma Astrophysics, Part I: Fundamentals and Practice*, by Boris V. Somov. Hardbound ISBN 0-387-34916-9, September 2006

Volume 339: *Cosmic Ray Interactions, Propagation, and Acceleration in Space Plasmas*, by Lev Dorman. Hardbound ISBN 1-4020-5100-X, August 2006

Volume 338: *Solar Journey: The Significance of Our Galactic Environment for the Heliosphere and the Earth*, edited by Priscilla C. Frisch. Hardbound ISBN 1-4020-4397-0, September 2006

For other titles see www.springer.com/astronomy

Short-Period Binary Stars: Observations, Analyses, and Results

Eugene F. Milone
University of Calgary, AB
Canada

Denis A. Leahy
University of Calgary, AB
Canada

David W. Hobill
University of Calgary, AB
Canada

Eugene F. Milone
University of Calgary, AB
Canada

Denis A. Leahy
University of Calgary, AB
Canada

David W. Hobill
University of Calgary, AB
Canada

ISBN 978-1-4020-6543-9 ISBN 978-1-4020-6544-6 (eBook)

Library of Congress Control Number: 2007935107

Cover figure: P. Marenfeld NOAO/AURA/NSF

Printed on acid-free paper

9 8 7 6 5 4 3 2 1

springer.com

Contents

Preface ... ix

List of Contributors ... xv

Part I Compact Relativistic Binary Systems

Black Hole Binaries: The Journey from Astrophysics to Physics
Jeffrey E. McClintock ... 3

Searches for Gravitational Waves from Binary Neutron Stars: A Review
Warren G. Anderson and Jolien D.E. Creighton ... 23

Observations of the Double Pulsar PSR J0737−3039A/B
I.H. Stairs, M. Kramer, R.N. Manchester, M.A. McLaughlin, A.G. Lyne, R.D. Ferdman, M. Burgay, D.R. Lorimer, A. Possenti, N. D'Amico, J.M. Sarkissian, G.B. Hobbs, J.E. Reynolds, P.C.C. Freire and F. Camilo ... 53

Gravitational Lensing in Compact Binary Systems
David W. Hobill, John Kollar, and Julia Pulwicki ... 63

Part II Accreting Neutron Star Binaries

Accreting Neutron Stars in Low-Mass X-Ray Binary Systems
Frederick K. Lamb and Stratos Boutloukos ... 87

Observations and Modeling of Accretion Flows in X-ray Binaries
D.A. Leahy ... 111

Part III Cataclysmic Variable Systems

Modeling the Hot Components in Cataclysmic Variables: Info on the White Dwarf and Hot Disk from GALEX, FUSE, HST and SDSS
Paula Szkody 137

The Cool Components in Cataclysmic Variables: Recent Advances and New Puzzles
Steve B. Howell 147

Models for Dynamically Stable Cataclysmic Variable Systems
Albert P. Linnell 161

Part IV Modeling Short-Period Eclipsing Binaries

Distance Estimation for Eclipsing X-ray Pulsars
R.E. Wilson, H. Raichur, and B. Paul 179

The Tools of the Trade and the Products they Produce: Modeling of Eclipsing Binary Observables
Eugene F. Milone and Josef Kallrath 191

The Closest of the Close: Observational and Modeling Progress
W. Van Hamme and R.E. Cohen 215

Part V Aspects of Short-Period Binary Evolution

Common Envelope Evolution Redux
Ronald F. Webbink 233

Index 259

Preface

This work had its genesis in a topical meeting on short-period binaries held during the 208th meeting of the American Astronomical Society in June, 2006. In spite of its origins it is not a meeting proceedings, but rather a series of contributions by experts in subfields of the discipline. Like the topical session, the contributions provided here are from researchers in various fields whose subject of interest is short–period binary stars. Some authors address the properties of short–period binary systems in general and others describe the behavior of specific systems.

The purpose of this contributed volume is to highlight the techniques and methodologies used in their respective studies. In addition the observational and theoretical state of knowledge of a broad realm of interacting binary stars, covering the gamut from unevolved binaries to black hole systems is presented in this volume.

1 Why are Binary Stars of Interest?

Many, if not most of the stellar objects in the Universe are members of multiple star systems. Indeed Cox (2000), §16.18, cites two sources to argue that the numbers of binaries are very impressive:

> "Indications are that some 40–60% of all stars are members of double or multiple systems (Herczeg, 1982), with some estimates running as high as 85% (Heintz 1969)."

It has long been known that novae are produced by interacting binary stars. Now that the frequency of hierarchies of multiple star systems has been found to be larger than previously thought, one can expect that phenomena once thought to be rather rare in occurrence may be more common and this provides justification for the search of such phenomena. One of the most common types of short–period eclipsing binaries is the W Ursae Majoris (W UMa) class. Recent ideas concerning these binaries are discussed here by Van Hamme and Wilson in this book. These objects are in direct contact, joined by a neck of material that varies in thickness from system to system, and for this reason, their physical configuration is referred to as "over–contact" (and by others merely as "contact"). [1] These objects are thought to be on their way to full merger. Such systems are very old — with ages of billions of years, judging from the galactic distribution of the systems outside of star clusters, their abundance in the very old globular clusters, and their relative paucity in the typically younger "open" clusters of the galactic plane. The contact or over–contact objects may have been born with relatively low amounts of angular momentum; why that should be the case has long been a problem in binary star research. A new study by Pribulla and Rucinski (2006) strongly suggests that all such systems may be triple star systems, in which a wide–orbit companion is in a longer–period orbit with the W UMa pair. In any case, a significant percentage of over–contact systems are involved in triple systems.

Aside from an argument based on numbers, however, binaries are important because their gravitational interactions provide precise information about the masses of the components. If we are lucky enough to observe eclipses, we gain also precise knowledge of the sizes and geometric shapes. Moreover, we can investigate the details of the surfaces, namely, the umbrae and plage regions, and, in special cases, prominences, surges, jets, and streams between or around the stars. Accurate and precise data from eclipsing and spectroscopic binaries provide the challenge needed to test theories of both stellar structure and evolution, and orbital variations can provide critical tests of basic physics.

2 Why *Short–period* Binaries?

The range of stellar masses is quite restricted, so a short orbital period implies component proximity. When the stars are sufficiently well–separated that there is no material link between them, they are referred to as "detached systems." Well–studied detached systems provide us with the most precise stellar parameters, the ultimate aim of the most refined light–curve analyses; the most recent techniques of such analyses are discussed by Milone and Kallrath in this volume. If the stars' radii are more than an order of magnitude smaller than their separation, the light curve of an eclipsing pair of stars will appear flat outside of the eclipses (unless the temperatures are disparate, in which case an increase in brightness around the eclipse of the cooler star due

[1] Astronomers who prefer the term "over–contact" reserve the term "contact" for the case where both objects are just in contact, that is, with an infinitesimally narrow neck of material connecting them; both sides concede that this latter case must be exceedingly rare, and, indeed, no definitive example is known.

to the "reflection effect" may be apparent). With decreasing separation, the stellar shapes become increasingly non–spherical, a feature that can be determined by light curve analysis. As stars age, and reach a stage where their central reserves of hydrogen become depleted, they undergo changes in size and temperature, so that their brightness and color change. On the Hertzsprung–Russell diagram, they evolve toward the regime of the cooler and more luminous stars as they enter the red giant phase. As such a swelling star in a short–period binary system approaches a limiting Lagrangian surface, a zero-potential surface, that effectively defines what belongs to the star and what does not, it loses both mass and angular momentum, certainly to its companion, but possibly also to the external universe. The mass loss can be effected by this "Roche–Lobe Overflow" and/or by stellar winds. A system in which one of the components has reached its critical lobe stage is known as a "semi-detached" system.

The mass–gaining star may gain so much additional mass from its shedding companion that it winds up with more mass than the donor star. If the separation is large enough, the envelope dissipates with only some orbital decay by the mass–gainer. In this circumstance we see a paradoxical condition: the more massive star appears to be the less evolved, and, indeed it is, because prior to Roche–lobe mass exchange, it had less mass than its companion. This is the *Algol paradox*, now readily understood because of studies of short–period binaries.

In some cases, the outer distended atmosphere of the more evolved component envelops the companion. The viscous drag contributes to orbital decay of the less massive component, that results in a spiraling inward, and, ultimately, collision. Ron Webbink takes a fresh look at this scenario in his paper.

Magnetic interactions between short–period binary components give rise to complex, large–scale and long–lasting spot activity cycles, which can be observed. Mass loss through Roche–lobe overflow and stellar winds (from both hot and cool stars), can be probed and studied with detailed Doppler modeling. Following the red–giant and asymptotic–giant phases, the system may wind up with a white dwarf as a component. Interacting systems involving a very–low mass (and thus very slowly evolving) component and a white dwarf can give rise to interesting interactions, leading to novae or even supernovae. Finally, in some cases, the supernova explosions may result not in merely a debris field, but a compact remnant — a neutron star, or a black hole. Determining the properties of objects in these systems is now an important priority of binary star research, as the discussion of the usefulness of X-Ray timings by Wilson, et al., in this volume indicates.

Binaries composed of such highly compact constituents (i.e. black holes, neutron stars and perhaps even quark stars), provide natural laboratories for testing the validity of general relativity and other alternative theories of gravity. Once again given the large percentage of objects found in multiple star systems provides us with the optimism that there are many more binary neutron star and hopefully binary black hole systems to be discovered.

Currently twenty-one X-ray sources can be identified with stellar mass black hole binary systems. The current state of knowledge concerning these objects is presented by Jeffrey McClintock, who also discusses what the future might hold in terms of our understanding of the fundamental physics associated with the dynamics of black holes interacting with their binary star companions.

Whereas the change in the period of the Hulse-Taylor system, PSR 1913+16, has provided indirect evidence of the existence gravitational radiation (and direct evidence that binary systems are of interest in Stockholm), the article by Anderson and Creighton describes the methods currently used to search for the gravitational waves produced by such systems.

The "double pulsar system", PSR J0737-3039AB, has been called the answer to many a relativist's dream. With both pulsar beams being observable, $\sin i \approx 1$ and an orbital period of 2.4^h, one could not ask for a better binary system with which to observe general relativistic effects. Ingrid Stairs and her colleagues, provide a description of this (so far) unique binary system. In the contribution by Hobill, *et al*, who introduce a method for studying the gravitational lensing of pulsar sources, that technique is applied specifically to the double pulsar system.

There has a been a recent surge of activity in the study of the behavior of accreting neutron stars, powered primarily by observations in the X-ray region of the electromagnetic spectrum. Much of this has been done with the Rossi X-ray Timing Explorer with studies of the luminous accreting low (companion) mass X-ray binary systems ($LMXB$s). Observations of kilohertz quasiperiodic oscillations (QPOs) and the latest theoretical work are described in the contribution by Lamb and Boutloukos. These are now combining to yield new inferences on the extreme conditions in the inner accretion disks around the neutron stars in these systems.

For the lower luminosity and higher companion mass X-ray pulsar binaries, analysis of disk precession light curves is now giving detailed information on the accretion disk geometry. Analysis of orbital light curves in X-ray, optical and ultraviolet traces different emission components in the binary system: companion; illuminated face of companion; accretion stream from companion to disk; accretion disk; inner disk edge; and accretion stream from inner disk to neutron star surface. The article by Leahy describes progress and methods in modeling various components of the accretion flow for X-ray pulsars.

An accreting white dwarf in close orbit with a low-mass secondary is known as a cataclysmic variable (CV). CVs display a variety of regularities in their light curves and spectra caused by regular geometrical viewing changes as the system orbit rotates the binary with respect to the observer. Recently, as the contribution by Szkody explains, new ultraviolet observation of the hot components in CVs are being physically modeled with spectacular success. Howell's article discusses the nature of the cool components in CVs, and the implications for the nature and evolution of CVs. As for much of astrophysics, spectra provide the most detailed observational information on physical conditions in *CV*s. The contribution by Linnell discusses the nature of *CV* accretion disk spectra and a general purpose program for calculating synthetic spectra and light curves for *CV*s.

A final argument for studying short–period binaries is that we see many cycles in short intervals of time. Somewhat analogous to the value of fruit flies to a genetics lab, *CV*s permit us to study the orbital and system changes from all sides of the orbit without having to wait a significant fraction of a human lifetime to collect the data! Short–period binaries can have exceedingly short orbital periods. For example X-ray binaries RX J0806.3+1527 and RX J1914.4+2456 have periods of 321^s and 569^s, respectively. Such ultra–short period binaries provide challenges to both theorists and observers in their attempt to understand the extreme nature of such systems. With orbital periods on the order of just a few minutes, the enormous amount of data accumulated over a relatively short interval is clearly advantageous.

3 More to Learn!

We still have much to learn about short–period objects, however. A recent paper (Pinsonneault and Stanek, 2006) promotes the notion that massive binaries are highly likely to involve "twin," components that are essentially identical in mass and other physical properties. These authors discovered in a sample of 21 detached systems in the Small Magellanic Cloud, with median mass ratio 0.87. They argue further that the frequency of twins may be generally as high as 50% in systems with periods less than 10^d, and possibly even as high as 35% for those with periods less than 1000^d. Because mass is the principal property determining rate of evolution, this finding has implications for production rates of white dwarf binaries, Type Ia supernovae, blue stragglers, binary neutron stars, and neutron star–black hole binaries.

Some of this volume is devoted to research on relatively unevolved binaries, and the state of the art of the analytical techniques for determining stellar parameters and system elements. Other contributions describe systems with highly evolved components. Study of all of these objects can benefit from an examination of techniques and results from other sub–field disciplines. Therefore it is our hope that this volume will provide an overview of the many aspects of research on short-period binaries that will be useful to all researchers and students.

Finally, it is a pleasure to thank the contributors for their interesting and timely contributions. We would also like to acknowledge both the assistance and patience of Sonja Japenga and Vaska Krabbe from Springer-Verlag who were instrumental in the completion of this project.

References

1. A.N. Cox, ed.: *Allen's Astrophysical Quantities* 4th ed. (AIP Press & Springer, New York), (2000).
2. W.D. Heintz: JRASC, **63**, 275, (1969).
3. T. Herczeg: *Landolt–Börnstein Tables* (Berlin: Springer–Verlag), **2**, 381 (1982).
4. M.H. Pinsonneault and K.Z. Stanek: ApJ, **639**, L67, (2006).
5. T. Pribulla and S.M. Rucinski: AJ, **131**, 2986, (2006).

Calgary, Alberta
May 2007

Eugene F. Milone
Denis A. Leahy
David W. Hobill

List of Contributors

Anderson, Warren Department of Physics University of Wisconsin – Milwaukee P.O. Box 413
Milwaukee, Wisconsin, 53201-0413
USA

Boutloukos, Stratos
Center for Theoretical Astrophysics and Department of Physics
University of Illinois
at Urbana-Champaign
1110 W Green Street
Urbana, Illinois 61801 USA

Burgay, M.
INAF - Osservatorio Astronomica di Cagliari
Loc. Poggio dei Pini, Strada 54
09012 Capoterra, Italy

Camilo, F.
Columbia Astrophysics Laboratory
Columbia University
550 West 120th Street
New York, NY 10027, USA

Cohen, R.E.
Department of Physics
Florida International University
Miami, Florida 33199, USA
and
Astronomy Department
University of Florida
Gainesville, Florida 32611, USA

Creighton, Jolien
Department of Physics
University of Wisconsin – Milwaukee
P.O. Box 413
Milwaukee, Wisconsin, 53201-0413
USA

D'Amico, N.
INAF - Osservatorio Astronomica di Cagliari
Loc. Poggio dei Pini, Strada 54
09012 Capoterra, Italy
and
Universita' degli Studi di Cagliari
Dipartimento di Fisica
SP Monserrato-Sestu km 0.7
09042 Monserrato (CA), Italy

Ferdman, R.D.
Department of Physics and Astronomy
University of British Columbia
6224 Agricultural Road
Vancouver, British Columbia
V6T 1Z1, Canada

Freire, P.C.C.
NAIC, Arecibo Observatory
HC03 Box 53995
Puerto Rico, 00612, USA

Hobbs, G.B.
Australia Telescope National Facility
CSIRO
P.O. Box 76
Epping NSW 1710, Australia

Hobill, David
Department of Physics
and Astronomy
University of Calgary
2500 University Drive, NW
Calgary, Alberta T2N 1N4, Canada

Howell, Steve B.
National Optical Astronomy
Observatory
950 North Cherry Avenue
Tuscon, Arizona, 85719, USA

Kallrath, Josef
Astronomy Department
University of Florida
Gainesville, Florida 32611, USA

Kramer, M.
University of Manchester
Jodrell Bank Observatory
Macclesfield, SK11 9DL, UK

Lamb, Frederick K.
Center for Theoretical Astrophysics
and Departments of Astronomy
and of Physics
University of Illinois
at Urbana-Champaign
1110 W Green Street
Urbana, Illinois 61801 USA

Leahy, Denis
Department of Physics
and Astronomy
University of Calgary
2500 University Drive, NW
Calgary, Alberta T2N 1N4, Canada

Linnell, Albert P.
Department of Physics
and Astronomy
University of Washington
Box 351580
Seattle, Washington, 98195-1580
USA

Lorimer, D.R.
Department of Physics, West
Virginia University
Morgantown, West Virginia 26506
USA

Lyne, A.G.
University of Manchester, Jodrell
Bank Observatory
Macclesfield, SK11 9DL, UK

Manchester, R.N.
Australia Telescope National Facility
CSIRO
P.O. Box 76
Epping NSW 1710, Australia

McClintock, Jeffrey E.
Harvard-Smithsonian Center
for Astrophysics
60 Garden St.
Cambridge, Massachusetts, 02138
USA

McLaughlin, M.A.
Department of Physics
West Virginia University
Morgantown, West Virginia 26506
USA

Milone, Eugene, F.
Department of Physics
and Astronomy
University of Calgary
2500 University Drive, NW
Calgary, Alberta T2N 1N4, Canada

Paul, B.
Raman Research Institute
Bangalore, India

Possenti, A.
INAF - Osservatorio Astronomica
di Cagliari
Loc. Poggio dei Pini, Strada 54
09012 Capoterra, Italy

Pulwicki, Julia
Department of Physics
and Astronomy
University of Calgary
2500 University Drive, NW
Calgary, Alberta T2N 1N4, Canada

Raichur, H.
Raman Research Institute
Bangalore, India

Reynolds, J.E.
Australia Telescope National Facility
CSIRO
P.O. Box 76
Epping NSW 1710, Australia

Sarkissian, J.M.
Australia Telescope National Facility
CSIRO
P.O. Box 76
Epping NSW 1710, Australia

Stairs, Ingrid H.
Dept. of Physics and Astronomy
University of British Columbia
6224 Agricultural Road
Vancouver, British Columbia
V6T 1Z1, Canada

Szkody, Paula
Department of Physics
and Astronomy
University of Washington
Box 351580
Seattle, Washington, 98195-1580
USA

Van Hamme, W.
Department of Physics, Florida
International University
Miami, Florida 33199, USA

Webbink, Ronald F.
Department of Astronomy
University of Illinois
at Urbana-Champaign
1002 W Green Street
Urbana, Illinois 61801 USA

Wilson, R.E.
Astronomy Department
University of Florida
Gainesville, Florida 32611, USA

Part I

Compact Relativistic Binary Systems

Black Hole Binaries: The Journey from Astrophysics to Physics

Jeffrey E. McClintock

Harvard-Smithsonian Center for Astrophysics, 60 Garden St., Cambridge, MA 02138, USA
jem@cfa.harvard.edu

1 Introduction

This paper is based on a talk presented at the 208th Meeting of the American Astronomical Society in the session on Short-Period Binary Stars. The talk (and this paper in turn) are based on a parent paper, which is a comprehensive review by Remillard and McClintock (2006; hereafter RM06) on the X–ray properties of binary stars that contain a stellar black-hole primary. We refer to these systems as black hole binaries. In this present paper, which follows closely the content of the talk, we give sketches of some of the main topics covered in RM06. For a detailed account of the topics discussed herein and a full list of references (which are provided only sketchily below), see RM06 and also a second review paper by McClintock & Remillard (2006; hereafter MR06). There is one subject that is treated in more detail here than in the two review papers just cited, namely, the measurement of black hole spin; on this topic, see McClintock et al. (2006) for further details and references.

There are a total of 21 stellar black holes with measured masses that are located in X–ray binary systems. Nearly all these systems are X–ray novae – transient systems that are discovered during their typically year-long outbursts. There are an additional ~20 binaries that likely contain stellar black holes based on their X–ray behavior; however, firm dynamical evidence is lacking in these cases, and we refer to these compact primaries as black hole candidates (RM06).

These black holes are the most visible representatives of an estimated population of ~300 million stellar-mass black holes that are believed to exist in the Galaxy. These stellar-mass black holes are important to astronomy in numerous ways. For example, they are one endpoint of stellar evolution for massive stars, and the collapse of their progenitor stars enriches the universe with heavy elements. Also, the measured mass distribution for even the small sample of 21 black holes featured here are used to constrain models of black hole formation and binary evolution. Lastly, some black hole binaries appear to be linked to the hypernova believed to power gamma–ray bursts (MR06).

E.F. Milone et al. (eds.), Short-Period Binary Stars: Observations, Analyses, and Results, 3–22.

In astronomy, for all practical purposes a black hole is completely specified in General Relativity (*GR*) by two numbers, its mass M and its specific angular momentum or spin $a = J/cM$, where J is the black hole angular momentum and c is the speed of light. The spin is usually expressed in terms of a dimensionless spin parameter, $a_* \equiv a/M = a/R_{\rm g}$, where $R_{\rm g} \equiv GM/c^2$. The spin is an important property of a black hole because it sets the geometry of space-time, whereas mass simply supplies a scale. The value of a_* is bounded to lie between 0 for a Schwarzschild hole and 1 for a maximally-rotating Kerr hole. The defining property of a black hole is its event horizon, the immaterial surface that bounds the interior region of space-time that cannot communicate with the external universe.

The event horizon, the existence of an innermost stable circular orbit (*ISCO*), and other properties of black holes are discussed in many texts (e.g., Shapiro & Teukolsky 1983). The radius of the event horizon of a Schwarzschild black hole ($a_* = 0$) is $R_{\rm S} = 2R_{\rm g} = 30$ km$(M/10M_\odot)$, the *ISCO* lies at $R_{\rm ISCO} = 6R_{\rm g}$, and the corresponding maximum orbital frequency is $\nu_{\rm ISCO} = 220$ Hz$(M/10M_\odot)^{-1}$. For an extreme Kerr black hole ($a_* = 1$), the radii of both the event horizon and the ISCO (prograde orbits) are identical, $R_{\rm K} = R_{\rm ISCO} = R_{\rm g}$, and the maximum orbital frequency is $\nu_{\rm ISCO} = 1615$ Hz$(M/10M_\odot)^{-1}$.

This paper is organized as follows. In the following section we describe and catalog the 21 black hole binaries. In §§3-5, we present a brief review of spectral and timing observations of these binaries, focusing on the three X–ray states of accretion defined by MR06 and RM06. In §6, we sketch a scenario for the potential impact of timing/spectral studies of accreting black holes on physics. In §§7-8, we discuss a current frontier topic, namely, the measurement of black hole spin.

2 The Twenty-one Black Hole Binaries

The names and some selected properties of the 21 black hole binaries are given in Table 1. The binaries are ordered by right ascension (column 1). Column 2 gives the common name of the source (e.g., LMC X–3) *or* the prefix to the coordinate name that identifies the discovery mission (e.g., XTE J, where a "J" indicates that the coordinate epoch is J2000). For X–ray novae, the third column gives the year of discovery and the number of outbursts that have been observed. The spectral type of the secondary star is given in column 4. Extensive optical observations of this star yield the key dynamical data summarized respectively in the last three columns: the orbital period, the mass function, and the black hole mass. Additional data on black hole binaries are given in Tables 4.1 & 4.2 of MR06.

An observational quantity of special interest is the mass function, $f(M) \equiv P_{\rm orb}K_2^3/2\pi G = M_1 \sin^3 i/(1+q)^2$ (see Table 1, column 6). The observables on the left side of this equation are the orbital period $P_{\rm orb}$ and the half-amplitude

Table 1. Twenty-one confirmed stellar black holes[a]

Coordinate Name	Common[b] Name/Prefix	Year[c]	Spec.	$P_{\rm orb}$ (hr)	f(M) ($M_\odot$)	M_1 ($M_\odot$)
0133+30[d]	M33 X-7	–	O	82.9	0.41±0.08	>8
0422+32	(GRO J)	1992/1	M2V	5.1	1.19±0.02	3.7–5.0
0538–641	LMC X–3	–	B3V	40.9	2.3±0.3	5.9–9.2
0540–697	LMC X–1	–	O7III	93.8[d]	0.13±0.05[e]	4.0–10.0:[f]
0620–003	(A)	1975/1[g]	K4V	7.8	2.72±0.06	8.7–12.9
1009–45	(GRS)	1993/1	K7/M0V	6.8	3.17±0.12	3.6–4.7:[e]
1118+480	(XTE J)	2000/2	K5/M0V	4.1	6.1±0.3	6.5–7.2
1124–684	Nova Mus 91	1991/1	K3/K5V	10.4	3.01±0.15	6.5–8.2
1354–64	(GS)	1987/2	GIV	61.1[g]	5.75±0.30	–
1543–475	(4U)	1971/4	A2V	26.8	0.25±0.01	8.4–10.4
1550–564	(XTE J)	1998/5	G8/K8IV	37.0	6.86±0.71	8.4–10.8
1650–500	(XTE J)	2001/1	K4V	7.7	2.73±0.56	–
1655–40	(GRO J)	1994/3	F3/F5IV	62.9	2.73±0.09	6.0–6.6
1659–487	GX 339–4	1972/10	–	42.1[j,k]	5.8±0.5	–
1705–250	Nova Oph 77	1977/1	K3/7V	12.5	4.86±0.13	5.6–8.3
1819.3–2525	V4641 Sgr	1999/4	B9III	67.6	3.13±0.13	6.8–7.4
1859+226	(XTE J)	1999/1	–	9.2 :[e]	7.4±1.1:[e]	7.6–12.0:[e]
1915+105	(GRS)	1992/Q[h]	K/MIII	804.0	9.5±3.0	10.0–18.0
1956+350	Cyg X–1	–	O9.7Iab	134.4	0.244±0.005	6.8–13.3
2000+251	(GS)	1988/1	K3/K7V	8.3	5.01±0.12	7.1–7.8
2023+338	V404 Cyg	1989/1[g]	K0III	155.3	6.08±0.06	10.1–13.4

[a]See Remillard & McClintock 2006 plus additional reference given below.
[b]A prefix to a coordinate name is enclosed in parentheses.
The presence/absence of a "J" indicates that the epoch of the coordinates is J2000/B1950.
[c]Year of initial X–ray outburst/total number of X–ray outbursts.
[d]Preliminary results based on a private communication by J. Orosz; also Pietsch et al. 2006.
[e]Period and f(M) corrections by AM Levine and D Lin, private communication.
[f]Colon denotes uncertain value or range.
[g]Additional outbursts in optical archives: A 0620 (1917) and V404 Cyg (1938, 1956).
[h]"Q" denotes quasi-persistent intervals (e.g., decades), rather than typical outburst.

of the velocity curve of the secondary K_2. On the right, the quantity of greatest interest is M_1, the mass of the black hole primary (given in column 7); the other parameters are the orbital inclination angle i and the mass ratio $q \equiv M_2/M_1$, where M_2 is the mass of the secondary. The value of $f(M)$ can be determined by simply measuring the radial velocity curve of the secondary star, and it corresponds to the absolute minimum allowable mass of the compact object.

An inspection of Table 1 shows that 15 of the 21 X–ray sources have values of $f(M)$ that require a compact object with a mass $\gtrsim 3\ M_\odot$. This is a widely agreed limit for the maximum stable mass of a neutron star in *GR* (e.g., Kalogera & Baym 1996). For the remaining five systems, some additional data

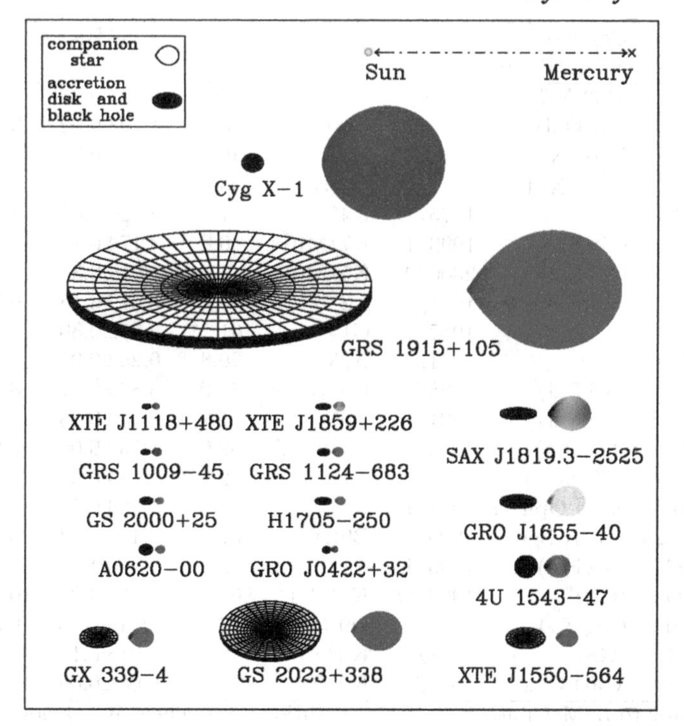

Fig. 1. Scale drawings of 16 black hole binaries in the Milky Way (courtesy of J. Orosz). The Sun–Mercury distance (0.4 AU) is shown at the top. The estimated binary inclination is indicated by the tilt of the accretion disk. The color of the companion star roughly indicates its surface temperature. Reprinted with permission from Volume 44 of Annual Reviews of Astronomy & Astrophysics (RM06).

are required to make the case for a black hole. Historically, the best available evidence for the existence of black holes is dynamical, and the evidence for these 21 systems is generally very strong, with two cautionary cases: LMC X–1 and XTE J1859+226 (see MR06). Thus, assuming that *GR* is valid in the strong-field limit, we choose to refer to these compact primaries as black holes, rather than as black hole candidates.

Figure 1 is a schematic sketch of 16 Milky Way black hole binaries with reasonably accurate dynamical data. Their diversity is evident: there are long-period systems containing hot and cool supergiants (Cyg X–1 and GRS 1915+105) and many compact systems containing K-dwarf secondaries. Considering all 21 black hole binaries (Table 1), only 4 are persistently bright X–ray sources (Cyg X–1, LMC X–1, LMC X–3 and M33 X–7). The 17 transient sources include 2 that are unusual. GRS 1915+105 has remained bright for more than a decade since its first known eruption in August 1992.

GX 339–4 undergoes frequent outbursts followed by very faint states, but it has never been observed to fully reach quiescence.

Nearly all black hole binaries are X–ray novae (Table 1) that are discovered when they first go into outburst. The X–ray light curves of nearly all of these black hole transients, which are frequently referred to as black hole X–ray novae, can be found either in MR06 or in a review paper on pre-*RXTE* X–ray novae by Chen et al. (1997). For X–ray outbursts that last between ~20 days and many months, the generally accepted cause of the outburst cycle is an instability that arises in the accretion disk. This model predicts recurrent outbursts; indeed, half of the black hole binaries are now known to recur on timescales of 1 to 60 years (Table 1). For details on the properties of X–ray novae, see Chen et al. (1997), MR06 and RM06.

3 X–ray Spectral and Timing Observations of Black Hole Binaries

It has been known for decades that the energy spectra of black hole binaries often exhibit a composite shape consisting of both a thermal and a nonthermal component. Furthermore, black hole binaries display transitions in which one or the other of these components may dominate the X–ray luminosity (e.g., Tanaka & Lewin 1995). The thermal component is well modeled by a multitemperature blackbody, which originates in the inner accretion disk and often shows a characteristic temperature near 1 keV (see §7). The nonthermal component is usually modeled as a power law (*PL*). It is characterized by a photon index Γ, where the photon spectrum is $N(E) \propto E^{-\Gamma}$. The *PL* generally extends to much higher Photon energies (E) than does the thermal component, and sometimes the *PL* suffers a break or an exponential cutoff at high energy. X–ray spectra of black hole binaries may also exhibit an Fe Kα emission line that is often relativistically broadened (§8.2.3). For further details and references, see MR06 and RM06.

An important resource for examining the near-vicinity of a black hole is the rapid variations in X–ray intensity that are so often observed (van der Klis 2005; MR06). The analysis tool commonly used for probing fast variability is the power–density spectrum (*PDS*). The continuum power in the *PDS* is of interest for both its shape and its integrated amplitude (e.g., 0.1–10 Hz), which is usually expressed in units of rms fluctuations scaled to the mean count rate. *PDS*s of black hole binaries also exhibit transient, discrete features known as quasi-periodic oscillations (*QPO*s) that may Range in frequency from 0.01 to 450 Hz. *QPO*s are generally modeled with Lorentzian profiles, and they are distinguished from broad power peaks using a coherence parameter, $Q = \nu/FWHM \gtrsim 2$ (van der Klis 2005).

4 A Quantitative Three-State Description for Active Accretion

In MR06, a new framework was used to define X–ray states (for an historical discussion of X–ray states, see MR06 and RM06). In Figure 2, we illustrate

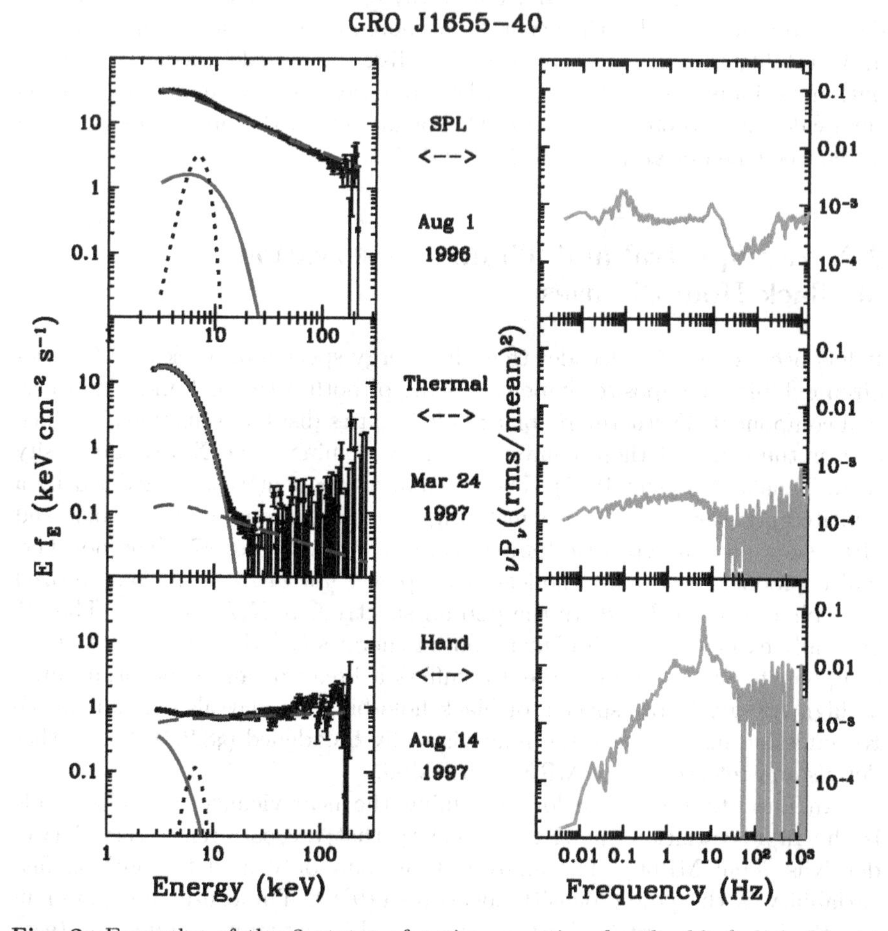

Fig. 2. Examples of the 3 states of active accretion for the black hole binary GRO J1655-40. Left panels show the energy spectra, with model components attributed to thermal-disk emission (red solid line), a power-law continuum (blue dashes) and a relativistically broadened Fe K–*alpha* line (black dotted). Power-law components for the *SPL* and hard states are distinguished by different values of the photon index (i.e. slope). The *PDS* (green solid lines) are shown in the right panels. A strong, band-limited continuum characterizes the hard state, while *QPO*s and the absence of the intense, broad continuum are usually seen in the *SPL* state. Reprinted with permission from Volume 44 of Annual Reviews of Astronomy & Astrophysics (RM06).

the character of each state by showing examples of *PDS*s and energy spectra for the black hole binary GRO J1655–40. The relevance of X–ray states fundamentally rests on the large differences in the energy spectra and *PDS*s that can be seen in a comparison of any two states.

4.1 The Thermal State

For extended periods of time during a transient outburst, the emission is observed to be dominated by thermal radiation from the inner accretion disk accompanied by a near-absence of complicating temporal variability. This well-defined "thermal" state (formerly "high/soft" state) is defined by the following three conditions: (1) the fraction f of the total 2–20 keV emission contributed by the accretion disk exceeds 75%, (2) there are no *QPO*s present with integrated amplitude above 0.5% of the mean count rate, and (3) the integrated power continuum is low, with rms power $r < 0.06$ averaged over 0.1–10 Hz.

For the thermal state there is a satisfactory paradigm, namely, thermal emission from the inner regions of an accretion disk. The best-known hydrodynamic model of a radiating gas orbiting in the gravitational potential of a compact object is the steady-state, thin accretion disk model (Shakura & Sunyaev 1973). This model leads to a temperature profile $T(R) \propto R^{-3/4}$ and the conclusion that the inner annulus in the disk dominates the thermal spectrum because $2\pi R dR\, \sigma T^4 \propto L(R) \propto R^{-2}$. This result has a striking observational consequence: X–ray astronomy is the window of choice for probing strong gravity near the horizon of an accreting stellar-mass black hole.

Our understanding of the thermal state is developing quickly as a result of observation, the development of fully-relativistic accretion disk models, and magnetohydrodynamic (*MHD*) simulations.

4.2 The Hard State

In the "hard" state, the accretion-disk component is either absent or it is modified in the sense of appearing comparatively cool and large. The hard state has been clearly associated with the presence of a steady type of radio jet. Transitions to either the thermal state or an *SPL* state effectively quench this radio emission. Thus, the presence of the jet is an important defining feature of this state. The definition of the hard state is based on three X–ray conditions: (1) $f < 0.2$, i.e. the power-law contributes at least 80% of the unabsorbed 2–20 keV flux, (2) $1.5 < \Gamma < 2.1$, and (3) the *PDS* yields $r > 0.1$.

Multiwavelength studies of XTE J1118+480 with its very low column density showed directly that in the hard state the thermal disk radiation is truncated at a large radius. Nevertheless, the physical condition of material within this radius remains somewhat uncertain. The presence of a hot advective flow that feeds the jet is one leading model. Both synchrotron and Compton components contribute to the broadband spectrum, with the Compton emission presumed to originate at the base of the jet. See RM06 for details and references.

4.3 The Steep Power Law State

The steep power law (*SPL*) component was first linked to the power-law "tail" found in the thermal state, and it was widely interpreted as inverse Compton radiation from a hot corona somehow coupled to the accretion disk. The picture became more complicated when X–ray *QPO*s were first detected with Ginga for two sources: GX 339-4 (6 Hz) and X–ray Nova Muscae 1991 (3–8 Hz). The *QPO*s, the high luminosity, and the strength of the power-law component prompted the interpretation that the *QPO*s signified a new black hole state, which was originally named the "very high" state (MR06). *RXTE* observations later showed that X–ray *QPO*s from black hole binaries are much more common than had been realized.

As noted above, *CGRO* observations have shown that the *SPL* may extend to photon energies as high as 800 keV). This forces consideration of nonthermal Comptonization models. The *QPO*s impose additional requirements for an oscillation mechanism that must be intimately tied to the electron acceleration mechanism (in the inverse Compton scenario), since the *QPO*s are fairly coherent ($\nu/\Delta\nu \sim 12$) and are strongest above 6 keV. Despite a wide range in *SPL* luminosities, the *SPL* tend to dominate as the luminosity approaches the Eddington limit. Furthermore, the occasions of high-frequency *QPO*s at 100-450 Hz in 7 black hole binaries almost always coincide with a strong *SPL* spectrum. Overall, the many fundamental differences between the thermal and *SPL* properties indicate that one cannot invoke some alternative state description that unifies thermal and *SPL* observations under a single "soft" state.

The physical origin of the *SPL* state remains one of the outstanding problems in high-energy astrophysics. It is crucial that we gain an understanding of this state, which is capable of generating *HFQPO*s, extremely high luminosity, and spectra that extend to $\gtrsim$1 MeV.

5 X–ray Quasi-Periodic Oscillations

X–ray *QPO*s are specialized and extraordinarily important avenues for the study of accreting black holes. They are transient phenomena associated with the nonthermal states and state transitions. For definitions of *QPO*s and analysis techniques, see van der Klis (2005). *QPO*s play an essential role in several key science areas, such as probing regions of strong field and defining the physical processes that distinguish X–ray states. In the following, we discuss in turn low-frequency *QPO*s (*LFQPO*s) and high-frequency *QPO*s (*HFQPO*s).

5.1 Low-Frequency Quasi–Periodic Oscillations

Low-frequency *QPO*s (*LFQPO*s; roughly 0.1–30 Hz) have been detected on one or more occasions for 14 of the 18 black hole binaries considered in Table 4.2 of MR06. They are important for several reasons. *LFQPO*s can have high amplitude (integrated rms/mean values of $a > 0.15$) and high coherence (often $Q > 10$), and their frequencies and amplitudes are generally correlated with the spectral parameters for both the thermal and *PL* components. With the exception of Cyg X–1, *QPO*s generally appear whenever the *SPL* contributes more than 20% of the flux at 2–20 keV, which is one component of the definition of the *SPL* state.

5.2 High-Frequency Quasi–Periodic Oscillations

High-frequency *QPO*s (*HFQPO*s; 40–450 Hz) have been detected in seven sources (5 black hole binaries and 2 black hole candidates). These oscillations are transient and subtle ($a \sim 0.01$), and they attract interest primarily because their frequencies are in the expected range for matter in orbit near the *ISCO* for a $\sim 10\ M_\odot$ black hole.

The entire sample of *HFQPO*s with strong detections ($>4\sigma$) is shown in Figure 3. Three sources have exhibited single oscillations. The other four sources display pairs of *HFQPO*s with frequencies that scale in a 3:2 ratio. Most often, these pairs of *QPO*s are not detected simultaneously. The four sources are GRO J1655–40 (300, 450 Hz), XTE J1550–564 (184, 276 Hz), GRS 1915+105 (113, 168 Hz), and H 1743–322 (165, 241 Hz). GRS 1915+105 also has a second pair of *HFQPO*s with frequencies that are not in a 3:2 ratio (41, 67 Hz).

*HFQPO*s are of further interest because they do not shift freely in frequency in response to sizable luminosity changes (factors of 3–4). There is evidence of frequency shifts in the *HFQPO* at lower frequency (referring to the 3:2 pairing), but such variations are limited to 15%. This is an important difference between these black hole binary *HFQPO*s and the variable-frequency kHz QPOs seen in accreting neutron stars, where both peaks can shift in frequency by a factor of two. Overall, black hole binary *HFQPO*s appear to be a stable and identifying "voice-print" that may depend only on the mass and spin of the black hole.

All of the strong detections ($>4\sigma$) above 100 Hz occur in the *SPL* state. In three of the sources that exhibit *HFQPO*s with a 3:2 frequency ratio, the $2\nu_0$ *QPO* appears when the *PL* flux is very strong, whereas $3\nu_0$ appears when the *PL* flux is weaker. Currently, there is no explanation for this result.

The commensurate frequencies of *HFQPO*s suggests that these oscillations are driven by some type of resonance condition. Abramowicz and Kluzniak (2001) proposed that orbiting blobs of accreting matter could generate the harmonic frequencies via a resonance between a pair of the coordinate frequencies given by *GR*. Earlier work had used *GR* coordinate frequencies and

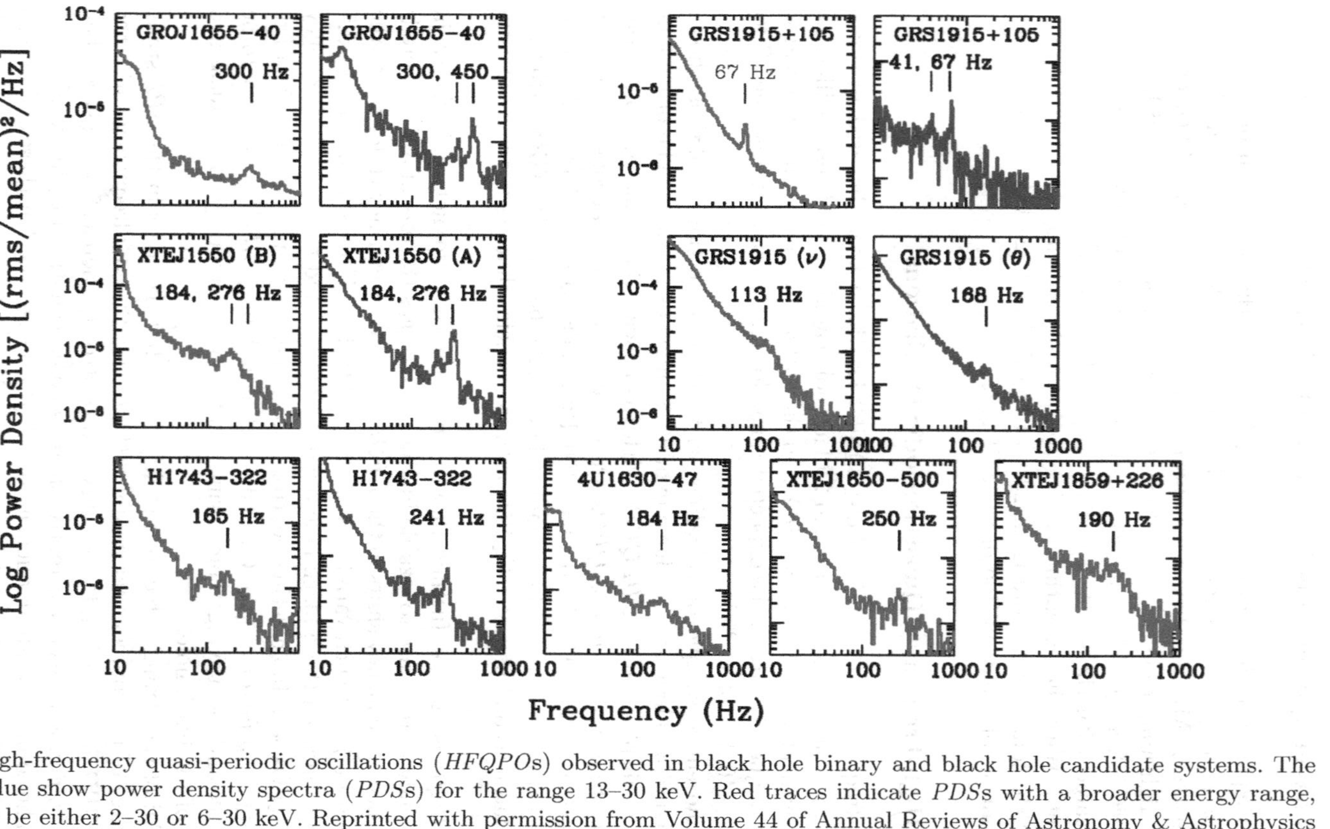

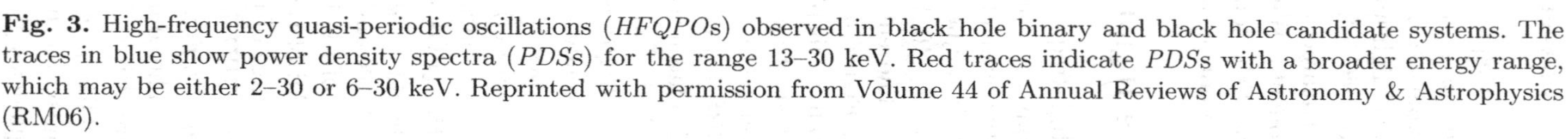
Fig. 3. High-frequency quasi-periodic oscillations (*HFQPOs*) observed in black hole binary and black hole candidate systems. The traces in blue show power density spectra (*PDSs*) for the range 13–30 keV. Red traces indicate *PDSs* with a broader energy range, which may be either 2–30 or 6–30 keV. Reprinted with permission from Volume 44 of Annual Reviews of Astronomy & Astrophysics (RM06).

associated beat frequencies to explain fast *QPO*s in both neutron-star and black hole systems (Stella et al. 1999), but without invoking a resonance condition. Current work on resonances as a means of explaining *HFQPO*s includes more realistic models for fluid flow in the Kerr metric. Resonance models are considered in more detail in §6.2.

6 Accreting Black Holes as Probes of Strong Gravity

The continuing development of gravitational wave astronomy is central to the exploration of black holes. In particular, we can reasonably expect that *LIGO* and *LISA* will provide us with intimate knowledge concerning the behavior of space-time under the most extreme conditions. Nevertheless, gravitational wave detectors are unlikely to provide us with direct information on the formation of relativistic jets, on strong-field relativistic *MHD* accretion flows, or on the origin of high-frequency *QPO*s or broadened Fe lines. Accreting black holes – whether they be stellar-mass, supermassive or intermediate mass – promise to provide detailed information on all of these topics and more. In short, accreting black holes show us uniquely how a black hole interacts with its environment. In this section, we first sketch a scenario for the potential impact of black hole binaries on physics, and we then discuss a current frontier topic, namely, the measurement of black hole spin.

6.1 The Journey from Astrophysics to Physics

Numerous examples can be given of how a discovery in astrophysics has impacted physics, such as Newton's and Einstein's theories of gravity and the quest to understand dark matter and dark energy. Likewise, studies of astrophysical black holes - the only type of black hole we are likely to ever know about - have the potential to revolutionize the foundations of frontier physics. We see a distinct possibility that the study of black hole binaries can make a significant contribution to such a revolution. In the following we provide an outline in five stages of how this development is presently unfolding and where it may ultimately lead.

Phase I—*Identify Dynamical Black Hole Candidates*: As discussed in §2, much has already been done during the past 20 years to secure the masses of many stellar black holes. This is extremely important because mass is the most fundamental parameter of a black hole. Obviously obtaining additional mass measurements and increasing the quality of the existing measurements is of great importance. However, the dynamical measurement of mass does not probe the space-time near the black hole, and we therefore curtail the discussion of this step.

Phase II—*Establish that the Candidates are True Black Holes*: The defining property of a black hole is its event horizon, and establishing its existence is the straight path to demonstrating that a black hole candidate is a bona fide black hole. Of course, it is quite impossible to detect any radiation from an event horizon, which is an immaterial surface of infinite redshift. Nevertheless, despite the complete absence of any emitted radiation, it is possible to marshal strong circumstantial evidence for the reality of the event horizon. One approach to this problem has been to compare black-hole X–ray binaries and neutron–star X–ray binaries under similar conditions and to show that there is a large difference in the luminosity or other basic observational property of the two systems, a difference that is most easily explained by invoking an event horizon in the candidate black holes. An initial and arguably the strongest evidence of this kind is based on X–ray observations of about 20 X–ray novae in quiescence with the Chandra X–ray Observatory that show that the black-hole candidate systems are about 100 times fainter than the nearly identical systems containing neutron stars (Garcia et al. 2001). The advection-dominated accretion flow model provides a natural explanation for this difference. In this model, the accreting gas reaches the compact object with a large amount of thermal energy. If the accretor is a black hole, the thermal energy will disappear through the event horizon, and the object will be very dim. On the other hand, if the accretor is a neutron star or any other object with a surface, the thermal energy will be radiated from That surface, and the object will be bright. In short, a black hole can "hide" most of its accretion energy behind its event horizon.

Also in quiescence, one observes that black holes lack a soft thermal component of emission that is very prevalent in the spectra of neutron stars and is widely agreed to be due to thermal emission from the surface of the star (McClintock, Narayan, & Rybicki 2004), again indicating that the black hole candidates lack a material surface. During outburst as well, the surface of a neutron star likewise gives rise to distinctive phenomena that are absent in the black hole binaries: (i) type I thermonuclear bursts (Narayan & Heyl 2002; Remillard et al. 2006); (ii) high-frequency timing noise (Sunyaev & Revnivtsev 2000), and (iii) a distinctive spectral component of emission from a boundary layer at the stellar surface (Done & Gierlinski 2003).

Of course, as stressed above, all attempts to confirm the existence of the event horizon can provide only indirect evidence. Nevertheless, unless one appeals to very exotic physics (e.g., McClintock, Narayan, & Rybicki 2004), the evidence considered here makes a strong case that the dynamical black hole candidates possess an event horizon and are therefore genuine black holes.

Phase III—*Measure the Spins of Black Holes*: As noted in §1, spin is one of the two fundamental parameters that completely define an astrophysical black hole. Moreover, it Is arguably the more important one in determining how the black hole interacts with its environment. Masses have already been determined for a good sample of black holes, and obviously the present challenge is to obtain reliable measurements of spin. Although there are several methods that may prove fruitful for estimating black hole spin, very few results thus far can be described as credible. The measurement of spin is presently at the frontier of black hole research and of first importance for modeling all the ways that a black hole interacts with its environment. In §6.2, we return to this central topic and discuss four approaches to measuring spin and some recent results that have been obtained for three black hole binaries.

Phase IV—*Relate Black Hole Spin to the Penrose Process and Other Phenomena*: Spin enlivens a black hole and gives it the potential to interact with its environment in ways that are not possible for a non-spinning black hole. A particularly important class of phenomena in this regard are the explosive and relativistic jets that have been observed for eight black hole binaries and black hole candidates. Large-scale X–ray jets have also been reported for two sources. Scientists have long speculated that these jets are powered by something akin to the Penrose (1969) process, which describes how in principle the spin energy of a black hole can be tapped. Many detailed models have been proposed for directing this spin energy into axial, relativistic jets of matter. Recently, some progress has been made on measuring black hole spin. If methods for measuring spin can be established and enough spin measurements can be amassed, it may soon be possible to test these models and to attack the jet-spin/Penrose-process connection in earnest.

Phase V—*Achieve Quantitative Tests of the Kerr Metric:* Arguably, the existence of an analytic solution that describes the space-time surrounding a spinning black hole using just two numbers is one of the "Seven Wonders" of 20th century physics. Testing this prediction is the most important contributions astrophysics can make to black hole physics. An obvious step toward this challenging goal is to amass many precise measurements of black hole masses and spins, as discussed above. All current attempts to measure spin assume the validity of the Kerr metric, which is a long way from establishing its validity. Nevertheless, we are confident that when a large number of black hole masses and spins have been precisely measured and related to the wide range of phenomena observed for accreting black holes, as well as coalescing black holes in vacuum environments, that astronomers will be strongly motivated to devise tests of the metric, a topic that is beyond the scope of this work.

7 Measuring Black Hole Spin: A Current Frontier

We now elaborate on Phase III by discussing three avenues for measuring black hole spin: (1) X–ray polarimetry, which appears very attractive but thus far has not been incorporated into any contemporary X–ray mission; (2) the Fe K line profile, which has yielded results, although the method is hampered by significant uncertainties; and (3) high-frequency *QPO*s (§5.2), which arguably offer the most reliable measurement of spin once a model is established. In §8, we consider in greater detail a fourth method, X–ray continuum fitting, which is already delivering plausibly reliable results.

7.1 Polarimetry

As pointed out by Lightman & Shapiro (1975) and Meszaros et al. (1988), polarimetric information (direction and degree) would increase the parameter space used to investigate compact objects from the current two (spectra and time variability) to four independent parameters that models need to satisfy. Such constraints are likely to be crucial in our attempts to model the hard state with its radio jet and the *SPL* state. However, because of the complexities of the accretion flows associated with these states it appears unlikely that their study will soon provide quantitative probes of strong gravity. We therefore focus on disk emission in the thermal state.

The polarization features of black hole disk radiation can be affected strongly by *GR* effects. The crucial requirement for a simple interpretation is that higher energy photons come from smaller disk radii, as they are predicted to do in conventional disk models (§7). If this requirement is met, then as the photon energy increases from 1 keV to 30 keV, the plane of linear polarization swings smoothly through an angle of about 40° for a $9M_{\odot}$ Schwarzschild black hole and 70° for an extreme Kerr black hole (Connors et al. 1980). The effect is due to the strong gravitational bending of light rays. In the Newtonian approximation, on the other hand, the polarization angle does not vary with energy, except for the possibility of a sudden 90° jump (Lightman & Shapiro 1976). Thus, a gradual change of the plane of polarization with energy is a purely relativistic effect, and the magnitude of the change can give a direct measure of a_*.

A model is now available in *XSPEC* that allows one to compute the Stokes parameters of a polarized accretion disk spectrum (Dovčiak et al. 2004). While the theoretical picture is bright, and very sensitive instruments can be built, unfortunately, results to date are meager and there are no mission opportunities on the horizon. Important advances in this promising area could be made by a relatively modest mission given that black hole binaries in the thermal state are bright.

7.2 Fe K Line Profile

The first broad Fe Kα line observed for either a black hole binary or an *AGN* was reported in 1985 in the spectrum of Cyg X–1 based on *EXOSAT* data. Since then, the line has been widely studied in the spectra of both black hole binaries and *AGN*. The Fe K fluorescence line is thought to be generated through the irradiation of the cold (weakly-ionized) disk by a source of hard X–rays (likely an optically-thin, Comptonizing corona). Relativistic beaming and gravitational redshifts in the inner disk region can serve to create an asymmetric line profile (for a review, see Reynolds & Nowak 2003).

The line has been modeled in the spectra of several black hole binaries. In some systems the inner disk radius deduced from the line profile is consistent with the $6R_{\rm g}$ radius of the *ISCO* of a Schwarzschild black hole, suggesting that rapid spin is not required (e.g., GRS 1915+105; V4641 Sgr). On the other hand, fits for GX 339–4 indicate that the inner disk likely extends inward to $(2-3)R_g$, implying $a_* \geq 0.8-0.9$. XTE J1650–500 is the most extreme case with the inner edge located at $\approx 2R_g$, which suggests nearly maximal spin. Large values of a_* have also been reported for XTE J1655–40 and XTE J1550–564. Sources of uncertainty in the method include the placement of the continuum, the model of the fluorescing source, and the ionization state of the disk (Reynolds & Nowak 2003). Also, thus far the analyses have been done using the *LAOR* model in *XSPEC*, which fixes the spin parameter at $a_* = 0.998$ (Laor 1991). A reanalysis of archival data using new *XSPEC* models that allow one to fit for a_* may prove useful. See RM06 and MR06 for references.

7.3 High Frequency Quasi-Periodic Oscillations

Arguably, *HFQPO*s (see §5.2) are likely to offer the most reliable measurement of spin once the correct model is known. Typical frequencies of these fast *QPO*s, e.g., 150–450 Hz, correspond respectively to the frequency at the *ISCO* for Schwarzschild black holes with masses of 15–5 $M_\odot$, which in turn closely matches the range of observed masses (Table 1). As noted in §6.2, these *QPO* frequencies (single or pairs) do not vary significantly despite sizable changes in the X–ray luminosity. This suggests that the frequencies are primarily dependent on the mass and spin of the black hole. Those black holes that show *HFQPO*s and have well-constrained masses are the best prospects for constraining the value of the black hole spin (a_*).

The four sources that exhibit harmonic pairs of frequencies in a 3:2 ratio (see §5.2) suggest that *HFQPO*s arise from some type of resonance mechanism (Abramowicz & Kluźniak 2001; Remillard et al. 2002a). Resonances were first discussed in terms of specific radii where particle orbits have oscillation frequencies in *GR* that scale with a 3:1 or a 3:2 ratio. Current resonance concepts now consider accretion flows in a more realistic context. For example, the "parametric resonance" concept (e.g., Török et al. 2005) describes oscillations rooted in fluid flow where there is a coupling between the radial and polar coordinate frequencies. As a second example, one recent *MHD* simulation provides evidence for resonant oscillations in the inner disk (Kato 2004); in this case, however, the coupling relation involves the azimuthal and radial coordinate frequencies. If radiating blobs do congregate at a resonance radius for some reason, then ray tracing calculations have shown that *GR* effects can cause measurable features in the X–ray power spectrum (Schnittman & Bertschinger 2004).

Other models utilize variations in the geometry of the accretion flow. For example, in one model the resonance is tied to an asymmetric structure (e.g., a spiral wave) in the inner accretion disk. In an alternative model, state changes are invoked that thicken the inner disk into a torus; the normal modes (with or without a resonance condition) can yield oscillations with a 3:2 frequency ratio. All of this research is still in a developmental state, and these proposed explanations for *HFQPOs* are basically dynamical models that lack radiation mechanisms and fail to fully consider the spectral properties of *HFQPOs* described in §5.2. For additional details and references, see RM06 and MR06.

8 Continuum Fitting

Among the several spectral states of accreting black holes, the thermal state (§4.1) is of particular interest. A feature of this state is that the X–ray spectrum is dominated by a soft blackbody-like component that is emitted by (relatively) cool optically-thick gas, presumed to be located in the accretion disk. A minor nonthermal (power-law) component in the spectrum, possibly from a hot optically thin corona, is energetically unimportant. The thermal state is believed to match very closely the classic thin accretion disk model of Shakura & Sunyaev (1973) and Novikov & Thorne (1973). This theoretical model has been widely studied for many decades and its physics is well understood.

The idealized thin disk model describes an axisymmetric radiatively efficient accretion flow in which, for a given black hole mass M, mass accretion rate $\dot{M}$ and black hole spin parameter a_* (§1), we can calculate precisely the total luminosity of the disk: $L_{\rm disk} = \eta \dot{M} c^2$, where η is a function only of a_*. We can also calculate precisely the local radiative flux $F_{\rm disk}(R)$ emitted at radius R by each surface of the disk. Moreover, the accreting gas is optically thick, and the emission is thermal and blackbody-like, making it straightforward to compute the spectrum of the emission. Most importantly, the inner edge of the disk is located at the innermost stable circular orbit (*ISCO*) of the black hole space-time, whose radius $R_{\rm ISCO}$ (in gravitational units) is a function only of the spin of the black hole: $R_{\rm ISCO}/(GM/c^2) = \xi(a_*)$, where ξ is a monotonically decreasing function of a_* (see Fig. 4). Thus, if we measure the radius of the disk inner edge, and if we know the mass M of the black hole, we can immediately obtain a_*. This is the principle behind the continuum-fitting method of estimating black hole spin, which was first described by Zhang et al. (1997; see also Gierlinski et al. 2001).

Before discussing how to measure $R_{\rm ISCO}$ of a disk, we remind the reader how one measures the radius R_* of a star. Given the distance D to the star, the radiation flux $F_{\rm obs}$ received from the star, and the temperature T of the continuum radiation, the luminosity of the star is given by

$$L_* = 4\pi D^2 F_{\rm obs} = 4\pi R_*^2 \sigma T^4, \qquad (1)$$

where σ is the Stefan-Boltzmann constant. Thus, from $F_{\rm obs}$ and T we immediately obtain the ratio R_*^2/D^2, the solid angle subtended by the star. Then, if we know the distance to the star, we obtain the stellar radius R_*. For accurate results we must allow for limb-darkening and other non-blackbody effects in the stellar emission by computing a stellar atmosphere model, a minor detail.

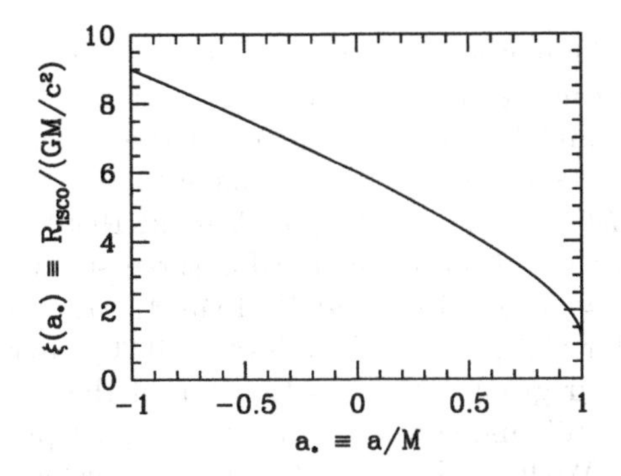

Fig. 4. Shows as a function of the black hole spin parameter, $a_* = a/M$, the variation of the radius of the *ISCO* $R_{\rm ISCO}$ in units of the black hole mass M. Negative values of a_* correspond to the black hole counter–rotating with respect to the orbit.

The same principle applies to an accretion disk, but with some differences. First, since $F_{\rm disk}(R)$ varies with radius, the radiation temperature T also varies with R. But the precise variation is known, so it is easily incorporated into the model. Second, since the bulk of the emission is from the inner regions of the disk, the effective area of the radiating surface is directly proportional to the square of the disk inner radius, $A_{\rm eff} = CR_{\rm ISCO}^2$, where the constant C is known. Third, the observed flux $F_{\rm obs}$ depends not only on the luminosity and the distance, but also on the inclination i of the disk to the line-of-sight. Allowing for these differences, one can write a relation for the disk problem similar in spirit to eq. (1). Therefore, in analogy with the stellar case, given $F_{\rm obs}$ and a characteristic T (from X–ray observations), one obtains the solid angle subtended by the *ISCO*: $\cos i\, R_{\rm ISCO}^2/D^2$. If we know i and D, we obtain $R_{\rm ISCO}$, and if we also know M, we obtain a_* [via $\xi(a_*)$, Fig. 1]. This is the basic idea of the method.

Note that $R_{\rm ISCO}$ varies by a factor of 6 between $a_* = 0$ and $a_* = 1$. This means that the solid angle subtended by the *ISCO* varies by a whopping factor of 36 (or even 81 if we include negative values of a_* to allow for counter-rotating disks, see Fig. 1). The method is thus potentially very sensitive. Also, all one asks the X–ray spectral data to provide are the characteristic temperature T of the radiation and the X–ray flux $F_{\rm obs}$ (or equivalently the normalization of the spectrum). These quantities can be obtained robustly from appropriately selected data. In contrast, other proposed approaches to estimate black hole spin require very intricate modeling of X–ray spectra (e.g., Brenneman & Reynolds 2006).

Zhang et al. (1997) first argued that the relativistic jets and extraordinary X–ray behavior of GRS 1915 are due to the high spin of its black hole primary. In their approximate analysis, they found that both GRS 1915 and GRO J1655–40 had high spins, $a_* > 0.9$. Subsequently, Gierliński et al. (2001) estimated the spin of GRO J1655–40 and LMC X–3. Recently, our group has firmly established the methodology pioneered by Zhang et al. and Gierliński et al. by constructing relativistic accretion disk models (Li et al. 2005; Davis et al. 2005) and by modeling in detail the effects of spectral hardening (Davis et al. 2005, 2006). We have made these analysis tools publicly available via *XSPEC* (*kerrbb* and *bhspec*; Arnaud 1996). Using this modern methodology, spins have been estimated for three stellar-mass black holes: GRO J1655–40 ($a_* = 0.65 - 0.75$) and 4U 1543–47 ($a_* = 0.75 - 0.85$; Shafee et al. 2006, hereafter S06), and LMC X–3 ($a_* < 0.26$; Davis et al. 2006).

Most recently, our group estimated the spin of the extraordinary microquasar GRS 1915 +105 and conclude that its compact primary is a rapidly-rotating black hole. We find a lower limit on the dimensionless spin parameter of $a_* > 0.98$. Our result is robust in the sense that it is independent of the details of the data analysis and insensitive to the uncertainties in the mass and distance of the black hole.

9 Critique of Methods for Measuring Spin

In short, there are four avenues for measuring spin – polarimetry, the Fe K line, *HFQPO*s and continuum fitting. Because spin is such a critical parameter, it is important to attempt to measure it by as many of these methods as possible, as this will provide arguably the best possible check on our results. The best current method, continuum fitting, has the drawback that its application requires accurate estimates of black hole mass (M), disk inclination (i), and distance. In contrast, assuming we have a well-tested model, *QPO* observations require knowledge of only M to provide a spin estimate. Broadened Fe K lines and polarimetry data do not even require M, although knowledge of i is useful in order to avoid having to include that parameter in the fit. On the other hand, the Fe-line and *HFQPO* methods are not well-enough developed to provide dependable results, and the required polarimetry data are not available, whereas the continuum method, despite its limitations, is already delivering results.

10 Importance of Measuring Black Hole Spin

A black hole's mass simply supplies a scale, whereas its spin fundamentally changes the geometry of space-time. Accordingly, in order to model the ways in which an accreting black hole can interact with its environment, one must know its spin. One key topic has already been discussed in §6.1 (Phase IV), namely, the relationship between relativistic jets and spin. This is one of the most intriguing unsolved problems in astrophysics. Clearly, in order to test the many beautiful jet models that have been devised, it will be necessary to obtain reliable values of spin for a good sample of black holes.

It is likely that the spins that have been estimated already for three black holes are not due to spin up by disk accretion, but rather are the natal spins that were imparted to the black holes during their formation; if true, this is obviously of major significance in building core-collapse models for SN and *GRBs* (McClintock et al. 2006).

The continuing development of gravitational wave astronomy is central to the exploration of black holes, and knowledge of black hole spin is fundamentally important to this effort. To detect the faint coalescence signal for two inspiralling black holes, one must compute the expected waveform and use it to filter the data. The first computation of such waveforms that includes the effects of spin were done by Campanelli et al. (2006) and were stimulated by the spin results reported by Shafee et al. (20060. In their abstract, Campanelli et al. conclude: "...the last stages of the orbital motion of black-hole binaries are profoundly affected by their individual spins."

11 Concluding Comments

We are at a special moment in the history of black hole astrophysics and the study of binary black holes in particular. The masses of 21 stellar-mass black holes have been measured or constrained. And just now the spin parameters a_* of four of these black holes have been estimated using a full general relativistic ray-tracing/indexgeneral relativistic ray–tracing model software model *KERRBB* and a state-of-the-art disk atmosphere model. It should be possible to obtain a dozen spin measurements over the next several years and to corroborate these results through modeling the Fe K line profiles and high-frequency *QPOs* that are observed for many of these sources. Since an astrophysical black hole is completely described by only two parameters, M and a_*, this advance would be a milestone. Apart from the accomplishment per se, the complete specification of these black holes would impact astronomy on many fronts.

12 Acknowledgments

Most of the ideas in this paper were generated collaboratively with Ron Remillard and Ramesh Narayan, and I am grateful for their permission to present them here. I also thank Annual Reviews of Astronomy and Astrophysics for permission to reprint three figures from Remillard and McClintock (2006).

References

1. Abramowicz, M. A., Czerny, B., Lasota, J. P., & Szuszkiewicz, E., ApJ, 332, 646 (1988)

2. Arnaud, K. A., in ASP Conf. Ser. 101, Astronomical Data Analysis Software and Systems V, ed. G. H. Jacoby & J. Barnes, San Francisco, ASP, p17 (1996)
3. Brenneman, L. W. & Reynolds, C. S., ApJ, in press, astro-ph/0608502 (2006)
4. Campanelli, M., Lousto, C. O., & Zlochower, Y., Phys.Rev. D 74, 041501 (2006)
5. Chen, W., Shrader, C. R., & Livio, M., ApJ, 491, 312 (1997)
6. Connors, P. A., Piran, R. F., & Stark, T., ApJ, 235, 224 (1980)
7. Davis, S. W., Blaes, O. M., Hubeny, I., & Turner, N. J., ApJ, 621, 372 (2005)
8. Davis, S. W., Done, C., & Blaes, O. M., ApJ, 647, 525 (2006)
9. Done, C., & Gierliński, M. MNRAS, 342, 1041 (2003)
10. Dovčiak, M., Karas, V., & Yaqoob, T., ApJS, 153, 205 (2004)
11. Garcia, M. R., McClintock, J. E., Narayan, R., Callanan, P., Barret, D., & Murray, S. S., ApJ, 553, L47 (2001)
12. Gierliński, M., Maciolek–Niedzwiecki, A., & Ebisawa, K., MNRAS, 325, 1253 (2001)
13. Kalogera, V., & Baym, G., ApJ, 470, L61 (1996)
14. Kato, Y., PASJ, 56, 931 (2004)
15. Laor, A., ApJ, 376, 90 (1991)
16. Li, L.-X., Zimmerman, E. R., Narayan, R., & McClintock, J. E., ApJS, 157, 335 (2005)
17. Lightman, A. P., & Shapiro, S. L., ApJ, 198, L73 (1975)
18. McClintock, J. E., Narayan R., & Rybicki, G. B., ApJ, 615, 402 (2004)
19. McClintock, J. E., & Remillard, R. A. 2006, in Compact Stellar X–ray Sources, eds. W. H. G. Lewin & M. van der Klis, Cambridge, CUP (2006) MR06
20. McClintock, J. E., Shafee, R., Narayan, R., Remillard, R. A., Davis, S. W., & Li, L.-X., ApJ, 652, 518 (2006)
21. Meszaros, P., Novick, R., Szentgyorgyi, A., Chanan, G. A., & Weisskopf, M. C., ApJ, 324, 1056 (1988)
22. Narayan, R., & Heyl, J. S., ApJ, 574, L139 (2002)
23. Novikov, I. D., & Thorne, K. S., in Black Holes, eds. C. Dewitt & B. DeWitt, Gordan & Breach, NY (1973)
24. Penrose, R., Riv. Nuovo Cim., 1, 252 (1969)
25. Pietsch, W., Haberl, F., Sasaki, M., Gaetz, et al. ApJ, 646, 420 (2006)
26. Remillard, R. A., Lin, D., Cooper, R., & Narayan, R., ApJ, 646, 407 (2006)
27. Remillard, R. A., & McClintock, J. E. 2006, ARAA, 44, 49 (2006) RM06
28. Remillard, R. A., Muno, M. P., McClintock, J. E., Orosz, J. A., ApJ, 580, 1030 (2002)
29. Reynolds, C. S., & Nowak, M. A., PhysRept, 377, 389 (2003)
30. Schnittman, J. D., & Bertschinger, E., ApJ, 606, 1098 (2004)
31. Shafee, R., McClintock, J. E., Narayan, R., Davis, S. W., Li, L.-X., & Remillard, R. A., ApJ, 636, L113 (2006)
32. Shakura, N. I., & Sunyaev, R. A., A&A, 24, 337 (1973)
33. Shapiro, S. L., & Teukolsky, S., Black Holes, White Dwarfs and Neutron Stars, New York, John Wiley & Sons (1984)
34. Stella, L., Vietri, M., & Morsink, S. M., ApJ, 524, L63 (1999)
35. Sunyaev, R., & Revnivtsev, M., A&A, 358 (2000)
36. Tanaka, Y., & Lewin, W. H. G., in X–ray Binaries, eds. W. Lewin, J. van Paradijs, & E. van den Heuvel, Cambridge, CUP, p.126 (1995)
37. Török, G., Abramowicz, M. A., Kluźniak, W., & Stuchlík, Z., A&A, 436, 1 (2005)
38. van der Klis, M. 2006, in Compact Stellar X–ray Sources, eds. W. H. G. Lewin & M. van der Klis, Cambridge, CUP (2006)
39. Zhang, S. N., Cui, W., & Chen, W., ApJ, 482, L155 (1997)

Searches for Gravitational Waves from Binary Neutron Stars: A Review

Warren G. Anderson[1] and Jolien D.E. Creighton[2]

[1] Department of Physics, University of Wisconsin – Milwaukee, P.O. Box 413, Milwaukee, Wisconsin, 53201-0413, U.S.A.
warren@gravity.phys.uwm.edu
[2] Department of Physics, University of Wisconsin – Milwaukee, P.O. Box 413, Milwaukee, Wisconsin, 53201-0413, U.S.A.
jolien@gravity.phys.uwm.edu

A new generation of observatories is looking for gravitational waves. These waves, emitted by highly relativistic systems, will open a new window for observation of the cosmos when they are detected. Among the most promising sources of gravitational waves for these observatories are compact binaries in the final minutes before coalescence. In this article, we review in brief interferometric searches for gravitational waves emitted by neutron star binaries, including the theory, instrumentation and methods. No detections have been made to date. However, the best direct observational limits on coalescence rates have been set, and instrumentation and analysis methods continue to be refined toward the ultimate goal of defining the new field of gravitational wave astronomy.

1 Introduction

It is no great exaggeration to say that the advent of a new type of astronomy is imminent. Gravitational wave astronomy is predicated on the observation of the cosmos not with a new band of the electromagnetic spectrum, but rather via a whole new spectrum, the spectrum of gravitational waves. As such it has the potential to revolutionize our understanding of the Universe because it will allow us to access phenomena which are electromagnetically dark or obscured [1]. The impetus for this revolution is the recent construction and operation of a new generation of gravitational wave detectors based on interferometry [2]. Within a decade, these interferometers will be reaching sensitivities at which gravitational wave observations should become routine.

There are four categories of gravitational wave signals which ground-based interferometers are currently trying to detect: quasi-periodic signals, such as

E.F. Milone et al. (eds.), Short-Period Binary Stars: Observations, Analyses, and Results, 23–52.

those expected from pulsars [3–7], stochastic background signals, such as remnant gravitational waves from the Big Bang [8–12], unmodeled burst signals, such as those that might be emitted by supernovae [13–19], and inspiral signals, such as those from neutron star or black hole binaries [20–26]. In this article, we will only be concerned with the last of these searches, and in particular the search for neutron star binaries [20, 23, 25, 26].

This article is organized as follows: Since many astronomers may not be very familiar with gravitational waves and the effort to use them for astronomy, Sect. 2 is devoted to background material. This includes a simple description of what gravitational waves are and how they are generated, a brief history of the instruments and efforts to detect them, and some background on the relevant aspects of neutron star binaries and the gravitational waves they produce. Section 3 describes in some detail how searches for gravitational waves from neutron star binaries have been conducted, including a description of the data, the data analysis methods employed, and coincidence vetoes.

Section 4 reviews the published upper limits that have been placed on the rate of neutron star binary coalescence in our galactic neighborhood by interferometric detectors. Also included is a description of the statistical analysis used to place these upper limits. In Sect. 5, we discuss prospects for better upper limits and discuss some of the astrophysics that might be done by interferometric gravitational wave detectors when they reach a sensitivity where gravitational waves from neutron star binaries are regularly observed. Concluding remarks are found in Sect. 6.

2 Background

2.1 Gravitational Waves

One of the many remarkable predictions of Einstein's general theory of relativity is the existence of gravitational waves (GWs) [27]. Einstein himself elucidated the theoretical existence of GWs as early as 1918 [28]. Today, however, GWs have still not been directly measured, although the measurements of the binary pulsar PSR 1913+16 [29–31], (discovered by Hulse and Taylor and for which they won the Nobel prize) leave little doubt that GWs do, in fact, exist.

The fundamental factor that has led to our failure to directly measure GWs is the exceptional weakness of the gravitational coupling constant. This causes GWs to be too feeble to detect unless produced under extreme conditions. In particular, gravitational waves are produced by accelerating masses (a more exact formulation of this statement appears in Sect. 2.3). Thus, for the highest GW amplitudes, we seek sources with the highest possible accelerations and masses. As a result, astrophysical objects are the most plausible sources of GWs [3].

Further restrictions on viable sources are imposed by our detectors. For instance, for Earth based instruments, even with the best seismic isolation technologies currently available [32, 33], noise from seismic vibrations limits large-scale precision measurements to frequencies above ~30 Hz. Through causality, this time-scale limitation implies a maximum length-scale at the source of 10^4 km. Given the Chandrasekhar limit on the mass of a white dwarf star [34,35] and the white dwarf mass-radius relationship this is approximately the minimum length-scale for white dwarf stars [36].

If searches are restricted to objects more compact than white dwarfs, then within the bounds of current knowledge, gravitational wave astronomy with ground-based interferometers is limited to black holes and neutron stars as sources. According to our current understanding of astrophysical populations, these objects are relatively rare. Thus, the probability of finding them in our immediate stellar neighborhood are small, and any realistic search must be sensitive to these sources out to extragalactic distances to have a reasonable chance of seeing them in an observation time measured in years. Is this a reasonable prospect? The answer is yes, but to understand why, it behooves us to first understand a bit more about what GWs are and how they might be measured.

Gravitational waves arriving at Earth are perturbations of the geometry of space-time. Heuristically, they can be understood as fluctuations in the distances between points in space. Mathematically, they are modeled as a metric tensor perturbation $h^\alpha{}_\beta$ on the flat spacetime background. Linearizing the Einstein field equations of general relativity in $h^\alpha{}_\beta$, we find that this symmetric four-by-four matrix satisfies the the wave equation,

$$\left(-\frac{\partial^2}{\partial t^2} + c^2\nabla^2\right) h^\alpha{}_\beta \;=\; 8\pi\, G\, T^\alpha{}_\beta, \tag{1}$$

in an appropriate gravitational gauge. Here ∇^2 is the usual Laplacian operator, G is the gravitational constant, and $T^\alpha{}_\beta$ is the stress energy tensor, another symmetric four-by-four matrix which encodes information about the energy and matter content of the spacetime.

From (1), it is obvious that GWs travel at c, the speed of light. To understand the *production* of gravitational waves by a source, we solve (1) with the stress-energy ($T^\alpha{}_\beta$) of that source on the right-hand-side. We will discuss this in more detail in Sect. 2.3. First, however, we wish to consider the propagation of gravitational waves.

[3] The exception to this statement is the big-bang itself, which should lead to a stochastic cosmological background of GWs, as mentioned in Sect. 1. Gravitational wave observations have just begun to bound previously viable theoretical models of this background [10]. We will not be considering this background further in this review.

For the propagation of gravitational waves, we require solutions to the homogeneous ($T^\alpha{}_\beta = 0$) version of (1). As usual, such solutions can be expressed as linear combinations of the complex exponential functions

$$h^\alpha{}_\beta = A^\alpha{}_\beta \exp(\pm i k_\mu x^\mu). \tag{2}$$

Here, $A^\alpha{}_\beta$ is a matrix of constant amplitudes, $k_\mu = (-\omega, \boldsymbol{k})$ is a four vector which plays the role of a wave vector in four dimensions, and $x^\mu = (t, \boldsymbol{x})$ are the spacetime coordinates. Above, and in what follows, we use the Einstein summation convention that repeated indices, such as the μ in $k_\mu x^\mu = -\omega t + \boldsymbol{k}\cdot\boldsymbol{x}$, indicate an implicit summation. It can be shown that for gauges in which (1) holds that

$$k_\mu h^\mu{}_\beta = 0. \tag{3}$$

In words, this means that the wave vector is orthogonal to the directions in which the *GW* distorts spacetime, i.e. the wave is transverse.

Since $h^\alpha{}_\beta$ is a four-by-four matrix, it has 16 components. However, because it is symmetric, only 10 of those components are independent. Further, (3) imposes four constraints on $h^\alpha{}_\beta$, reducing the number of free components to six. One can use remaining gauge freedom to impose four more conditions on $h^\alpha{}_\beta$. There are therefore only two independent components of the matrix $h^\alpha{}_\beta$. Details can be found in any elementary textbook on General Relativity, such as [37].

The two independent components of $h^\alpha{}_\beta$ are traditionally called h_+ and $h_\times$. These names are taken from the effect that the components have on a ring of freely moving particles laying in the plane perpendicular to the direction of wave propagation, as illustrated in Fig. 1. The change in distance between particles, $\delta\ell$, is proportional to the original distance between them, ℓ, and the amplitude of the gravitational waves, $A^\alpha{}_\beta$. For a point source, which all astrophysical sources will effectively be, the amplitude decreases linearly with the distance from the source. For astrophysical source populations from which gravitational wave emission have been estimated, the typical gravitational wave strain, $h \sim 2\delta\ell/\ell$, at a detector at Earth would be expected to be less than or of the order of 10^{-21} [1]. Through interferometry, it is possible to measure $\delta\ell \sim 10^{-18}$ m. Thus, interferometers of kilometer scales are required to have any chance of measuring these sources.

It might seem that the challenge of attempting to measure gravitational waves is so daunting as to call into question whether it is worthwhile at all. However, there are several factors which make the measurement of gravitational waves attractive. First, astrophysical gravitational wave sources include systems, such as black hole binaries, which are electromagnetically dark. Gravitational waves may therefore be the best way to study such sources. Second, since gravitation couples weakly to matter, gravitational waves propagate essentially without loss or distortion from their source to the detector. Thus, sources obscured by dust or other electromagnetically opaque media may still

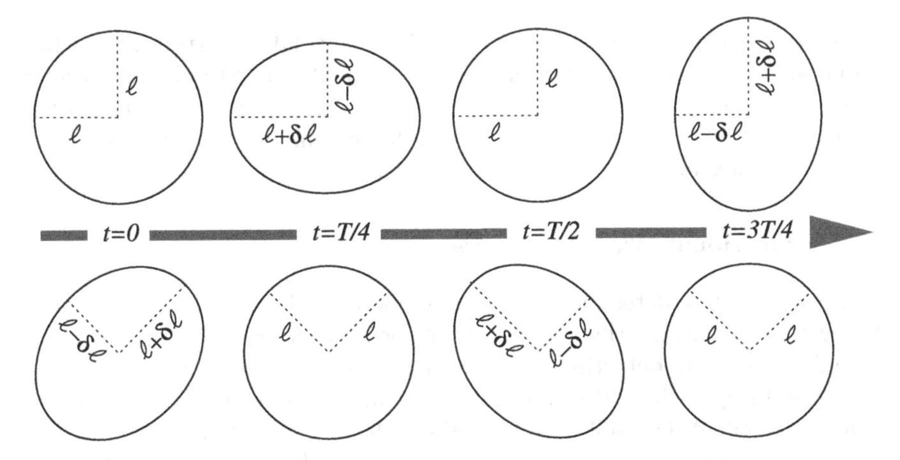

Fig. 1. Distortion of a ring of freely falling dust as a gravitational wave passes through. The wave is propagating into the page. From left to right are a series of four snapshots of the distortion of the ring. The top row are distortions due to h_+. The bottom row are distortions due to $h_\times$. The snapshots are taken at times $t = 0$, $t = T/4$, $t = T/2$ and $t = 3T/4$ respectively, where T is the period of the gravitational wave. The relative phase between h_+ and $h_\times$ corresponds to a circularly polarized gravitational wave.

be observed with gravitational waves. Also, interferometers behave as amplitude sensing devices for *GW*s (like antennas), rather than energy gathering devices (like telescopes), leading to a $1/r$ fall-off with distance, rather than the more usual $1/r^2$ fall-off [2].

But perhaps the most compelling reason for pursuing the measurement of gravitational waves is that they constitute an entirely new medium for astronomical investigation. History has demonstrated that every time a new band of the electromagnetic spectrum has become available to astronomers, it has revolutionized our understanding of the cosmos. What wonders, then, await us when we start to see the Universe through the lens of gravitational waves, (which will surely begin to happen within the next decade as *GW* detectors continue their inevitable march toward higher sensitivities)? Only time will tell, but there is every reason to be optimistic.

In this article, we concentrate on one of the many sources of gravitational waves for which searches are ongoing – binary neutron star (*BNS*) systems. In particular, the sensitive frequency band of ground-based interferometers, which is approximately 40 Hz to 400 Hz, dictates that we should be interested in neutron star binaries within a few minutes of coalescence [38]. These sources hold a privileged place in the menagerie of gravitational waves sources that interferometers are searching for. The Post-Newtonian expansion, a general relativistic approximation method which describes their motion, gives us expected waveforms to high accuracy [39–41]. They are one of the few sources for which such accurate waveforms currently exist, and they are therefore amenable to the most sensitive search algorithms available. Furthermore, while the population of neutron star binaries is not well understood, there are at least observations of this source with which to put some constraints on the population [42]. These two factor give *BNS* systems one of the best (if not *the* best) chance of discovery in the near future.

2.2 Gravitational Wave Detectors

Gravitational wave detectors have been in operation for over forty years now. However, it is only in the last five years or so that detectors with a non-negligible chance of detecting gravitational waves from astrophysical sources have been in operation. The first gravitational wave detector to operate was built by Joseph Weber [43]. It consisted of a large cylinder of aluminum, two meters long and a meter in diameter, with piezoelectric crystals affixed to either end.

The fundamental idea for such detectors is that if a strong enough gravitational wave was to pass by, that it would momentarily reduce the interatomic distances, essentially compressing the bar, and setting it ringing, like a tuning fork. The ringing would create electric voltages in the piezoelectric crystals which could then be read off. Of course, as with a tuning fork, the response of the apparatus is greatly increased if it is driven at its resonant frequency (about 1660 Hz, for Weber's bars).

Weber's bar was isolated from seismic and electromagnetic disturbances and housed in a vacuum. He attributed the remaining noise in his instrument to thermal motion of the aluminum atoms. This noise limited Weber's measurements to strains of $h{\sim}10^{-16}$, about five orders of magnitude less sensitive than the level now believed necessary to make the probability of detection non-negligible. Nonetheless, by 1969, Weber had constructed two bars and had observed coincident events in them although they were housed approximately 1000 miles apart. He calculated that his noise would create some of these events at rates as low as one per thousand years, and subsequently published his findings as "good evidence" for gravitational waves [44].

Many groups followed up on this and subsequent claims of gravitational wave detections made by Weber. No other group was ever able to reproduce these observations, and it is now generally agreed that Weber's events were spurious. Nonetheless, this launched the era of gravitational wave detectors, as resonant mass detectors (as Weber-like bars are now called) began operating in countries around the globe. Today, there is a network of these detectors operated under the general coordination of the International Gravitational Event Collaboration [45]. Technical advances have led to considerable improvements in sensitivity over the past decades. They continue to be rather narrow band detectors, however, and they are not the most sensitive instruments in operation today.

That distinction belongs to interferometric gravitational wave detectors [2]. The basic components of these detectors are a laser, a beam splitter to divide

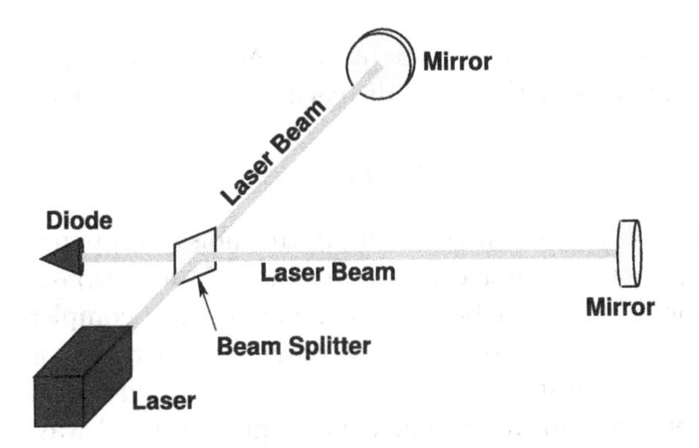

Fig. 2. Schematic diagram of an interferometric gravitational wave detector. The beam splitter is coated to allow half the light to be transmitted to one of the mirrors, and the other to be reflected to the other mirror. Real interferometric gravitational wave detectors are much more sophisticated, including frequency stabilization of the laser, a second mirror on each arm between the beam splitter and the end mirror to create Fabry-Perot cavities, and control feedback loops which "lock" the interferometer onto an interference fringe. Thus, rather than measuring the current from the photo-diode directly, the gravitational wave signals are encoded in the feedback loop voltages needed to maintain the lock of the interferometer. These and many other enhancements are necessary to reach the required sensitivity level.

the laser light into two coherent beams, hanging mirrors to reflect the laser beams, and a light sensing diode, as shown in Fig. 2. The mirrors act essentially as freely moving particles in the horizontal directions. If a sufficiently strong gravitational wave with a vertical wave-vector component impinges, it shortens the distance between the beam splitter and one of the mirrors, and lengthens the distance between the beam splitter and the other mirror. This is registered as a shifting interference pattern by the light sensing diode, thus detecting the gravitational wave.

More specifically, what is measured is a quantity proportional to the strain on the detector, $s := (\delta x - \delta y)/\ell$, where δx and δy are the length changes in the two equal length arms of the interferometer, traditionally called the x and y arms, and ℓ is the unperturbed length of each arm. If there were no noise, then in the special case that h_+ or $h_\times$ was aligned with the arms, we would have $\delta x = -\delta y = \delta\ell$ and the measured quantity would be proportional to $2\delta\ell/\ell$. For a more general alignment, but still in the absence of noise, some linear combination of h_+ and $h_\times$ is measured

$$h(t) \;:=\; F_+(\theta,\phi,\psi)\, h_+(t) + F_\times(\theta,\phi,\psi)\, h_\times(t), \tag{4}$$

where F_+ and $F_\times$ are called *beam pattern functions.* They project the gravitational wave components, h_+ and $h_\times$, onto a coordinate system defined by the detector, and are functions of the Euler angles (θ,ϕ,ψ) which relate this coordinate system to coordinates which are aligned with the propagating GW.

In the case where there is noise, δx and δy are sums of the gravitational wave displacements and the noise displacements, so that the detector strain is

$$s(t) = h(t) + n(t) \tag{5}$$

where $n(t)$ is the noise contribution. If the noise component can be kept from dominating the signal component, there is a reasonable chance that the gravitational wave can be detected. For a more complete and detailed account of interferometric detection of gravitational waves, as well as many other aspects of gravitational wave physics, we recommend [1].

The idea of using an interferometer as a gravitational wave detectors was explored repeatedly but independently by several different researchers over approximately 15 years [46–49]. The first prototype of an interferometric gravitational wave detector was built by one of Weber's students, Robert Forward, and collaborators in 1971 [50, 51]. The advantages of this idea were immediately understood, but, as mentioned above, an understanding also emerged that kilometer-scale interferometry would be needed. Thus, these detectors needed to be funded, built and operated as coordinated efforts at the national or international level.

To date, there have been six large-scale (100 m plus) interferometers operated at five sites. Three of the these are located in the United States and constitute the Laser Interferometer Gravitational-wave Observatory (*LIGO*) [52, 53]. There is one four kilometer instrument at each of the *LIGO*-Hanford site in Washington State [54] and *LIGO*-Livingston site in Louisiana [55] (H1 and L1 respectively), and a two kilometer instrument, housed in the same enclosure, at the *LIGO*-Hanford Observatory (H2). The other interferometers are the three kilometer Virgo instrument, built in Italy by a French-Italian collaboration [56], the 600 meter German-English Observatory (*GEO600*) in Germany [57], and the *TAMA* observatory, a 300 meter instrument located in and funded by Japan [58].

Other large-scale interferometers are in various stages of planning, although funding has not been secured for them. One of the most interesting is *LISA*, a planned joint *NASA-ESA* mission, which would consist of two independent million-kilometer-scale interferometers created by three satellites in solar orbit [59, 60]. The larger scale and freedom from seismic noise will make this instrument sensitive at much lower frequencies than its Earth-based brethren.

Because they are currently the most sensitive gravitational wave detectors in the world, and the ones with which we are most familiar, this article will often use the particular example of *LIGO* and its methods to illustrate our discussion. *LIGO* began construction in 1994 and was commissioned in 1999. It began taking scientific data on 23 August, 2002. That data taking run, called Science Run One (or S1 for short), ended 9 Sept., 2002. There have been four subsequent science runs: S2 from 14 Feb. to 14 Apr. 2003, S3 from 31 Oct. 2003 to 9 Jan. 2004, S4 from 22 Feb. to 23 Mar. 2005, and finally S5 has been ongoing from Nov. 2005 and is expected to end in fall, 2007. At the conclusion of S5, the *LIGO* interferometers are scheduled for component enhancements which are expected to double the sensitivity of the instrument [61, 62]. The upgrade process is scheduled to last approximately a year.

As with all the new interferometers, initial runs were short and periods between them were long because scientists and engineers were still identifying and eliminating technical and environmental noise sources which kept the instruments from running at their design sensitivity. As designed, interferometer sensitivity is expected to be bounded by three fundamental noise sources [49]. Below approximately 40 Hz, seismic noise transmitted to the mirrors through housing and suspension dominates. Between approximately 40 and 200 Hz, thermal vibrations in the suspension system for the mirrors dominates. Finally, above approximately 200 Hz, the dominant contribution comes from photon shot noise associated with counting statistics of the photons at the photodiode. These three fundamental noise regimes, which contribute to the "noise floor" of a detector, are described in the caption of Fig. 3.

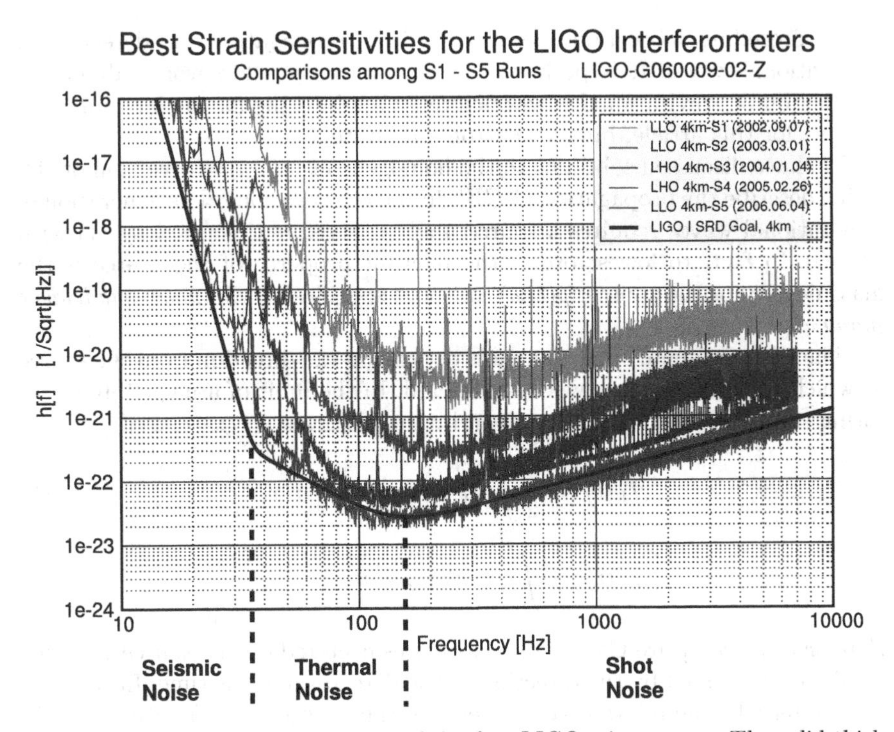

Fig. 3. Interferometer noise in each of the five *LIGO* science runs. The solid thick black curve is design goal of *LIGO*. There are two changes of slope indicated by dashed thick black lines. These correspond to changes in the dominant noise source – the three noise regimes are marked. Over four years of commissioning, the noise floor was reduced by approximately three orders of magnitude.

Apparent in Fig. 3 is the remarkable progress that has been made in lowering the noise floor and the other noises superimposed upon it in going from S1 to S5. Indeed, at S5, *LIGO* matches or exceeds its design specification over most frequency bands, including the most sensitive region from approximately 100–300 Hz. When S5 is over and S5 data is analyzed, it will provide by far the best opportunity to date to detect gravitational waves.

The degree to which improvements in the noise floor such as those in Fig. 3 correspond to increase detection probabilities depend on the shape of the noise floor and the signal being sought. For neutron star binary inspiral signals, one can devise a figure of merit called the inspiral range. This is the distance, averaged over all sky positions and orientations, to which an instrument can expect to see a 1.4-1.4 $M_\odot$ binary with signal-to-noise ratio (defined below) of 8. For S2, typical inspiral ranges for L1 were just below 1 Mpc. For the *LIGO* design curve, the inspiral range is approximately 14 Mpc [63].

2.3 Neutron Star Binaries as Gravitational Wave Emitters

In the preceding subsections, we discussed the propagation and detection of gravitational waves. It is perhaps useful now to say a few words about the production of gravitational waves before diving into the particular sources of interest for this article, binary neutron star systems.

We have discussed solutions to the homogeneous ($T^\alpha{}_\beta = 0$) version of (1), which describe the propagation of *GWs* in the far wave zone. The generation of gravitational waves requires a source, and is therefore described by (1) with $T^\alpha{}_\beta \neq 0$. In fact, to lowest order, the relevant property of the source is the mass density of the source, $\rho(t, \boldsymbol{x})$, or more specifically, the *mass quadrupole moments.*

The relevant quadrupole moments depend on the direction from the source at which the gravitational wave is detected. In the limit of a negligibly gravitating source, the moments are given by the integrals

$$\mathcal{J}_+(t) := \frac{1}{2} \int \rho(t, \boldsymbol{x}) \left[x^2 - y^2\right] d^3x, \tag{6}$$

$$\mathcal{J}_\times(t) := \int \rho(t, \boldsymbol{x})\, x\, y\, d^3x. \tag{7}$$

Here, $\boldsymbol{x} = \{x, y, z\}$ are Cartesian coordinates centered on the source with the z-axis being defined by the direction to the detector and t is time. In terms of these integrals, the relevant components of the solution to (1) with source is

$$h_+(t) = \frac{2}{r}\frac{G}{c^4}\frac{\partial^2}{\partial t^2}\mathcal{J}_+(t - r), \tag{8}$$

and likewise for $h_\times(t)$. Here r is the distance between the source and the detector.

While higher order multipole moments of the mass distribution can contribute to the radiation, for most systems the quadrupole will dominate. Further, the mass monopole and dipole moment will not contribute any gravitational waves. Thus, such events as a spherically symmetric gravitational collapse and axially symmetric rotation do not emit any gravitational radiation. On the other hand, a rotating dumbell is an excellent emitter of gravitational waves, making binary systems potentially amongst the brightest emitters of gravitational waves in the Universe.

In general, the better the information about the gravitational wave signal, the better will be our chances of detecting it. This, in turn, prods us to find the best possible model of the dynamics of a *GW* emitting system. For the gravitational wave sources suitable for Earth-based interferometric detectors, this proves to be difficult in general. For instance, the physics of core-collapse supernovae is not understood at the level of detail needed to accurately predict gravitational waveforms. This is also true of colliding black holes or neutron stars, which should emit large amounts of energy in gravitational waves. In fact, there are only two sources for ground-based interferometers, inspiralling binary neutron stars and perturbed intermediate mass ($\sim$100 $M_\odot$) black holes, for which the physics is well enough understood that it is believed that the waveforms calculated provide a high degree of confidence for detection[4].

Of these two potential sources, we at least know with great certainty that neutron star binaries, like the Hulse-Taylor binary, exist [42,64,65]. According to current thinking, these binaries are formed from the individual collapse of binaries of main sequence stars [66]. If one can establish a population model for our galaxy, therefore, one should be able to deduce the population outside the galaxy by comparing the rate of star formation in our galaxy to elsewhere. One measure of star formation is blue light luminosity of galaxies (corrected for dust extinction and reddening) [67,68]. Thus, once a galactic population model is established, given a noise curve for an instrument it is possible to estimate the rate at which gravitational waves from binary neutrons star (*BNS*) systems out to extragalactic distances will be seen.

To develop a population model, astrophysicists use Monte Carlo codes to model the evolution of stellar binary systems [69, 70]. To cope with uncertainties in the evolutionary physics of these systems, a great many ($\sim$30) parameters must be introduced into the models, some of which can cause the predicted rates to vary by as much as two orders of magnitude. This can

[4] To be fair, although the source mechanics of rotating neutron stars, like pulsars, are not well understood, the gravitational waves they produce if there is some asymmetry about the axis of rotation will be quasi-periodic. We therefore do not need to model these sources well to search for their gravitational waves. Similarly, although a stochastic background of gravitational waves in intrinsically unknowable in detail, a search for these waves can be optimized if their statistical properties are known.

be reduced somewhat by feeding what is known about *BNS* systems from observation into the models [71]. However, only seven *BNS* systems have been discovered in the galactic disk. Furthermore, the most relevant for detection rates with interferometric *GW* detectors are the four *BNS* which are tight enough to merge within 10 Gyrs, since, as mentioned above, only those within minutes of merger can be observed. Thus, there is not much information to feed into these models, and estimates can still vary widely. The current best estimate is that *LIGO* should now be able to observe of the order of one *BNS* system approximately every hundred years, although uncertainties extend this from a few per thousand years to almost one every ten years [66]. An improvement in *LIGO*'s ability to detect *BNS* systems by a factor of three would therefore raise the most optimistic case to almost one *BNS* signal per year.

With projected rates this low, it is essential to search for *BNS* systems with the highest possible detection efficiency. Since, in general, the more information that can be fed into the detection algorithm, the more efficient it will be, it is important to have a high accuracy theoretical prediction for the waveform. In the case of *BNS* systems, this prediction is provided by the restricted post-Newtonian approximation [39–41]. The post-Newtonian formalism uses an expansion in orbital velocity divided by c, the speed of light. Since this is a slow motion approximation, and the orbital velocity of the binary constantly increases during inspiral, this approximation becomes worse as the binary evolves. However, the accuracy is still good for *BNS* systems when they are in the interferometric detection band. Furthermore, we need only deal with circular orbits because initially eccentric orbits are circularized through the *GW* emission process [72]. Finally, one can safely ignore spin terms [73] and finite size effects [74]. Figure 4 shows the post-Newtonian prediction for a neutron star binary in the sensitive frequency band of an interferometric detector.

The general features of this waveform are easily understood. From (7), it is apparent that the mass quadrupole moment will be proportional to the square of the orbital radius, a^2. The second time derivative of this quadrupole moment therefore goes as $a^2\omega^2$, and from (8) we have that $h{\sim}a^2\omega^2$. For a Keplerian orbit, which relativistic orbits approximate when the Post-Newtonian approximation is valid, $\omega^2{\sim}a^{-3}$ and therefore $h{\sim}a^{-1}$. Thus, when the binary emits *GW*s, which carry away orbital energy, it experiences a progressive tightening, leading to an increased frequency and amplitude. If the two neutron stars are of equal mass, then the quadrupole configuration repeats itself every twice every orbital period, and the gravitational wave frequency is twice the orbital frequency. Because the two masses are comparable in every BNS system, the dominant frequency component is always at twice the orbital frequency.

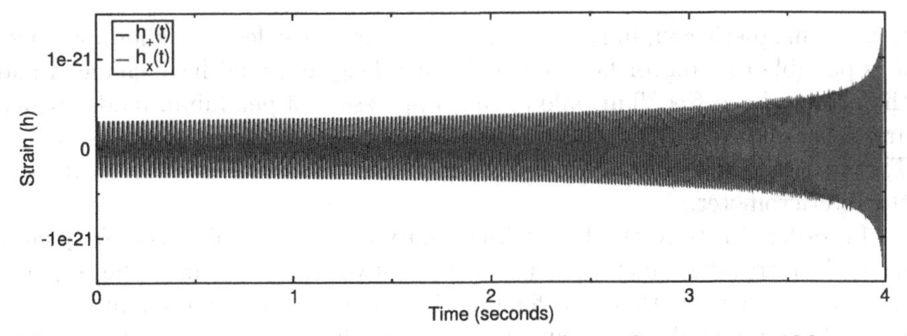

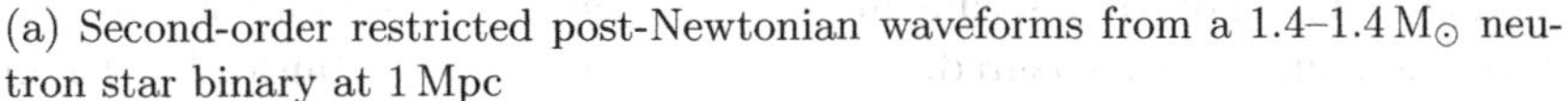
(a) Second-order restricted post-Newtonian waveforms from a 1.4–1.4 $M_\odot$ neutron star binary at 1 Mpc

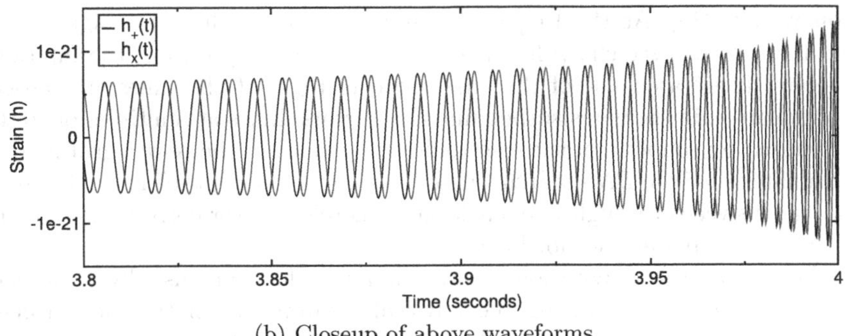

(b) Closeup of above waveforms

Fig. 4. Second-order restricted post-Newtonian h_+ and $h_\times$ waveforms from an optimally located and oriented 1.4–1.4 $M_\odot$ neutron star binary system at 1 Mpc. The top panel, (a), shows approximately the portion (with frequency sweeping through $\sim$ 40–400 Hz) of the waveform visible to a ground-based interferometric *GW* detector. The bottom panel, (b), shows a closeup of the last 0.2 s of panel (a). From the closeup, we see that the waveforms are sinusoids sweeping upward in both frequency and amplitude as the binary companions inspiral toward one another. The two phases of the gravitational wave, h_+ and $h_\times$, are 90° out of phase. Thus, this is a circularly polarized gravitational wave.

3 Search Method

3.1 Interferometric Data

As was mentioned in the caption of Fig. 2, gravitational wave interferometers are not the simple Michelson interferometers portrayed in that figure. Rather, they have a number of important and sophisticated refinements to that basic configuration, all designed to increase instrument sensitivity and reliability [2, 75]. For the purpose of understanding the data, the most important enhancement is that, when the *IFO*s are operational, their arm-lengths are held fixed, or *locked.* This is done by measuring the light at the antisymmetric port (the diode in Fig. 2) and applying feed-back controls to the mirrors through magnetic couplings [76,77]. By measuring the feed-back loop voltage, it is possible to monitor how much the arm-lengths would have changed had the mirrors been free. This allows the suppression of pendulum modes which are either spurious or which might damage the optics if driven at resonance. The feedback control voltage is therefore the gravitational wave channel of the interferometer.

In order to convert this voltage into detector strain, the frequency dependent transfer function between these two quantities must be applied. The overall shape of this transfer function in the frequency domain is determined by a model of the instrument, however, the overall scaling of the transfer function must be measured. This can be done, for instance, by driving the mirrors at specific frequencies and measuring the response of the control loop voltage [78]. At the frequencies at which these calibration signals are injected, one can see sharp line features in the noise spectra of the interferometers. These are some of the lines found in Fig. 3. Other lines are caused by resonances in the wires suspending the mirrors. Lines at multiples of 60 Hz are caused by coupling of the electronic subsystems to the power grid mains. Fortunately, for the *BNS* search, these lines are not very problematic because the signal sweeps through frequencies and therefore never corresponds to one of these noisy frequencies for long.

As well as the gravitational wave channel, there are usually a myriad of other channels monitoring the physical environment of the interferometer (seismic, magnetic, etc) and the internal status of the instrument (mirror alignment, laser power, etc). These can be used to veto epochs of gravitational wave data which are unreliable. Furthermore, if one has access to simultaneous data from multiple interferometers (like *LIGO* does with its three interferometers, all of which have similar alignment[5] and two of which are co-located), most false alarms can be eliminated by requiring coincidence of received signals in all interferometers.

Although noise from an interferometer that is functioning well is primarily Gaussian and stationary, there are occasional noise excursions, called *glitches*. The causes of some of these glitches are not apparent in the auxiliary channels. These constitute the primary source of background noise in the search for signals from neutron star binaries.

3.2 Matched Filtering and Chi-Squared Veto

When the expected signal is known in advance and the noise is Gaussian and stationary, the optimal linear search algorithm is *matched filtering* [79]. The idea behind matched filtering is to take the signal, and data segments of the same length as the signal, and treat them as members of a vector space. As with any two vectors in a vector space, the degree to which the signal and a data vector overlap is calculated using an inner product.

[5] The detectors in Washington are somewhat misaligned with the detector in Louisiana due to the curvature of the Earth.

To be more precise, consider detector strain $s(t)$ and a signal $h(t)$ that lasts for a duration of T. If the signal arrives at the detector at time t_0, then the detector strain can be written

$$s(t) = \begin{cases} h(t - t_0) + n(t), & t_0 < t < t_0 + T \\ n(t), & \text{otherwise} \end{cases} \tag{9}$$

where $n(t)$ is the detector noise. For this paragraph, we will assume that, apart from being stationary and Gaussian, $n(t)$ is white (same average power at all frequencies) for simplicity. Then, the matched filter output, $\zeta(t)$, is given by

$$\zeta(t) = 2 \int_0^T h(\tau)\, s(t+\tau)\, d\tau. \tag{10}$$

At time $t = t_0$, we have

$$\zeta(t_0) = 2 \int_0^T h^2(\tau)\, d\tau + 2 \int_0^T h(\tau)\, n(t_0 + \tau)\, d\tau. \tag{11}$$

Let us denote the first and second integrals in (11) by I_1 and $I_2(t_0)$ respectively. Clearly, the integrand of I_1 is deterministic and positive everywhere. However, the integrand of I_2 is stochastic. The average of I_2 over all noise realizations vanishes. In other words, on average $\zeta(t_0) = I_1$ when there is a signal starting at time t_0. On the other hand, when there is no signal, $\zeta(t) = I_2(t)$. Denoting the standard deviation of I_2 over all noise realizations by σ, we define the *signal-to-noise ratio* (*SNR*) for the data to be

$$\varrho(t) := |\zeta(t)|/\sigma. \tag{12}$$

Clearly, at time t_0 the expected value of the *SNR* is $\varrho(t_0) = I_1/\sigma$. Thus, if the signal is strong enough that I_1 is several times larger than σ, there is a high statistical confidence that it can be detected.

In practice, it is preferable to implement the matched filter in the frequency domain. Thus, rather than a stretch of data $s(t)$, one analyzes its Fourier transform

$$\tilde{s}(f) = \int_{-\infty}^{\infty} e^{-2\pi i f t}\, s(t)\, dt, \tag{13}$$

where f labels frequencies. This has several advantages: first, it allows for the non-white noise spectrum of interferometers (cf. Fig. 3) to be more easily handled. Second, it allows the use of the stationary phase approximation to the restricted post-Newtonian waveform [1,80], which is much less computationally intensive to calculate, and accurate enough for detection [81]. Third, it allows one to easily deal with one of the search parameters, the unknown phase at which the signal enters the detector's band.

In the frequency domain, the matched filter is complex and takes the form

$$z(t) = x(t) + iy(t) = 4 \int_0^{\infty} \frac{\tilde{s}^*(f)\tilde{h}(f)}{S_n(f)}\, e^{2\pi i f t}\, df, \tag{14}$$

where $S_n(f)$ is the one-sided noise strain power spectral density of the detector and the $*$ superscript denotes complex conjugation. It can be shown that the variance of the matched filter due to noise is

$$\sigma^2 = 4 \int_0^\infty \frac{\tilde{h}^*(f)\tilde{h}(f)}{S_n(f)}\, df. \tag{15}$$

In terms of $z(t)$, the *SNR* is given by

$$\varrho(t) = |z(t)|/\sigma. \tag{16}$$

Note that σ and $z(t)$ are both linear in their dependence on the signal template h. This means that the *SNR* is independent of an overall scaling of $h(t)$, which in turn means that a single template can be used to search for signals from the same source at any distance. Also, a difference of initial phase between the signal and the template manifests itself as a change in the complex phase of $z(t)$. Thus, the *SNR*, which depends only on the magnitude of the matched filter output, is insensitive to phase differences between the signal and the template.

Equations (14-16) tell us how to look for a signal if we know which signal to look for. However, in practice, we wish to look for signals from any neutron star binary in the last minutes before coalescence. Because, as mentioned above, finite-size effects are irrelevant, a single waveform covers all possible equations of state for the neutron stars. Likewise, as stated above, the spinless waveform will find binaries of neutron stars with any physically allowable spin. Further, as just discussed, a single template covers all source distances and initial signal phases. However, a single template does not cover all neutron star binaries because it does not cover all masses of neutron stars.

Population synthesis models for neutron star binaries indicate that masses may span a range as large as ~1–3 $M_\odot$. Since mass is a continuous parameter, it is not possible to search at every possible mass for each of the neutron stars in the binaries. However, if a signal is "close enough" to a template, the loss of *SNR* will be small. Thus, by using an appropriate set of templates, called a *template bank*, one can cover all masses in the 1–3 $M_\odot$ range with some predetermined maximum loss in *SNR* [82, 83]. The smaller the maximum loss in *SNR*, the larger the number of templates needed in the bank. Typically, searches will implement a template bank with a maximum *SNR* loss of 3%, which leads to template banks containing of the order of a few hundred templates (the exact number depends on the noise spectrum because both $z(t)$ and σ do, and therefore the number of templates can change from epoch to epoch).

When the noise is stationary and Gaussian, then matched filtering alone gives the best probability of detecting a signal (given a fixed false alarm rate). However, as mentioned earlier, gravitational wave interferometer noise generically contains noise bursts, or glitches, which provide a substantial noise background for the detection of binary inspirals. It is possible for strong glitches to cause substantial portions of the template bank to simultaneously yield high *SNR* values. It is therefore highly desirable to have some other way of distinguishing the majority of glitches from true signals.

The method which has become standard for this is to use a *chi-squared* (χ^2) veto [84]. When a template exceeds the trigger threshold in *SNR*, it is then divided into p different frequency bands such that each band should yield $1/p$ of the total *SNR* of the data if the high *SNR* event were a signal matching the template. The sum of the squares of the differences between the expected *SNR* and the actual *SNR* from each of the p bands, that is the χ^2 statistic, is then calculated. The advantage of using the χ^2 veto is that glitches tend to produce large (low probability) χ^2 values, and are therefore distinguishable from real signals. Thus, only those template matches with low enough χ^2 values are considered triggers.

If the data were a matching signal in Gaussian noise, the χ^2 statistic would be χ^2 distributed with $2p-2$ degrees of freedom [84]. However, it is much more likely that the template that produces the highest *SNR* will not be an exact match for the signal. In this case, denoting the fractional loss in *SNR* due to mismatch by μ, the statistic is distributed as a non-central chi-squared, with non-centrality parameter $\lambda \leq 2\varrho^2\mu$. This simply means that the χ^2 threshold, χ^*, depends quadratically on the measured *SNR*, ϱ, as well as linearly on μ.

In practice, the number of bins, p, and the parameters which relate the χ^2 threshold to the *SNR* , as well as the *SNR* threshold ϱ^* which an event must exceed to be considered a trigger are determined empirically from a subset of the data, the *playground data.* A typical playground data set would be ~10% of the total data set, and would be chosen to be representative of the data set as a whole. Playground data is not used in the actual detection or upper limit analysis, since deriving search parameters from data which will be used in a statistical analysis can result in statistical bias. Values for these parameters for the *LIGO* S1 and S2 *BNS* analyses are given in Table 1.

Finally, let us say a few words about clustering. As discussed earlier, when a glitch occurs, many templates may give a high *SNR*. This would also be true for a strong enough signal. It would be a misinterpretation to suppose that there might be multiple independent and simultaneous signals – rather, it is preferable to treat the simultaneous events as a cluster and then try to determine the statistical significance of that cluster as a whole. The simplest

Table 1. Search algorithm parameters for S1 and S2 *BNS* searches. These parameters were determined using playground data extracted from the S1 and S2 data sets respectively. Note that the χ^2 threshold, χ^*, is different for the Louisiana and Washington instruments in the S2 run.

Data Set	ϱ^*	p	L1 χ^*	H1/H2 χ^*
S1	6.5	8	$5\,(p+0.03\varrho^2)$	$5\,(p+0.03\,\varrho^2)$
S2	6.0	15	$5\,(p+0.01\varrho^2)$	$12.5\,(p+0.01\,\varrho^2)$

strategy, and the one used thus far, is to take the highest *SNR* in the cluster and perform the χ^2 using the corresponding template. Another possibility might have been to take the template with the lowest χ^2 value as representative, or some function of ϱ and χ^2. In fact, there is reason to believe that the last option may be best [23].

3.3 Coincidence and Auxiliary Channel Veto

Although the χ^2 veto reduces the rate of triggers from glitches, some glitch triggers survive. However, there are further tests that can be used to eliminate them. Most importantly, if more than one interferometer is involved in the search, one can require consistency between their triggers. *LIGO* is especially well designed in this regard. Because *LIGO*'s three detectors are almost co-aligned, they should all be sensitive to the same signals (although the 2km H2 will only be half as sensitive to them as the 4km instruments). Thus, any signal that appears in one should appear in all. On the other hand, there is no reason for glitches in one instrument to be correlated with glitches in another, especially between a detector located in Washington state and the Louisiana instrument. Thus, one way to distinguish between triggers generated by actual gravitational waves and those generated by glitches is to demand coincidence between triggers in different instruments [85, 86].

The most fundamental coincidence is coincidence in time. The timing precision of the matched filtering for properly conditioned interferometric data is $\sim$1 ms. The larger effect is the time it could take the gravitational wave to traverse the distance between detectors (i.e. the *light travel time* between them). The maximum time delay for this is also measured in ms (e.g. 10 ms between the Washington site and the Louisiana site for *LIGO*). Thus, triggers at one site which are not accompanied by triggers at another site within the light travel time plus 1 ms are likely not gravitational waves and can be discarded. For triggers from instruments which are sufficiently well aligned, there are several other quantities for which one could required coincidence. Of these, the only one which had been applied to date as a trigger veto is the template which generated the trigger – for the *LIGO* S2 search the same template was required to have generated all coincident triggers (or represent all coincident clusters of triggers) or they were discarded.

The final hurdle that a trigger may have to overcome to remain viable is that it not be associated with a known instrumental disturbance. Auxiliary channels which monitor the instruments and their environment contain information about many potential disturbances. Those channels most likely to correspond to spurious disturbances which would be manifest in the gravitational wave channel have been studied intensively. To date, studies have revealed that channels which allows for safe and useful auxiliary channel vetoes are not forthcoming for most instruments. However, for *LIGO*'s second science run it was discovered that a channel which measures length fluctuations in a certain optical cavity of the L1 interferometer had glitches which were highly correlated with glitches in that instrument's gravitational wave channel [87]. Thus, triggers which occurred within a time window 4 seconds earlier to 8 seconds later than a glitch in this auxiliary channel in Louisiana were also discarded for that analysis.

Finally, triggers which survive all of these cuts must be examined individually to determine if they are genuine candidates for gravitational wave signals. The flow chart for the procedure we have just describe is shown in Fig. 5. A more comprehensive and detailed discussion of such an algorithm can be found in [88]. Details of the pipelines used by *LIGO* for the S1 and

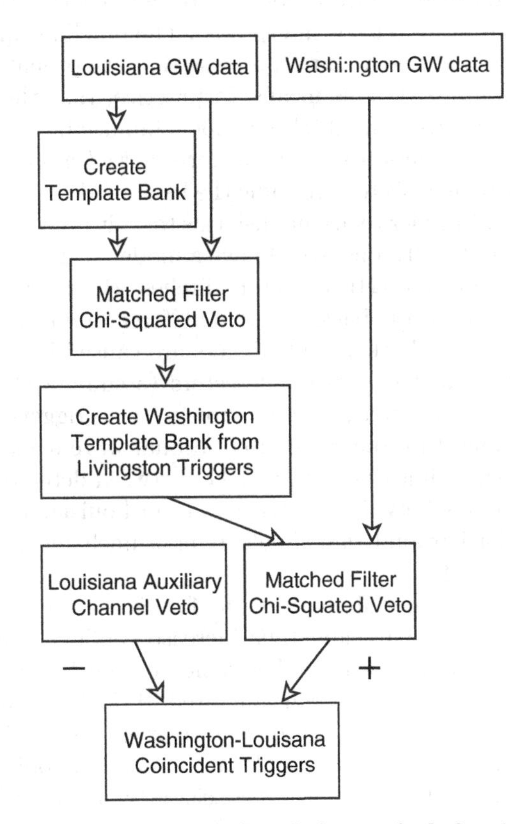

Fig. 5. A diagram showing a typical work-flow to look for signals from neutron star binaries. This is a two instrument work-flow for *LIGO* involving one detector from Washington and the Louisiana detector. It is similar to the one used for the S2 analysis. There would a slightly more complicated work-flow diagram when all three instruments were used. Note that the Washington data is only analyzed at times and for binary templates that correspond to a trigger in Louisiana, thus minimizing both false alarms and computing time. The + and – signs by the two bottom-most arrows indicated that the addition and subtraction of coincident triggers respectively – i.e. passing the *SNR* and χ^2 thresholds adds coincident triggers, while occurring during an auxiliary channel glitch removes them.

S2 analyses can be found in [89] and [90] respectively. In the next section, we discuss how the surviving candidates are analyzed and how upper limits are set from them.

4 Statistics and Results

4.1 Background

Once the method outlined above is completed, one is left with zero or more coincident triggers that need to be interpreted. The pipeline, and in particular the search parameters ϱ^* and χ^*, are chosen so that the probability of missing a real signal are minimized. This means, in practice, that the probability of having coincident triggers from glitches is not minimized.

These coincident triggers from glitches form a *background* against which one does a statistical analysis to determine the likelihood that there are signals (foreground events). In order to accomplish this task, it is useful to understand the background rate. Ideally, one would have a model for the background, but there is no such model for *GW* detector glitches. However, there is a fundamental difference between background and foreground triggers that can be exploited – coincident background triggers are caused by random glitches which happened to coincide between detectors (assuming that the glitches in detectors are not correlated), where as foreground triggers are coincident because they are caused by the same gravitational wave signal. Thus, if one introduces a large enough artificial time delay between detector data streams (e.g. greater than 11 ms between the triggers from Louisiana and those from Washington), the background rate should remain unchanged while the foreground rate must vanish.

A large number of time shifts (e.g. 40 for S2) are used to provide a great deal of background data and thus increase confidence in the inferred background rate. Both the mean and variance of the background rate are estimated. The foreground rate is then compared. If the foreground rate is significantly higher than the mean background rate, it is possible that some of the coincident triggers are from gravitational waves. A careful look at each coincident trigger then ensues. Criteria such as consistency of *SNR* values, problems with auxiliary channels, and behaviour of the gravitational wave channel are invoked to try to eliminate surviving background events. In particular, it is known that poor instrument behavior often leads to many consecutive glitches, and coincident triggers are much more likely during these glitchy times.

4.2 Upper Limits and the Loudest Event

If no gravitational wave candidate is identified amongst the coincident triggers, then the best one can do is set upper limits on the rate at which such signals occur. Clearly, it is meaningless to quote a rate unless one specifies the minimum *SNR* one is considering. Since triggers can have different *SNR*'s in each instrument, for coincident searches it is convenient to consider a measure of the combined *SNR*.

If the noise in all instruments were Gaussian, then it would be appropriate to combine *SNR*'s in quadrature, e.g. $\varrho_C^2 = \varrho_H^2 + \varrho_L^2$, where ϱ_H and ϱ_L are the *SNR*'s at Washington and Louisiana respectively. This would still be appropriate for non-Gaussian noise if the rate and *SNR*s of glitches were the same for instruments at both sites. However, this is not guaranteed to be the case, and if it is not, the noisier instrument can dominate the the upper limit statistics. To avoid this problem, the combined *SNR* needs to be modified slightly to

$$\varrho_C^2 = \varrho_H^2 + \alpha \varrho_L^2, \tag{17}$$

where α needs to be determined empirically.

Now, let us denote the mean rate at which signals arrive at a set of detectors with *SNR* $\varrho_C > \varrho_C^*$ by $\mathcal{R}$. If we model the arrival of such signals as a Poisson process, then the probability of detecting such a signal within time T is given by

$$P(\varrho_C > \varrho_C^* \,;\, \mathcal{R}) = 1 - e^{\mathcal{R}T\varepsilon(\varrho_C^*)}, \tag{18}$$

where $\varepsilon(\varrho_C^*)$ is the detection efficiency, i.e. the ratio of detected signals to incident signals at the threshold ϱ_C^*. This is not the probability of observing a trigger with $\varrho_C > \varrho_C^*$ however, because it does not account for background triggers. If the probability of having no background event with $\varrho_C > \varrho_C^*$ is P_b, then the probability of observing at least one trigger with $\varrho_C > \varrho_C^*$ is

$$P(\varrho_C > \varrho_C^* \,;\, \mathcal{R}, b) = 1 - P_b \, e^{\mathcal{R}T\varepsilon(\varrho_C^*)}. \tag{19}$$

The loudest event (i.e. the event with the maximum combined *SNR*, $\varrho_{\max}$) sets the scale for the upper limit on the rate. More precisely, to find the 90% frequentist upper limit, one needs to determine the value of $\mathcal{R}$ such that there is a 90% chance that no combined *SNR* would exceed $\varrho_{\max}$ over the course of the run. In other words, one needs to solve $0.9 = P(\varrho_C > \varrho_{\max} \,;\, \mathcal{R}, b)$ for $\mathcal{R}$. Doing so, we find

$$\mathcal{R}_{90\%} = \frac{2.303 + \ln P_b}{T\varepsilon(\varrho_{\max})}. \tag{20}$$

There are two quantities in (20) which need to be ascertained. The first is the efficiency of the detection algorithm to signals with combined *SNR* $\varrho_{\max}$. This is evaluated through Monte Carlo simulations where simulated signals from the target population of neutron star binaries (using the population models discussed in Sect. 2.3) are injected into the data.

The second is the background probability, P_b. The most straightforward approach might be to estimate P_b using the background events resulting from the timeshifts as described in Sect. 4.1. However, these background rates are known to be subject to significant variation depending on the details of search

– e.g. reasonable changes to the event clustering criteria lead to different background rate estimates [23]. Since for detections one would follow up with detailed investigation of coincident triggers anyway, the background estimates are used in a more qualitative manner, and this variation is not an issue for detection. For an upper limit, however, one needs a quantitative result for the background rates, and if there is uncertainty in the value obtained, it also must be quantified. Failing to do so could lead to undercoverage, i.e. a 90% upper limit which is below the actual rate that can be inferred from the data. Undercoverage is considered a "cardinal sin" in frequentist analyses.

Therefore, rather than try to get a quantitative estimate, the standard practice is to simply use $P_b = 1$ in (20). Note that this maximizes $\mathcal{R}_{90\%}$ with respect to P_b. It therefore gives a conservative upper limit – the actual 90% confidence rate is certainly lower. While undesirable, this is considered a "venial sin" in frequentist statistics, and therefore far preferable to undercoverage. Furthermore, since P_b is the probability that no background coincidences will occur with *SNR* above $\varrho_{\max}$, and since $\varrho_{\max}$ is the *SNR* of the loudest background coincidence that actually did occur, it is statistically unlikely that the actual $\mathcal{R}_{90\%}$ would be very much below the one obtained by this method. This statistic is known as the *loudest event statistic*, and is discussed at some length in [91].

Finally, *BNS* rate limits are typically quoted in units of "per Milky Way equivalent galaxy (*MWEG*) per year". This results from expressing the efficiency in terms of effective number of Milky Way equivalent galaxies to which the search was sensitive, N_G. The conversion between $\varepsilon(\varrho_{\max})$ and N_G is

$$N_G := \varepsilon(\varrho_{\max}) \left(\frac{L_{\text{pop}}}{L_G} \right), \tag{21}$$

where L_{pop} is the effective blue-light luminosity of the target population and $L_G = 9 \times 10^9 \, L_\odot$ is the blue-light luminosity of the Milky Way galaxy. In terms of N_G, the 90% frequentist rate upper limit is written

$$\mathcal{R}_{90\%} = 2.303 \times \left(\frac{1\,\text{yr}}{T} \right) \times \left(\frac{1}{N_G} \right) \text{yr}^{-1} MWEG^{-1}, \tag{22}$$

where T has units of years and $P_b = 1$ has been used. This, then, is the most common form of the upper limit that is quoted in the literature, and in this article.

4.3 Results

To date, there have been five published searches for *BNS* coalescence by large scale (>100 m) interferometric detectors[6]. The first was performed by *TAMA*

[6] One other was done using *TAMA300* and *LISM*, a 20 m prototype interferometer [24] – we will not consider this one here since we feel that the results are essentially subsumed by the *TAMA* 2000-2004 result.

Table 2. Results from five searches for neutron star binaries involving data from large scale (>100 m) interferometric gravitational wave detectors. T is the observation time used for setting the upper limit, i.e. the total observation time minus playground data set length minus time periods lost due to vetoes. In the first and last entry, *TAMA* did not convert their efficiency to N_G. The results quoted in these rows, therefore, are not in units of $MWEG^{-1}$ yr^{-1}, but rather in "per population observed yr^{-1}". Note that where N_G is used, the quoted rates are those from the lower bound for N_G, thus giving the most conservative upper limit.

Data Set	T (hrs)	N_G	$\mathcal{R}_{90\%}$ (MWEG^{-1}yr^{-1})
TAMA DT2	6	-	5170
LIGO S1	236	$0.60^{+0.12}_{-0.10}$	170
LIGO S2	339	$1.34^{+0.06}_{-0.07}$	47
LIGO S2/*TAMA* DT8	584	$0.76^{+0.05}_{-0.06}$	49
TAMA 2000-2004	2075	-	20

with a single instrument [26] on the 1999 DT2 (Data Taking 2) data. Subsequent multi-detector searches were performed by *LIGO* based on S1 [25] and S2 [23] data and jointly by *LIGO* and *TAMA* [20]. Most recently, *TAMA* has analyzed cumulatively data taken over many Data Taking runs between 2000 and 2004. No detections have been claimed for any analysis thus far. The results therefore have all been observational upper limits on coalescence rates. They are summarized in Table 2.

All of these rates are substantially above the theoretically predicted rates for these searches [92]. Nonetheless, these are the best direct observational limits on neutron star binaries to date. Furthermore, even at these sensitivities, given how little is known about these gravitational wave sources, there is a real chance, however marginal, of a serendipitous discovery. Finally, these studies were used as a testing ground to develop the analysis tools and methodologies that will maximize the possibility of discoveries in future analyses.

5 Future Prospects

5.1 Interferometers Now and Future

As mentioned in the introduction, to date (Spring 2007), *LIGO* has completed four science runs, S1 from 23 Aug–9 Sep 2002, S2 from 14 Feb–14 Apr 2003, S3 from 31 Oct 2003–9 Jan 2004, and S4 22 Feb– 23 Mar 2005. These short science runs were interspersed with periods of commissioning and have yielded dramatic improvements in sensitivity (see Fig. 3). *LIGO* is now operating with its design sensitivity. The fifth science run, S5, began in Nov 2005, and has the goal of collecting a full-year of coincident data at the *LIGO* design sensitivity. This goal is expected to be achieved by late summer 2007 [62]. The scientific reach of S5 – in terms of the product of the volume of the Universe surveyed times the duration of the data sample – will be more than two orders of magnitude greater than the previous searches.

A period of commissioning following the S5 run will hopefully improve the sensitivity of *LIGO* by a factor of ∼2 [61,62]. An anticipated S6 run with the goal of collecting a year of data with this "enhanced" *LIGO* interferometer is projected to commence in 2009. If S6 sensitivity is indeed doubled, enhanced *LIGO* will survey a volume eight times as great as the current *LIGO* sensitivity.

Following S6, in 2011, the *LIGO* interferometers are slated for decommissioning in order to install advanced interferometers [62, 93]. These advanced *LIGO* interferometers are expected to operate with 10 times the current sensitivity (a factor of 1000 increase in the volume of the Universe surveyed) by ∼2014.

In addition to *LIGO*, the Virgo and *GEO600* observatories are also participating in the S5 science run and are expected to undergo upgrades that will improve their sensitivity along with *LIGO*. In Japan, *TAMA* will hopefully be replaced by the Large Scale Cryogenic Gravitational-wave Telescope (*LCGT*), a subterranean cryogenic observatory with similar performance to advanced *LIGO* [62]. Having several detectors operating at the time of a gravitational wave event will allow better determination of an observed system's parameters.

5.2 Future Reach and Expected Rates

It is possible, even before doing analyses, to get estimates of how instruments will perform in future analyses given their noise curves. For instance, since S2 a sequence of improvements to *LIGO*'s sensitivity have been made (see Fig. 3) that have resulted in an order of magnitude increase in the range to which it is sensitive to neutron star binary inspiral. Galaxy catalogs can be used to enumerate the nearby galaxies in which *BNS* inspirals could be detected. The relative contribution to the overall rate of *BNS* inspirals for each galaxy is determined by its relative blue light luminosity compared to that of the Milky Way. Monte Carlo methods are used to determine the fraction of *BNS* signals that would be detectable from each galaxy during a particular science run. As described above, this procedure yields a figure for sensitivity during any science run: the number of Milky Way Equivalent Galaxies (*MWEG*) that are visible.

In the case of the second *LIGO* science run, S2, the search was sensitive to 1.34 *MWEG*. The S2 run produced a total of 339 hours (around 0.04 years) of analyzed data. The inverse of the product of the number of galaxies to which a search was sensitive times the livetime is a measure of the scientific reach of the search. For S2 this was $\sim 2\times 10^7\,\mathrm{Myr}^{-1}\,MWEG^{-1}$. For initial *LIGO* sensitivity (which has been achieved during the current S5 science run) Nutzman et al. [94] estimate the effective number of *MWEG* surveyed to be

$\sim$600 $MWEG$, though this number depends on the actual sensitivity achieved during the search. Assuming that S5 produces one year of analyzed data, the scientific reach of this search is $\sim$2000 Myr^{-1} $MWEG^{-1}$. These numbers should be compared to estimated *BNS* merger rates in the Milky Way. For example, the "reference model" (model 6) of Kalogera et al. [92] quotes a Galactic rate of between $\sim$20 and $\sim$300 Myr^{-1} $MWEG^{-1}$ with a most likely rate of $\sim$80 Myr^{-1} $MWEG^{-1}$.

Extrapolating to the future, the enhancements that are expected to double the range of *LIGO* before the sixth science run could increase the scientific reach of *LIGO* to $\sim$200 Myr^{-1} $MWEG^{-1}$, which would now begin to probe the expected range of *BNS* coalescence rates. Since the Advanced *LIGO* is expected to improve the sensitivity by a factor of 10 compared to the current sensitivity, given a few years of operation a scientific reach of $\sim$1 Myr^{-1} $MWEG^{-1}$ should be achieved. It is therefore likely that detection of *BNS* inspirals will become routine during the operation of Advanced *LIGO*.

5.3 BNS Astrophysics with GWs

As discussed above, we expect that future *GW* observations will begin to probe the interesting range of *BNS* inspiral rates over the next several years; when advanced detectors like Advanced *LIGO* begin running at the anticipated sensitivity, we expect *BNS* inspirals to be routinely detected, which will give a direct measurement of the true *BNS* merger rate as well as observed properties (such as the distribution of masses of the companions) of the population. Such constraints on the population of *BNS* can then be compared to the predictions from population synthesis models, which then can give insight into various aspects of the evolution of binary stars [71].

Design differences between current and future interferometers might also open new avenues of investigation. One intriguing possibility involves making measurements of neutron star equations of state using *GW* observations of *BNS* mergers. It is possible that advanced interferometers will be able to be tuned to optimize the sensitivity for a particular frequency band (while sacrificing sensitivity outside that band). This is already being discussed as a feature for Advanced *LIGO*. This raises the possibility of tuning one of the *LIGO* interferometers to be most sensitive to gravitational waves at high frequencies, around 1 kHz, where effects due to the size of the neutron stars is expected to be imprinted in the gravitational waveform. Advanced interferometers could then make a direct measurement of the ratio of neutron star mass to radius and thereby constrain the possible neutron star equations of state [95].

Finally, recent evidence suggests that short hard gamma-ray bursts (*GRB*s) could be associated with *BNS* mergers or neutron-star/black-hole binaries. Once *GW* interferometers are sufficiently sensitive, a search for inspiral waveforms in conjunction with observed *GRB*s could confirm or refute the role of binaries as *GRB* progenitors if a nearby short *GRB* were identified. If the binary progenitor model is accepted then gravitational wave observations of the associated inspiral could give independent estimates or limits on the distance to the *GRB*. While most *GRB*s will be too distant for us to hope to detect an associated inspiral, a few may occur within the observed range.

6 Concluding Remarks

We have reviewed here the recently published results of searches for gravitational waves from neutron star binaries. No detections have been made in analyses published thus far, but this is hardly surprising given the gap between current sensitivities and those required to reach astrophysically predicted rates. Nonetheless, steady progress is being made in refining instrumentation and analyses in preparation for that time in the not-too-distant future when this gap has been closed.

It should be obvious to the reader that this has hardly been a comprehensive description of any aspect of this search. Indeed, to provide such a description would require at least an entire volume in itself. Notably, our description of the mathematical theory of gravitational waves, the theoretical and computational underpinnings of population synthesis modeling and our discussion of interferometric design were all sketchy at best. Nor have we delved at all into the important topics of error analysis and pipeline validation.

Nonetheless, we have attempted to provide a birds-eye view of the relevant aspects of searches for gravitational waves from coalescing neutron star binaries with enough references to the literature to provide a starting point for readers interested in any particular aspect. And there is good reason to believe that, in time, gravitational wave searches will become increasingly interesting and relevant to a ever broadening group of astronomers and astrophysicists outside the gravitational wave community. Already, viable (if somewhat marginal) theoretical models have been constrained [10], and greater volumes of more sensitive data are in hand. It seems that it is only a matter of time before GW detectors begins seeing the Universe through gravitational waves. We look forward to the opportunity to review the findings when they do.

7 Acknowledgments

We would like to thank Patrick Brady, Duncan Brown, Matt Evans, Gabriela Gonzáles, Patrick Sutton and Alan Weinstein for their timely readings and comments on this manuscript. Our understanding and appreciation of gravitational wave physics has been enriched by our participation in the *LIGO* Scientific Collaboration, for which we are grateful. The writing of this article was supported by NSF grant PHY-0200852 from the National Science Foundation. This document has been assigned *LIGO* Document number P070053-02.

References

1. K. S. Thorne: in *300 Years of Gravitation*, ed by S. W. Hawking, W. Israel (Cambridge University Press, 1987).
2. P. R. Saulson: *Fundamentals of Interferometric Gravitational Wave Detectors* (World Scientic, Singapore 1994).
3. B. Abbott et al. (LIGO Scientific Collaboration), M. Kramer, A. G. Lyne: *Upper limits on gravitational wave emission from 78 radio pulsars* (2007), gr-qc/0702039.
4. B. Abbott et al. (LIGO Scientific Collaboration): *Coherent searches for periodic gravitational waves from unknown isolated sources and Scorpius X-1: Results from the second LIGO science run* (2006), gr-qc/0605028.
5. B. Abbott et al. (LIGO Scientific Collaboration): Phys. Rev.D **72**, 102004 (2005).
6. B. Abbott et al. (LIGO Scientific Collaboration): Phys. Rev. Lett. **94**, 181103 (2005).
7. B. Abbott et al. (LIGO Scientific Collaboration): Phys. Rev. D **69**, 082004 (2004).
8. B. Abbott et al. (LIGO Scientific Collaboration and Allegro Collaboration): *First cross-correlation analysis of interferometric and resonant-bar gravitational-wave data for stochastic backgrounds* (2007), gr-qc/0703068.
9. B. Abbott et al. (LIGO Scientific Collaboration): *Upper limit map of a background of gravitational waves* (2007), astro-ph/0703234.
10. B. Abbott et al. (LIGO Scientific Collaboration): *Searching for a stochastic background of gravitational waves with LIGO* (2006), astro-ph/0608606.
11. B. Abbott et al. (LIGO Scientific Collaboration): Phys. Rev. Lett. **95**, 221101 (2005).
12. B. Abbott et al. (LIGO Scientific Collaboration): Phys. Rev. D **69**, 122004 (2004).
13. B. Abbott et al. (LIGO Scientific Collaboration): *Search for gravitational wave radiation associated with the pulsating tail of the SGR 1806-20 hyperflare of 27 December 2004 using LIGO* (2007),astro-ph/0703419.
14. B. Abbott et al. (LIGO Scientific Collaboration): Class. Quantum Grav. **23**, S29-S39 (2006).
15. B. Abbott et al. (LIGO Scientific Collaboration), T. Akutsu et al. (TAMA Collaboration): Phys. Rev. D **72**, 122004 (2005).
16. B. Abbott et al. (LIGO Scientific Collaboration): Phys. Rev. D **72**, 062001 (2005).
17. B. Abbott et al. (LIGO Scientific Collaboration): Phys. Rev. D **72**, 042002 (2005).
18. M. Ando et al. (the TAMA collaboration): Phys. Rev. D **71**, 082002 (2005).
19. B. Abbott et al. (LIGO Scientific Collaboration): Phys. Rev. D **69**, 102001 (2004).
20. B. Abbott et al. (LIGO Scientific Collaboration), T. Akutsu et al. (TAMA Collaboration): Phys. Rev. D **73**, 102002 (2006).
21. B. Abbott et al. (LIGO Scientific Collaboration): Phys. Rev. D **73**, 062001 (2006).
22. B. Abbott et al. (LIGO Scientific Collaboration): Phys. Rev. D **72**, 082002 (2005).

23. B. Abbott et al. (LIGO Scientific Collaboration): Phys. Rev. D **72**, 082001 (2005).
24. H. Takahashi et al. (TAMA Collaboration and LISM Collaboration): Phys. Rev. D **70**, 042003 (2004).
25. B. Abbott et al. (LIGO Scientific Collaboration): Phys. Rev. D **69**, 122001 (2004).
26. H. Tagoshi et al. (TAMA Collaboration): Phys. Rev. D **63**, 062001 (2001).
27. A. Einstein: Sitzungsberichte der Königlich Preussischen Akademie der Wissenschaften Berlin, 688696 (1916).
28. A. Einstein: Sitzungsberichte der Königlich Preussischen Akademie der Wissenschaften Berlin, 154167 (1918).
29. J. H. Taylor, J. M. Weisberg: Astrophys. J. **253**, 908 (1982).
30. J. H. Taylor, J. M. Weisberg: Astrophys. J. **345**, 434 (1989).
31. J. M. Weisberg, J. H. Taylor: in *Radio Pulsars*, ed by M. Bailes, D. J. Nice, S. Thorsett (ASP. Conf. Series, 2003).
32. J. Giaime, P. Saha, D. Shoemaker, L. Sievers: Rev. Sci. Inst. **67**, 208 (1996).
33. J. Giaime et al: Rev. Sci. Inst. **74**, 218 (2003).
34. S. Chandrasekhar: MNRAS **91**, 456 (1931).
35. S. Chandrasekhar: MNRAS **95**, 207 (1935).
36. J. L. Provencal, H. L. Shipman, E. Høg, P. Thejll: Astrophys. J. **494**, 759 (1998).
37. B. F. Schutz: *A First Course in General Relativity* (Cambridge University Press, 1985).
38. C. Cutler et al: Phys. Rev. Lett. **70**, 2984 (1993).
39. L. Blanchet, T. Damour, B. R. Iyer, C. M. Will, A. G. Wiseman: Phys. Rev. Lett. **74**, 3515 (1995).
40. L. Blanchet, B. R. Iyer, C. M. Will, A. G. Wiseman: Class. Quantum Grav. **13**, 575 (1996).
41. T. Damour, B. R. Iyer, B. S. Sathyaprakash: Phys. Rev. D **63**, 044023 (2001).
42. I. H. Stairs: Science **304**, 547 (2004).
43. J. Weber: Phys. Rev. Lett. **20**, 1307 (1968).
44. J. Weber: Phys. Rev. Lett. **22**, 1320 (1969).
45. Z. A. Allen et al. (International Gravitational Event Collaboration): Phys. Rev. Lett. **85**, 5046 (2000).
46. F. A. E. Pirani: Acta Physica Polonica **15**, 389 (1956).
47. M. E. Gertsenshtein, V. I. Pustovoit: Sov. Phys. JETP **14**, 433 (1962).
48. J. Weber: unpublished.
49. R. Weiss: Quarterly Progress Report of the Research Laboratory of Electronics of the Massachusetts Institute of Technology **105**, 54 (1972).
50. G. E. Moss, L. R. Miller, R. L. Forward: Appl. Opt. **10**, 2495 (1971).
51. R.L. Forward: Phys. Rev. D **17**, 379 (1978).
52. A. Abramovici et al: Science **256**, 325 (1992).
53. B. C. Barish, R. Weiss: Phys. Today **52** (Oct), 44 (1999).
54. See `http://www.ligo-wa.caltech.edu/`.
55. See `http://www.ligo-la.caltech.edu/`.
56. F. Acernese et al. (Virgo Collaboration): Class. Quantum Grav. **19**, 1421 (2002).
57. B. Willke et al. (GEO): Class. Quantum Grav. **19**, 1377 (2002).
58. H. Tagoshi et al. (TAMA): Phys. Rev. D **63**, 062001 (2001).
59. See `http://lisa.nasa.gov/`.
60. See `http://www.esa.int/esaSC/120376_index_0_m.html`.

61. R. Adhikari, P. Fritschel, S. Waldman: *Enhanced LIGO*, LIGO-T060156-01-I. http://www.ligo.caltech.edu/docs/T/T060156-01.pdf.
62. Jay Marx: *Overview of LIGO*, G060579-00-A, presented at the *23rd Texas Symposium on Relativistic Astrophysics.* http://www.texas06.com/files/presentations/Texas-Marx.pdf.
63. L. S. Finn: in *Astrophysical Sources For Ground-Based Gravitational Wave Dectectors*, ed by J. M. Centrella (Amer. Inst. Phys., Melville, N.Y., 2001), gr-qc/0104042.
64. R. A. Hulse: Rev. Mod. Phys. **66**, 699 (1994).
65. J. H. Taylor: Rev. Mod. Phys. **66**, 711 (1994).
66. V. Kalogera, K. Belczynski, C. Kim, R. OShaughnessy, B. Willems: *Formation of Double Compact Objects*, (2006) astro-ph/0612144.
67. V. Kalogera, R. Narayan, D. N. Spergel, J. H. Taylor: Astrophys. J. **556**, 340 (1991).
68. E. S. Phinney: Astrophys. J. **380**, L17 (1991).
69. K. Belczynski, V. Kalogera, T. Bulik: Astrophys. J. **572**, 407 (2002).
70. K. Belczynski et al: *Compact Object Modeling with the StarTrack Population Synthesis Code*, 2005 astro-ph/0511811.
71. R. O'Shaughnessy, C. Kim, V. Kalogera, K. Belczynski: *Constraining population synthesis models via observations of compact-object binaries and supernovae*, (2006) astro-ph/0610076.
72. P. C. Peters: Phys. Rev. **136**, B1224 (1964).
73. T. A. Apostolatos: Phys. Rev. D **52**, 605 (1995).
74. L. Bildsten, C. Cutler: Astrophys. J. **400**, 175 (1992).
75. B. Abbott et al. (LIGO Scientic Collaboration): Nucl. Instrum. Methods **A517**, 154 (2004).
76. P. Fritschel et al: Appl. Opt. **37**, 6734 (1998).
77. P. Fritschel et al: Appl. Opt. **40**, 4988 (2001).
78. R. Adhikari, G. González, M. Landry, B. OReilly: Class. Quantum Grav. **20**, S903 (2003).
79. L. A. Wainstein, V. D. Zubakov: *Extraction of signals from noise* (Prentice-Hall, Englewood Cliffs, NJ, 1962).
80. B. S. Sathyaprakash, S. V. Dhurandhar: Phys. Rev. D **44**, 3819 (1991) .
81. S. Droz, D. J. Knapp, E. Poisson, B. J. Owen: Phys. Rev. D **59**, 124016 (1999).
82. B. J. Owen: Phys. Rev. D **53**, 6749 (1996).
83. B. J. Owen, B. S. Sathyaprakash: Phys. Rev. D **60**, 022002 (1999).
84. B. Allen: Phys. Rev. D **71**, 062001 (2005).
85. E. Amaldi et al: Astron. Astrophys. **216**, 325 (1989)
86. P. Astone et al: Phys. Rev. D **59**, 122001 (1999).
87. N. Christensen, P. Shawhan, G. González: Class. Quantum Grav. **21**, S1747 (2004).
88. B. Allen, W. G. Anderson, P. R. Brady, D. A. Brown, J. D. E. Creighton: *FINDCHIRP: An algorithm for detection of gravitational waves from inspiraling compact binaries*, (2005) gr-qc/0509116.
89. D. A. Brown et al: Class. Quantum Grav. **21**, S1625 (2004).
90. D. A. Brown (for the LIGO Scientific Collaboration: Class. Quantum Grav. **22**, S1097 (2005).
91. P. R. Brady, J. D. E. Creighton, A. G. Wiseman: Class. Quantum Grav. **21**, S1775 (2004).

92. V. Kalogera et al: Astrophys. J. **601**, L179 (2004). [Erratum: ibid. **614**, L137 (2004)].
93. See `http://www.ligo.caltech.edu/advLIGO/`.
94. P. Nutzman, V. Kalogera, L. S. Finn, C. Hendricksen, K. Belczynski: Astrophys. J. **612**, 364 (2004).
95. J. A. Faber, P. Grandclément, F. A. Rasio, Phys. Rev. Lett. **89**, 231102 (2002).

Observations of the Double Pulsar PSR J0737−3039A/B

I.H. Stairs[1], M. Kramer[2], R.N. Manchester[3], M.A. McLaughlin[4], A.G. Lyne[2], R.D. Ferdman[1], M. Burgay[5], D.R. Lorimer[4], A. Possenti[5], N. D'Amico[5,6], J.M. Sarkissian[3], G.B. Hobbs[3], J.E. Reynolds[3], P.C.C. Freire[7], and F. Camilo[8]

[1] Dept. of Physics and Astronomy, University of British Columbia, 6224 Agricultural Road, Vancouver, BC V6T 1Z1, Canada, stairs@astro.ubc.ca
[2] University of Manchester, Jodrell Bank Observatory, Macclesfield, SK11 9DL, UK
[3] Australia Telescope National Facility, CSIRO, P.O. Box 76, Epping NSW 1710, Australia
[4] Department of Physics, West Virginia University, Morgantown, WV 26506, USA
[5] INAF - Osservatorio Astronomica di Cagliari, Loc. Poggio dei Pini, Strada 54, 09012 Capoterra, Italy
[6] Universita' degli Studi di Cagliari, Dipartimento di Fisica, SP Monserrato-Sestu km 0.7, 09042 Monserrato (CA), Italy
[7] NAIC, Arecibo Observatory, HC03 Box 53995, PR 00612, USA
[8] Columbia Astrophysics Laboratory, Columbia University, 550 West 120th Street, New York, NY 10027, USA

The double pulsar J0737−3039A/B consists of a recycled 22-millisecond pulsar and a young 2.7-second pulsar in a highly relativistic 2.4-hour orbit. This system provides an unprecedented laboratory for testing the timing predictions of strong-field gravitational theories and for investigating the interactions of the emissions of the two pulsars through eclipse and profile studies. We review the recent observations of this remarkable object, including the most precise test to date of general relativity in the strong-field regime.

1 Introduction

When a massive star undergoes a supernova explosion at the end of its fuel-burning lifetime, its core of about $1.5\,M_{\odot}$ may contract down to a compact object of about 10 km in radius: a neutron star. Many of these are highly magnetized, and some fraction produce beams of radio waves that impinge on the Earth each time the star spins: these "pulsating" objects (Hewish et al., 1968) are known as "pulsars". The extreme compactness of these stars

E.F. Milone et al. (eds.), Short-Period Binary Stars: Observations, Analyses, and Results, 53–62.

means that they are strongly self-gravitating and produce significant general-relativistic distortions of the space-time in their vicinities. This raises the exciting possibility of using the radio emission from these objects to probe these distortions and verify the predictions of gravitational theories in the strong field. Specifically, this is best done with pulsars in short-period eccentric orbits around other compact objects – second neutron stars, white dwarfs or (perhaps in the future) black holes.

The first double-neutron-star (*DNS*) system was PSR B1913+16, discovered more than 30 years ago (Hulse & Taylor, 1975) and its importance for practical tests of general relativity (*GR*) and other gravitational theories was quickly understood (e.g., Wagoner, 1975; Eardley, 1975; Barker & O'Connell, 1975). Despite the discovery and timing of other, similar systems, such as PSRs B1534+12 (Wolszczan, 1991) and B2127+11C (Prince et al., 1991), B1913+16 has provided the most famous self-consistency tests of *GR* (Taylor & Weisberg, 1989).

A *DNS* system is generally thought to evolve from a binary that contained two initially very massive ($>8\,M_\odot$) stars; see, for example Bhattacharya & van den Heuval (191) for a review. After the first supernova explosion, nuclear evolution of the secondary star would cause it to expand and likely engulf the neutron star, resulting in common-envelope evolution, the expulsion of the secondary's envelope, and the "spinning-up" or "recycling" of the neutron star to a period of tens of milliseconds and a comparatively low magnetic field of $\sim 10^9$ G. The second supernova explosion would then leave behind a neutron star which might be a normal "young" pulsar, with a spin period of about a second and a larger magnetic field of about 10^{12} G. Until recently, only the recycled pulsars were ever observed in *DNS* systems. As young pulsars have much shorter active radio lifetimes than recycled pulsars, it was generally considered unlikely that we would ever have the good fortune to catch both neutron stars as active pulsars, especially once radio beaming fractions were taken into account.

It was therefore a very welcome surprise for the pulsar community in 2003 when first a *DNS* system was discovered with a recycled pulsar (hereafter "A") in an eccentric and very short 2.4-hour orbit (Burgay et al., 2003) and later its companion star was discovered also to be an active radio pulsar (hereafter "B") – bright only at certain orbital phases, but indubitably a young, high-magnetic-field pulsar (Lyne et al., 2004). This pairing of a recycled and young pulsar was perfectly in line with the predictions of the standard evolutionary picture described above, and also opened up entirely new methods of testing the predictions of *GR* (Lyne et al., 2004). The system produced further surprises in the form of strong electromagnetic interactions between the two stars.

2 Interactions in the Double Pulsar System

The two pulsars, separated by less than 3 light-seconds, show strong evidence that the emission of each influences the observed radio waves from the other, permitting the investigation of many plasma physics phenomena. Here we summarize the various effects and their interpretations, and refer the reader to the original papers for full details.

2.1 Eclipses of A

One of the most astonishing features of the double-pulsar system is that its orbit is inclined so nearly parallel to our line of sight that the emission from A is actually extinguished in an eclipse for about 30 seconds when A passes behind B (Lyne et al., 2004). The eclipse is asymmetric in time, with a slow ingress and a steep egress, and shows only a weak dependence on observing frequency, being slightly longer in duration at low frequencies (Kaspi et al., 2004). Detailed investigations of the flux density of A as a function of time show that the eclipse is in fact not smooth nor always total, but is strongly modulated at half the spin period of the B pulsar for most of its overall duration (McLaughlin et al., 2004c, see Figure 1). An elegantly simple geometric model has been put forth to explain these variations (Lyutikov & Thompson, 2005). In brief, B's plasma-filled magnetosphere is represented as dominated by a dipolar component, misaligned with B's spin axis. As B spins, the dipole field lines move in and out of a configuration in which they can intercept A's radio beam and extinguish it via a form of synchrotron absorption. This model leads to estimates of the angles B's spin axis makes to the orbital angular momentum and to its own dipole axis (Lyutikov & Thompson, 2005).

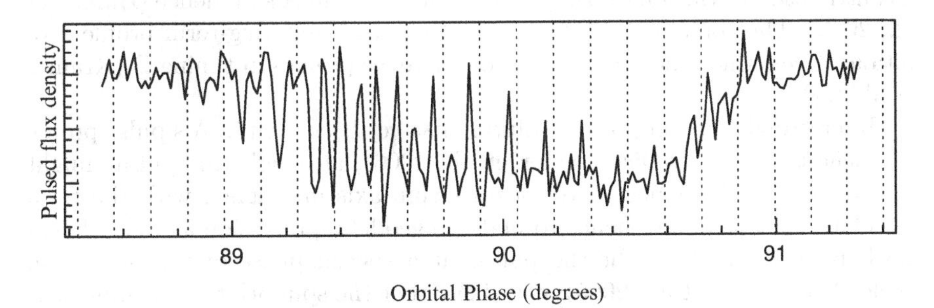

Fig. 1. The pulsed flux density of A in arbitrary units, as a function of orbital phase (true anomaly re-referenced to the ascending node), and averaged over three orbits. The vertical dashed line indicate the arrival times of pulses from B. This plot clearly shows the overall dip in observed A flux due to the eclipse, as well as the modulation of the flux at twice the spin frequency of the B pulsar. Figure adapted from McLaughlin et al., (2004c).

2.2 Orbital-phase and Long-term Modulation of B's Profile

The vast majority of pulsars show very stable "integrated profiles" when several hundred to several thousand pulses are summed. From the start, however, it was clear that B's average profile displayed very different shapes at different orbital phases; moreover the pulsar was only bright at two orbital phases as observed with the 64-m Parkes telescope (Lyne et al., 2004). More sensitive observations with the 100-m Green Bank Telescope (*GBT*) showed that B was in fact visible at all orbital phases, although with much reduced flux (Ransom et al., 2005). There are also severe timing systematics for B (Ransom et al., 2005), particularly during the two bright phases; in fact, in the most recent published timing solution, the brightest orbital phases are left out of the timing analysis altogether (Kramer et al., 2006). The orbital phase modulation is likely due to the effects of A's strong particle wind interacting with the magnetosphere of B. The simple dipole model used to explain the A eclipse (Lyutikov & Thompson, 2005), and distorted based on the geometry of the Earth's magnetosphere, has been able to replicate approximately the orbital phases at which B is bright and dim (Lyutikov, 2005).

B also shows strong profile shape evolution over secular (long-term) timescales (Burgay et al., 2005), with the two bright phases each shrinking in duration, and the average profile shapes also changing. This is likely interpretable as being due to geodetic precession (e.g., Barker & O'Connell, 1975; Weisberg et al., 1989). After the second supernova explosion, it is likely that B's spin axis was misaligned with the total angular momentum vector (this is further supported by the dipole model of the modulation of A's eclipse (Lyutikov & Thompson, 2005)), and its precession about the total angular momentum vector would naturally lead to different observed pulse shapes with time (Burgay et al., 2005). It is however difficult to completely rule out a contribution to the long-term profile evolution from A's influence (Burgay et al., 2005). The combined orbital-phase-dependent and long-term profile evolutions necessitate a detailed matrix of reference profiles in timing B (Kramer et al., 2006).

Interestingly, no long-term evolution has yet been seen in A's pulse profile (Manchester et al., 2005; Kramer et al., 2006), although this pulsar might also reasonably be expected to have its spin axis misaligned with the total angular momentum and hence to undergo geodetic precession as well. Unless we happen to have caught the pulsar at a special phase of the precession cycle (Manchester et al., 2005), it is likely that the spin-orbit misalignment is actually small for this pulsar. This would be consistent with recent modeling of the properties of the supernova explosion that created B (e.g. Willems et al., 2006; Piran & Shaviv, 2005; Stairs et al., 2006).

2.3 Short-term Modulation of B's Profile

The profile variation of B on orbital timescales contains a further short-term direct contribution from the interaction with A: over a small range of orbital phase, individual pulses of B show a "drifting subpulse" phenomenon in which a strong pulse feature migrates earlier in time with each successive B pulse (McLaughlin et al., 2004b). The drift timescale corresponds exactly to the beat frequency of the A and B pulses. These modulations are seen only near one of the orbital phases at which the line between A and B is in the plane of the sky. The physics behind the modulation is presently unclear, but the interaction must be induced by the radiation at the 44 Hz spin frequency of A (McLaughlin et al., 2004b).

2.4 Unpulsed Emission

A comparison of the initial radio observations of J0737−3039A/B with the Parkes single-dish telescope and the Australia Telescope Compact Array interferometer suggested that the combined flux densities of the pulsed emission from A and B could not account for the full flux density observed with the interferometer (Burgay et al., 2003), and that there might therefore be significant unpulsed radio emission, perhaps arising in the region where the winds of the two pulsars collide (Lyne et al., 2004). This unpulsed emission has however not been seen in follow-up observations with the Very Large Array (Chatterjee et al., 2005) and could therefore have been spurious or due to a problem in the data analysis.

Meanwhile, there is excellent evidence for X-ray emission from the system (McLaughlin et al., 2004a; Pellizzoni et al., 2004; Campana et al., 2004). While several models have proposed to account for the existence of X-ray emission (e.g., Zhang & Loeb, 2004; Granot & Mészáros, 2004), the origin of the observed flux is unclear: in particular, there is no evidence yet for modulation at either the orbital period or either of the two pulsar spin periods. More sensitive observations should soon yield a solution to this problem; meanwhile we note that X-ray emissions from the *DNS* PSR B1534+12 show modulation with orbital phase and likely arise from the interaction of the (recycled) pulsar's wind with that of its companion (Kargaltsev et al., 2006).

3 Pulsar Timing and Tests of Strong-field Gravity

Pulsar timing makes use of the stability of integrated pulse profiles: by cross-correlating an observed profile with a high signal-to-noise standard profile, a precise phase offset can be obtained (e.g., Taylor, 1992) and this can be added to the recorded start time of the observation to yield a "Time of Arrival" (*TOA*) of a representative pulse from the observation. Typically these *TOA*s are referenced to an accurate long-term atomic clock standard, such as Coordinated Universal Time or Terrestrial Time. *TOA*s may be transformed from the telescope reference frame to that of the Solar System Barycentre (which is approximately inertial relative to the centre-of-mass frame of the pulsar system) using a good Solar System ephemeris, such as the JPL DE405 standard (Standish, 2004). From this point, pulsar timing consists of fitting a model enumerating every rotation of the the pulsar, accounting for spin frequency and derivatives, astrometric parameters, Keplerian orbital parameters, and any further binary orbit corrections that may be needed. For detailed descriptions of the pulsar timing procedure, consult Lorimer & Kramer (2005) or Edwards et al. (2006).

We have recently published the timing models for PSRs J0737−3039 A and B based on nearly three years of data with the Parkes, Jodrell Bank 76-m Lovell, and *GBT* telescopes (Kramer et al., 2006). Here we present an overview of these results and their implications.

Data were taken with various instruments (filterbanks and coherent dedispersers) at the three telescopes. While Parkes provided the bulk of the data, the highest-precision *TOA*s came from the more sensitive *GBT*. *TOA*s were derived for A at intervals of 30 seconds, a compromise between acceptable signal-to-noise levels and the need to sample the 2.4-hour orbit as finely as possible. As mentioned above, only *TOA*s from the comparatively dim orbital phases of B were used, to minimize the systematics due to profile evolution. Because of the limited orbital phase coverage and poor *TOA* uncertainties of B, only the spin parameters and the projected orbital semi-major axis were fit for this pulsar, with A providing the rest of the parameters.

As in all eccentric binary pulsar systems, five Keplerian parameters are needed to describe the orbit, since the available information is similar to that of a spectroscopic binary. These five parameters are the orbital period P_b, the projected semi-major axis x, the eccentricity e and the longitude ω and time T_0 of periastron passage. Further "post-Keplerian" (*PK*) parameters are then necessary to account for relativistic corrections to the orbit; these include the advance of periastron $\dot{\omega}$, the combined time-dilation and gravitational redshift parameter γ, the orbital period decay rate $\dot{P}_b$ and the range r and shape s of the Shapiro delay. These *PK* parameters are fit in a way that makes no assumptions about the validity of *GR* or any other theory of gravity (Damour & Deruelle, 1985, 1986) and can therefore be used directly for self-consistency tests of any such theory. Finally, because the size of the projected orbit x is measured separately for both A and B, we have the mass ratio R. This parameter is unique to the double pulsar among *DNS* systems, and is furthermore also expected to be the same in all theories of gravity, at least in the first-order strong-field expansions (Damour & Schäfer, 1988; Damour & Taylor, 1992).

The test of any strong-field gravity theory then consists of calculating the dependence of each *PK* parameter (and R) on the two (otherwise unknown) stellar masses, and verifying that the allowed families of solutions all intersect in a common region. If this does not occur, then the theory must be discarded or modified. In *GR*, the dependence of the *PK* parameters (as measured for A) on the masses m_A and m_B are (Damour & Deruelle, 1986; Taylor & Weisberg, 1989; Damour & Taylor, 1992).

$$\dot{\omega} = 3 \left(\frac{P_b}{2\pi}\right)^{-5/3} (T_\odot M)^{2/3} (1-e^2)^{-1}$$

$$\gamma = e \left(\frac{P_b}{2\pi}\right)^{1/3} T_\odot^{2/3} M^{-4/3} m_B (m_A + 2m_B)$$

$$\dot{P}_b = -\frac{192\pi}{5} \left(\frac{P_b}{2\pi}\right)^{-5/3} \left(1 + \frac{73}{24}e^2 + \frac{37}{96}e^4\right) (1-e^2)^{-7/2} T_\odot^{5/3} m_A m_B M^{-1/3}$$

$$r = T_\odot m_B$$

$$s = x \left(\frac{P_b}{2\pi}\right)^{-2/3} T_\odot^{-1/3} M^{2/3} m_B^{-1}$$

where $M = m_A + m_B$ and $T_\odot \equiv GM_\odot/c^3 = 4.925490947\,\mu$s is the mass of the Sun in seconds. In *GR*, we have $s \equiv \sin i$, where i in the inclination angle of the orbit relative to the plane of the sky.

The allowed curves in the m_A–m_B plane for each *PK* parameter and for R are plotted in Figure 2. There is a small region allowed by all the curves simultaneously; this shows us that *GR* does in fact provide an acceptable description of the observed system (Kramer et al., 2006).

The absolute quality of our *GR* test can be derived by considering the most precisely measured parameters. We use our measured values and uncertainties of $\dot{\omega}$ and R (actually x_B) in a Monte Carlo simulation to predict the our best estimates of the two neutron-star masses ($m_A = 1.3381 \pm 0.0007\,M_\odot, m_B = 1.2489 \pm 0.0007\,M_\odot$) and hence all the other *PK* parameters, assuming *GR* is correct. We may then make direct comparisons to our measured values. Our next most precisely measured parameter is the Shapiro delay shape s, and it agrees with the *GR*-predicted value to within 0.05%, incorporating all the 1-σ uncertainties (Kramer et al., 2006). This is the most precise test to date of *GR* in the strong-field regime, an important complement to the more stringent tests conducted in the comparatively weak gravitational field of our Solar System (Bertotti et al., 2003).

We note that in the space of less than three years, we already have a measurement of the orbital period derivative $\dot{P}_b$ at the 1.4% level. This parameter agrees well with its GR-predicted value, and we have estimated that the kinematic corrections necessary to the observed value will only start to become a systematic problem once the measurement precision increases to the 0.02% level (Kramer et al., 2006). Thus this parameter will ultimately provide a very stringent indirect proof of the existence of quadrupolar gravitational radiation (c.f., Taylor & Weisberg, 1989). The system will merge in about 85 million years, and the existence of this relativistic binary has already been used to infer a larger predicted event rate for gravitational-wave observatories (Burgay et al., 2003; Kalogera et al., 2004).

In the future, a precise measurement of $\dot{P}_b$ combined with s has the potential to provide the means to identify a departure of $\dot{\omega}$ from the value predicted in the first-order ("1PN") expansion of the relativistic orbit (Lyne et al., 2004;

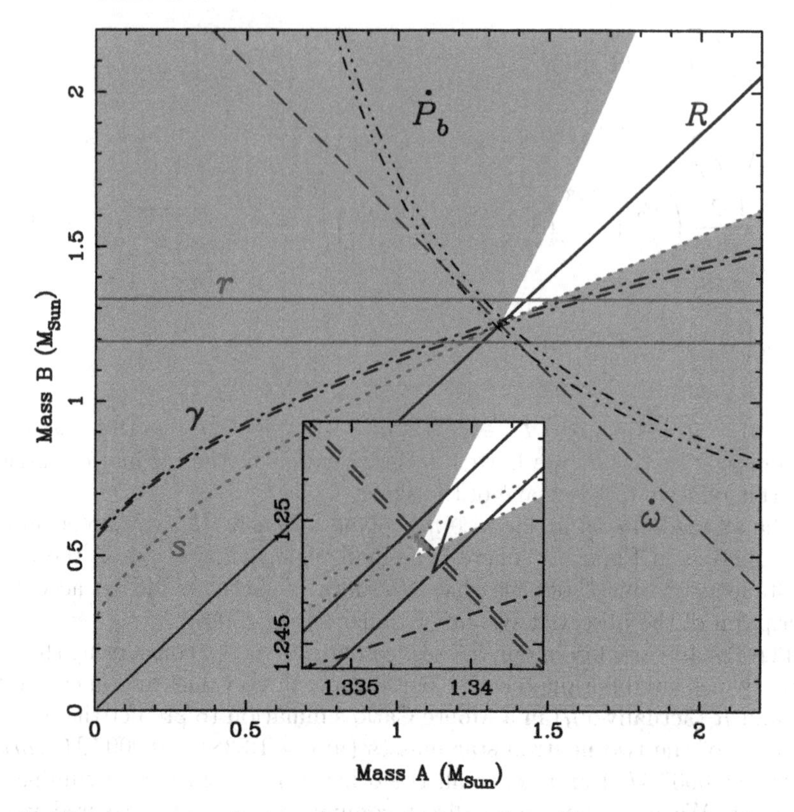

Fig. 2. The mass-mass diagram for the double pulsar. The measured values and uncertainties for each of the post-Keplerian (*PK*) parameters and the mass ratio R have been used to predict the curves indicating the contraints on the two masses according to general relativity (*GR*). All six finite-width curves intersect in one small region, indicating that *GR* satisfies the self-consistency requirement imposed by this system on strong-field theories of gravity. The shaded regions indicate mass-mass parameter space disallowed by the mass functions of the two pulsars and the requirement that the sine of the orbital inclination angle be less than or equal to 1. Figure adapted from (Kramer et al., 2006).

Kramer et al., 2006). The most important second-order contribution is expected to arise from the spin-orbit coupling of the A pulsar to the orbit (this is actually a 1PN effect that enters numerically at the 2PN level (Damour & Schäfer, 1988; Wex, 1997)). Successful identification of this effect would permit the measurement of the moment of inertia of the A pulsar and could provide important constraints on acceptable neutron-star equations of state (see e.g. Lattimer & Prakash, 2001). Further observations of this most exciting system should prove very rewarding indeed.

References

1. Barker, B. M. & O'Connell, R. F. 1975, ApJ, 199, L25
2. Bertotti, B., Iess, L., & Tortora, P. 2003, Nature, 425, 374
3. Bhattacharya, D. & van den Heuvel, E. P. J. 1991, Phys. Rep., 203, 1
4. Burgay, M., et al. 2003, Nature, 426, 531
5. Burgay, M., et al. 2005, ApJ, 624, L113
6. Campana, S., Possenti, A., & Burgay, M. 2004, ApJ, 613, L53
7. Chatterjee, S., Goss, W. M., & Brisken, W. F. 2005, ApJ, 634, L101
8. Damour, T. & Deruelle, N. 1985, Ann. Inst. H. Poincaré (Physique Théorique), 43, 107
9. Damour, T. & Deruelle, N. 1986, Ann. Inst. H. Poincaré (Physique Théorique), 44, 263
10. Damour, T. & Schäfer, G. 1988, Nuovo Cim., 101, 127
11. Damour, T. & Taylor, J. H. 1992, Phys. Rev. D, 45, 1840
12. Eardley, D. M. 1975, ApJ, 196, L59
13. Edwards, R. T., Hobbs, G. B., & Manchester, R. N. 2006, MNRAS, 1152, in press.
14. Granot, J. & Mészáros, P. 2004, ApJ, 609, L17
15. Hewish, A., Bell, S. J., Pilkington, J. D. H., Scott, P. F., & Collins, R. A. 1968, Nature, 217, 709
16. Hulse, R. A. & Taylor, J. H. 1975, ApJ, 195, L51
17. Kalogera, V., et al. 2004, ApJ, 601, L179
18. Kargaltsev, O., Pavlov, G. G., & Garmire, G. P. 2006, ApJ, 646, 1139
19. Kaspi, V. M., Ransom, S. M., Backer, D. C., Ramachandran, R., Demorest, P., Arons, J., & Spitkovskty, A. 2004, ApJ, 613, L137
20. Kramer, M., et al. 2006, Science, 314, 97
21. Lattimer, J. M. & Prakash, M. 2001, ApJ, 550, 426
22. Lorimer, D. R. & Kramer, M. 2005, Handbook of Pulsar Astronomy (Cambridge University Press)
23. Lyne, A. G., et al.2004, Science, 303, 1153
24. Lyutikov, M. 2005, MNRAS, 362, 1078
25. Lyutikov, M. & Thompson, C. 2005, ApJ, 634, 1223
26. Manchester, R. N., et al. 2005, ApJ, 621, L49
27. McLaughlin, M. A., et al. 2004a, ApJ, 605, L41
28. McLaughlin, M. A., et al. 2004b, ApJ, 613, L57
29. McLaughlin, M. A., et al. 2004c, ApJ, 616, L131
30. Pellizzoni, A., De Luca, A., Mereghetti, S., Tiengo, A., Mattana, F., Caraveo, P., Tavani, M., & Bignami, G. F. 2004, ApJ, 612, L49
31. Piran, T. & Shaviv, N. J. 2005, Phys. Rev. Lett., 94, 051102
32. Prince, T. A., Anderson, S. B., Kulkarni, S. R., & Wolszczan, W. 1991, ApJ, 374, L41
33. Ransom, S. M., Demorest, P., Kaspi, V. M., Ramachandran, R., & Backer, D. C. 2005, in Binary Radio Pulsars, ed. F. Rasio & I. H. Stairs (San Francisco: Astronomical Society of the Pacific), 73
34. Stairs, I. H., Thorsett, S. E., Dewey, R. J., Kramer, M., & McPhee, C. A. 2006, MNRAS, 373, L50.
35. Standish, E. M. 2004, A&A, 417, 1165
36. Taylor, J. H. 1992, Philos. Trans. Roy. Soc. London A, 341, 117

37. Taylor, J. H. & Weisberg, J. M. 1989, ApJ, 345, 434
38. Wagoner, R. V. 1975, ApJ, 196, L63
39. Weisberg, J. M., Romani, R. W., & Taylor, J. H. 1989, ApJ, 347, 1030
40. Wex, N. 1997, A&A, 317, 976
41. Willems, B., Kaplan, J., Fragos, T., Kalogera, V., & Belczynski, K. 2006, Phys. Rev. D, 74, 043003
42. Wolszczan, A. 1991, Nature, 350, 688
43. Zhang, B. & Loeb, A. 2004, ApJ, 614, L53

Gravitational Lensing in Compact Binary Systems

David W. Hobill, John Kollar, and Julia Pulwicki

Department of Physics and Astronomy, University of Calgary, Calgary, Alberta, Canada, T2N 1N4
hobill@crag.ucalgary.ca

The theory of microlensing is modified for sources producing pulsed, directional light beams in order to analyze the changes in pulse profiles that arise due to time delays and magnification factors occurring as a result of gravitational lensing by other compact objects. The results are applied to the double pulsar PSR J0737-3039 to determine whether or not anomalous enhancements observed during the eclipse of pulsar A by pulsar B have the signature of gravitational lensing. Predictions regarding the eclipse of pulsar B by pulsar A are also provided.

1 Introduction

Short period binary systems composed of compact stars present a number of challenges to both observational and theoretical astrophysicists. If the systems are composed of very compact stellar objects that have or have nearly completed their evolution, the question of how such systems might have been formed remains unanswered with any certainty. In spite of such an important question, binary systems composed of two mutually bound neutron stars or black holes provide laboratories for testing general relativity and other gravitational theories. The recent discovery of the double pulsar system PSR J0737-3039 consisting of two neutron stars, both of which are seen as pulsars Earth based observers, has provided an unprecedented example of a compact binary system that is able to act as a high precision test of general relativity [4]. With an inclination angle close to 90°, a small eccentricity, and orbital period of only 2.5 hours, this system is the fulfillment of many a relativist's dream.

For binary pulsar systems with a large inclination angle such that the orbital plane is viewed nearly edge on, gravitational lensing of one pulsar's signals by the other becomes a real possibility. An understanding of the effects of the lensing of the pulsar signals by the second compact object in the system

E.F. Milone et al. (eds.), Short-Period Binary Stars: Observations, Analyses, and Results, 63–84.

requires more than just an application of standard "microlensing" formulae [6,7,9] where it is assumed that compact objects might act as gravitational lenses for background sources, both of which are assumed to be found in our local galactic environment.

If the source of photons being lensed is a pulsar then one must recognize that those photons are both pulsed (and therefore have an intensity that varies rapidly in time) and are anisotropically distributed (and therefore must be directed at a very specific angle from the source in order that the beam is able to reach the observer on Earth). The theory of microlensing assumes that the source is radiating continuously and isotropically (or nearly isotropically) in all directions. When one considers that the pulses emitted from the pulsar source are of short duration and are narrow in width, certain modifications must be made to account for this. These modifications will require an analysis of time delays suffered by the beams as well as the changes in both the magnification and pulse profiles that might occur. If the gravitational lensing of such compact systems is observable then the observed pulse profiles (and their changes) might be able to provide independent measurements of the mass of the lensing object and other orbital parameters without having to make use of the orbital measurements themselves.

In this paper we shall review the methods that have been developed to analyze the effects of gravitational lensing by a very compact object such as a black hole or a neutron star where the radius of the star is much less than the distance of closest approach of the photon. Alternatively, for weak gravitational lensing this also means that the stellar radius is much smaller than the impact parameter of the photon orbit because in a first order perturbation analysis of the null geodesics in a curved spacetime the equivalence of the impact parameter and the distance of closest approach is obeyed. This means that at this level we do not consider the spin of the lensing object which itself may have observable consequences if the angular momentum parameter is large or if the multipole structure of the lens is such that it makes significant contributions to the vacuum gravitational field surrounding the lens.

Therefore we will assume that the lens is slowly rotating and that it is also close to being spherically symmetric. In this approximation it can be assumed that the Schwarzschild solution provides a good representation of the spacetime surrounding the lensing companion of the pulsar. This means that the results will be applicable to both black holes and neutron stars with small spin angular momentum. Any effects due to higher rotation rates would require two separate analyses: using the Kerr spacetime if the lens is a rotating black hole and using a Lens-Thirring metric if the lens is a material object with spinning multipole moments which would depend upon a specific equation of state. Such extensions along with the possible effects of electromagnetic fields will be analyzed later and will require at least a second-order perturbation analysis in order to account for such contributions.

This paper will review the modifications of "microlensing" theory that are required in order to account for directional, pulsed light beams from a lensed source. It will be shown that if the lensing object is a black hole with a mass on the order of tens of solar masses, significant distortions of the pulse profiles may arise due to the combined effects arising from magnification factors and pulse arrival times.

Using the measurements of the orbital parameters of the double pulsar system, the method will be applied to that particular system to determine whether or not the gravitational lensing by one of the neutron stars of the pulsar beam of the other is observable. While currently the "eclipse" of pulsar A by pulsar B has been observed, the application presented will also analyze the gravitational lensing of pulsar B by pulsar A. In both cases the actual pulse profiles associated with each pulsar will be used to make such predictions.

Strictly speaking, for an orbital inclination of close to 90°, a binary system consisting of a pulsar source with a black hole or neutron star where both objects are considered "clean" (i.e. without material or fields extending outside of their surfaces) one would not expect any eclipsing to occur since there will always be a beam that is bent around the companion in such a way that it will reach the distant observer. However in the case of PSR J0737-3039, it would appear that the atmosphere surrounding pulsar B is of sufficient extent as to provide an attenuation of the light from pulsar A. This too is taken into account in modeling the gravitational lensing that might occur in the double pulsar system.

2 Gravitational Lensing of Pulsar Sources

The assumptions that are made in order to develop a model of gravitational lensing in a compact binary system consisting of at least one observable pulsar are the following:

1. The light rays pass through empty (or nearly empty) space. This allows one to use the equations describing motion along null (zero-mass particle) geodesics of a vacuum spacetime to compute bending angles, time delays and magnification factors for the light beams as they traverse the region from source to observer.

2. The light beams pass by the lens at large distances compared to the size of the source and the deviations from straight line motion are small. This allows one to use weak lensing conditions with small angle approximations and a perturbation analysis where the geodesic equations are expanded in powers of $\ell_0/r \ll 1$ where ℓ_0 is some appropriate scale length (mass of the black hole in geometric units, the radius of the star, or the impact parameter) and r is the instantaneous radial position of the photon.

3. The spin angular momentum of the lens is small, as is its translational motion. This means that $v/c \ll 1$ and $a/r \ll 1$ where v is the translational velocity of the lens, c is the speed of light, a is the spin angular momentum parameter of the star or black hole (again in geometric units).

4. Electromagnetic effects are small. That is, there are no residual charges on either the source or lens $q \approx 0$, magnetic fields are small $B^2r^2 \ll M/r$ where B is the magnetic field strength and M is the mass of the lens (again as measured using geometric units).

5. The multipole moments associated with the distributions of mass in the lens are small i.e. the lens is nearly spherically symmetric. This and assumptions 1, 3, and 4 mean that one can use the vacuum Schwarzschild solution to approximate the motion of the photons.

6. The light source produces narrow, pulsed, directional beams of photons. This will require the modifications to the standard microlensing theory [9], where the sources emit isotropically and continuously in time.

In this section the lensing produced by a Schwarzschild-type object will first be reviewed which will then be followed by a introduction of the effects resulting from the addition of the last assumption.

To begin, the geometry of the Schwarzschild lens having a mass M is shown in Figure 1 along with the various angles and distances that will be used to compute the gravitational lensing effects.

The position of the source is labelled by S, the location of the observer is labelled by O, and the location of the lens is labelled L. It will be assumed that the optical axis of the system is defined by the straight line OL connecting the lens with the observer. Two planes perpendicular to the optical axis are then defined by the source position S (called the source plane) and the position of L (the lens plane). The following distance measurements are now introduced: D_s is the distance between the observer and source plane, D_d is the distance

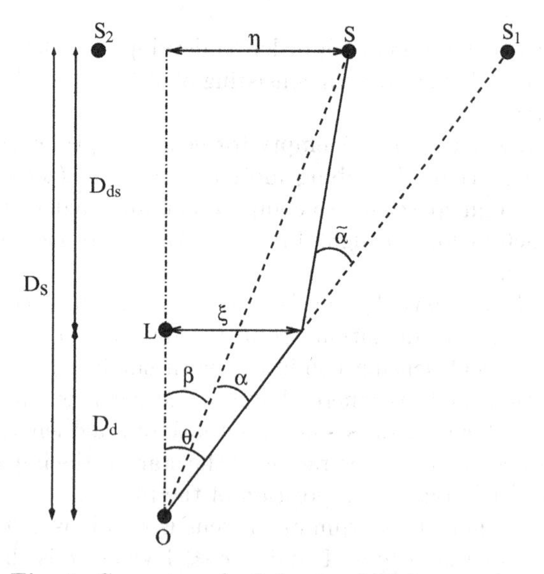

Fig. 1. Geometry of a Schwarzschild Lens system

between the observer and lens plane and D_{ds} is the distance from the lens plane to the source plane. In the simple case of a Euclidean metric $D_s = D_{ds} + D_d$.

One now defines two other distances that are measured orthogonal to the optical axis. The distance η measures the the position of the source with respect to the optical axis and ξ measures the distance of closest approach of the bent light ray as it passes by the lens. Clearly from the definition of the source and lens planes, these measurements are restricted to lie within their respective planes.

In addition, the following angular measurements (assumed to be small) are introduced: β is the angular separation of the source position from the lens, θ measures the angle between the incoming light ray and the line OL (this produces an image of the source located at S_1 in the plane containing the source and perpendicular to OL). The angle $\tilde{\alpha}$ is the the deflection angle for the light ray while the angle $\alpha = \theta - \beta$ is the angular separation of the image from the source.

At conjunction, when $\eta = 0$ ($\beta = 0$) there is no longer a preferred plane defined by OL and OS. This results in a characteristic angular deviation known as the "Einstein angle", α_0, (thereby producing a ring - the "Einstein Ring" - of images) at an angular separation of:

$$\alpha_0 = \sqrt{\frac{4Gm}{c^2}\frac{D_{ds}}{D_d D_s}} \tag{1}$$

from the optical axis. Here G is the Newtonian gravitational constant, m is the physical mass and c is the speed of light). The characteristic length $\xi_0 \approx \alpha_0 D_d$ is the radius if the Einstein ring in the lens plane:

$$\xi_0 = \sqrt{2R_s \frac{D_d D_{ds}}{D_s}}$$

The geometric mass M, the physical mass m, and the Schwarzschild radius of the lens are related by $M = Gm/c^2 = R_s/2$.

Using the relation that: separation = angle × distance, one can easily see that the geometry of Figure 1 leads to the so-called "lens equation",

$$\theta D_s = \beta D_s + \tilde{\alpha} D_{ds} \tag{2}$$

or re-written in terms of angular separations:

$$\beta = \theta - \alpha(\theta) \tag{3}$$

For every angular position of the source β, the geometry of Figure 1 and equation (3) leads to the relation:

$$\beta = \theta - \frac{\alpha_0^2}{\theta}$$

and this has the solutions:

$$\theta_{1,2} = \frac{1}{2}\left(\beta \pm \sqrt{\beta^2 + 4\alpha_0^2}\right) \tag{4}$$

where the subscript '1" refers to the solution with the plus sign i.e. image S_1 in Figure 1 and the and the subscript "2" refers to the solution with the minus sign (image S_2). The two solutions are associated with the light rays that pass on either side of the lens. The angular separation between the images is then determined by

$$\Delta\theta = \theta_1 - \theta_2 = \sqrt{4a_0^2 + \beta^2}$$

which for sources located close to the galactic centre leads to a separation on the order of micro-arcseconds.

In order to simplify the computations to follow, one now defines the source and image angular positions normalized with respect to the Einstein angle:

$$\tilde{\theta} = \theta/\alpha_0 \qquad \tilde{\beta} = \beta/\alpha_0$$

then the relationship equation (4) becomes

$$\tilde{\theta}_{1,2} = \frac{1}{2}\left(\tilde{\beta} \pm \sqrt{\tilde{\beta}^2 + 4}\right). \tag{5}$$

One can use these normalized angles to define the magnification factor associated with each image. This may be determined using a method recognizing that not only does the gravitational field bend light rays but it can also change the cross-sectional area of a bundle of light rays. In what follows it is assumed that the specific intensity I_ν of the light remains constant along a light ray since gravitational bending does not affect the emission or absorption processes. In addition if the lens and source are nearly static then it may be assumed that no frequency shifts occur. Certainly in the case of short period compact binaries where the orbital time scales are on the order of hours, these time scales are long in comparison with the pulsar period, which is on the order of a few seconds or less. Therefore the surface brightness I for a image is equal to the surface brightness of the source. The photon flux of an image produced by an infinitesimal source is simply the product of the surface brightness with the solid angle, $\Delta\Omega$ that the image subtends on the sky.

One therefore defines the magnification factor for an image as the ratio of the flux received from the image to the flux that would be received directly from the source:

$$\mu = \left|\frac{\Delta f}{\Delta f_s}\right| = \left|\frac{I\Delta\Omega}{I\Delta\Omega_S}\right| = \left|\frac{\Delta\Omega}{\Delta\Omega_S}\right| \tag{6}$$

where $\Delta\Omega_S$ denotes the undeflected value of the subtended solid angle of the source. It will be assumed that the source is located at a normalized angular position $\tilde{\beta}$ and subtends a (normalized) solid angle $\Delta\tilde{\Omega}_S \approx \Delta\tilde{\beta}\tilde{\beta}\Delta\varphi$ while the images which are located at angles $\tilde{\theta}_i$ $(i = 1, 2)$ subtend normalized solid angles $\Delta\tilde{\Omega}_i \approx \Delta\tilde{\theta}_i\tilde{\theta}_i\Delta\varphi$. The change in azimuthal angle, $\Delta\varphi$, is the same for both source and images as a result of spherical symmetry. The magnification factors for the images then becomes:

$$\mu_i = \left| \frac{\Delta\tilde{\theta}_i \cdot \tilde{\theta}_i}{\Delta\tilde{\beta} \cdot \tilde{\beta}} \right|$$

Using the expressions for the normalized angles in equation (5) the magnification of each image is then given by:

$$\mu_{1,2} = \frac{1}{4}\left[\frac{\tilde{\beta}}{\sqrt{\tilde{\beta}^2 + 4}} + \frac{\sqrt{\tilde{\beta}^2 + 4}}{\tilde{\beta}} \pm 2\right] \tag{7}$$

which obeys the relation

$$\mu_1 - \mu_2 = 1. \tag{8}$$

If the source is emitting continuously and isotropically in all directions, then the total magnification factor is the sum of the two individual expressions in equation (7) and this forms the basis for the search for dark compact objects (*MACHOS*, black holes, etc.). The idea is that brightness enhancements of luminous sources will result from lensing by dark compact objects with gravitational fields large enough to produce magnification factors many times larger than unity.

However for pulsed, directional beams such as those emitted by pulsar sources, the finite time interval during which the beam can be seen by the observer must be taken into account. The magnification factors are then time dependent. If the arrival times for the two beams do not occur simultaneously then the total magnification is not simply the sum of the two factors. Delays in arrival times will occur due to path length differences, Shapiro delay differences and differences in the initial beam directions for each image. The pulse profiles are also time dependent and each of these effects need to be analyzed properly in order to compute the time dependent behaviour of each magnification factor and their contributions to the total overall magnification.

Such a situation requires a modification of the microlensing magnification equation that takes into account the pulse profile and the Shapiro delays suffered by each image. While the source is close to superior conjunction, situations can arise where the delay between the pulses is sufficient to prevent them from overlapping. In this case the pulse profile, the magnification of each beam and the delay between the pulses must be modeled accurately in order to obtain a more realistic prediction of the effect of the lens on the overall pulse profiles. In addition, as stated above, the directional property of the pulsar beam must also be taken into account. The rotational period of the pulsar will also contribute to the overall delay in arrival times since the orientation of the initial direction of the two beams does not occur simultaneously but requires some fraction of the pulsar period so that the two beams are positioned correctly in order for the pulses to reach the Earth bound observer.

2.1 Microlensing for Orbiting Pulsar Sources

In order to develop a method for the analysis of the gravitational lensing of pulsar beams, we need to introduce some new parameters (based upon the both orbital motions of the pulsar and the pulsar rotational period). These parameters then will be used to describe the the time dependent position of the pulsar in the source plane and the orientation of the beams associated with the creation of the two lensed images. By relating these parameters to the source angular position β, one can then return to standard lensing theory to reconstruct image positions, time delays and magnification factors. It must be remembered that each image requires a specific initial alignment of the pulsar beam in order for it to reach the observer. Since two different alignments are required for the two lensed images, a "geometric time delay" will occur as the beam rotates from one initial angle to the other. This time delay required for the pulsar beam to change from one direction to the other will simply be a fraction of the total pulsar rotation period and will be added to the difference in the Shapiro and travel time delays associated with each beam.

The geometry associated with the motion of a pulsar in orbit around a compact lensing object is shown in Figure 2. The notation used in Figure 2 is the same as that introduced in Figure 1 but now the orbit of the pulsar (shown here as circular but higher eccentricity orbits can also be analyzed) is included along with the angles that will be used to locate the position of the pulsar with respect to the lens plane.

Assuming that the portion of the pulsar orbit of interest forms part of a circular arc with a radius R, an angle ψ is introduced and this will measure the pulsar position with respect the lens. Here $\psi = 0$ is defined to be the angular position of the source when it is on the optical axis OL. The angle ψ has a time dependence determined by the orbital period of the source. Since we are only interested in measurements made over short time scales, this approximation will be adequate over a few pulse periods. However if the orbit is significantly elliptic with a relativistic shift in periastron, then a more complicated mapping of the orbital radial and angular positions to the lensing source angle β and source position D_s will be required, especially if the lensing observations are to take place over many orbits of the pulsar.

The angles γ_1 and γ_2 which represent the specific directions of the primary and secondary pulsar beams (measured with respect to the radial vector pointing from the lens position to the position of the source) required for the beam to create the images S_1 and S_2 as seen by the observer. These are shown in Figure 3 which represents a detailed picture of the region between the source located at the left vertex and the lens. The sum of these two angles $\gamma = \gamma_1 + \gamma_2$ yields the angle between the two beams that produce the images S_1 and S_2. The time for the the pulsar to rotate through this angle gives the "geometric time delay" between the two beams.

The small angle deviation requirements can be translated into lower limits on the distances of closest approach:

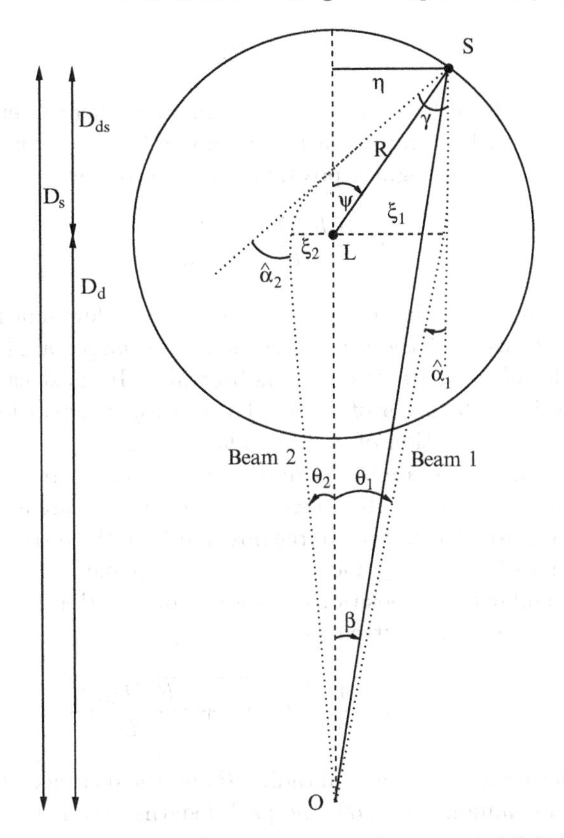

Fig. 2. Geometry of a compact binary system with a central Schwarzschild Lens

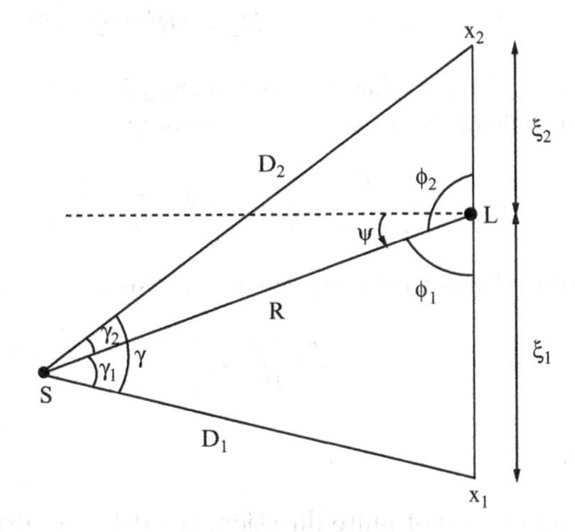

Fig. 3. Detailed geometry of a pulsar source which emits two light beams that pass by the gravitational lens at distances $\xi_{1,2}$.

$$\xi_{1,2} \geq 10R_s$$

which is equivalent to a "weak-field" approximation where ξ remains outside of the region defined by a radius of ten Schwarzschild horizon radii. This in turn places limits on the angular position ψ of the source:

$$\psi \leq \frac{10R_s D_s}{D_d R} \pm \frac{1}{5}\frac{D_{ds}}{R}$$

In practice the angle ψ of interest will never exceed a few tenths of a radian since the magnification of the secondary image for larger angles is to small (i.e. on the order of $\mu_2 < 10^{-6}$) to be unobservable. In most cases where the mass of the lens is on the order of a few solar masses, ψ .005 or less is required for a significant magnification of the secondary image.

Having introduced the angles γ_1 and γ_2 these define (to first order) the distances D_1 and D_2 respectively. These distances (see Figure 3) are the paths the light rays take from the pulsar source position S, to the respective of closest approach positions (x_1 and x_2) located in the lens plane.

From the angular image positions $\theta_{1,2}$ given in equation (4) the distances of closest approach can be written as:

$$\xi_{1,2} = \frac{D_d}{2}\beta \pm \frac{1}{2}\sqrt{(D_d\beta)^2 + 8\frac{R_s D_{ds} D_d}{D_s}} \tag{9}$$

Since these together with the orbital radius R and the distances $D_{1,2}$ form two triangles with a common edge (R), the path lengths traveled by the photons can be computed easily:

$$D_1 = (R^2 + \xi_1^2 - 2\xi_1 R \sin\psi)^{1/2} \qquad D_2 = (R^2 + \xi_2^2 + 2\xi_2 R \sin\psi)^{1/2}$$

Finally the "cosine law" for the large triangle $x_1 S x_2$ formed by the two smaller ones discussed above leads to an evaluation of $\cos\gamma$:

$$\cos\gamma = \frac{1}{2}\left[\frac{D_1}{D_2} + \frac{D_2}{D_1} - \frac{(\xi_1 + \xi_2)^2}{D_1 D_2}\right] \tag{10}$$

which to lowest order leads to the expected small angle result:

$$\gamma \approx \frac{(\xi_1 + \xi_2)}{R}. \tag{11}$$

2.2 Time Delays

Since the pulsar signals are of finite duration, the delay resulting in different arrival times for the two pulses needs to be computed. These will be used to compute a predicted pulse profile during conjunction. The are a number of contributions to the differences in arrival times that must be considered when computing the overall time delays. These are:

[**1**.] Differences in the path lengths traversed by the two pulses.

The Euclidean path lengths of rays that form the images S_1 and S_2 are respectively

$$ct_1 = \ell_1 = D_1 + (D_d^2 + \xi_1^2)^{1/2} \tag{12}$$

and

$$ct_2 = \ell_2 = D_2 + (D_d^2 + \xi_2^2)^{1/2} \tag{13}$$

These distances simply represent the flat space distance from the source to the closest approach positions added to the distance from the closest approach positions to the observer position.

[**2**.] Differences in the gravitational Shapiro Delays.

These delays are due to the photons passing through the gravitational potential of the lens and account for the first order gravitational curvature. They are therefore computed to first order in the geometric mass $M = Gm/c^2$ for a Schwarzschild spacetime and are given by the expressions:

$$ct_{S1} = 2M \ln\{[D_1 + R][(D_d^2 + \xi_1^2)^{1/2} + D_d]/\xi_1^2]\} - M[D_1/R + (D_d^2 + \xi_1^2)^{1/2}/D_d] \tag{14}$$

and

$$ct_{S2} = 2M \ln\{[D_2 + R][(D_d^2 + \xi_2^2)^{1/2} + D_d]/\xi_2^2]\} - M[D_2/R + (D_d^2 + \xi_2^2)^{1/2}/D_d] \tag{15}$$

The time delay for the beam with the smallest distance of closest approach will be larger of the two since to this level of approximation the difference between D_1 and D_2 is negligible.

[**3**.] Differences in the initial angular orientation of the pulse.

Finally there is a delay associated with the difference in orientation of the beam at the source position so that it forms the two images on the sky. The angular difference in the beam orientations is given by γ as shown in Figure 3 which represents a fraction of the total pulsar period:

$$\delta t = \frac{\gamma}{2\pi} T_p \tag{16}$$

where T_p is the rotational period of the pulsar.

The total time delay between the two beams is therefore:

$$\Delta t = \delta t + ct_1 - ct_2 + ct_{S1} - ct_{S2} \tag{17}$$

With the time delays, magnification factors and the angular positions of the images, one can now create a picture of what the overall effect of gravitational lensing might be when the source is pulsed and directional. First if the lens is located within our galaxy the distances D_d are on the order of a few hundred parsecs. The distances from the lens plane to the image(source) plane is of the order of the orbital radius R which can be approximated from the period of the Keplerian orbit of the pulsar about the lens.

3 Applications to Compact Binary Systems

Two applications of this analysis will be presented here. The first will determine the pulse profiles that can be expected to emerge from the gravitational lensing process for a pulsar in orbit around a black hole. This provides a good demonstration of the differences that arise between pulsed, directional sources and continuous isotropically emitting sources. The second application will be aimed at determining whether or not anomalous enhancements in the light curves of the double pulsar system PSR J0737-3039 during the eclipse phase are a result of the lensing of the light from pulsar A by the gravitational field of pulsar B.

In what follows we will assume a particular pulse profile for the source that is representative of a particular pulsar. In general the pulse profiles are assumed to not vary greatly from one pulse to the next. However, one can add a stochastic noise to the overall pulse profile to make a more realistic comparison with pulse-to-pulse differences. The motion of the pulsar (at least during conjunction) is assumed to take place at a constant radial distance R from the lens. For the systems studied the angular position of the source ψ with respect to the lens location is so small that the entire motion of the source is confined to a single source plane at a distance $D_{ds} \approx R$ from the lens. That is, the analysis of gravitational lensing will only lead to significant results when the source is close to superior conjunction. In this case the magnification factors of both beams are larger than unity and only then will they contribute to an enhancement of the over-all amplitude of the pulse profiles.

The motion of the source is obtained from a knowledge of its orbital period and the mass M of the lens. This in turn (assuming a nearly circular Keplerian orbit in the gravitational field of a geometric mass M) gives the radius R. The position and motion of the source is given by the angle ψ as a function of time. The radius of the orbit, the mass of the lens and the distance to the lens from the observer then determine the Einstein angle associated with the particular binary system.

Since ψ is measured with respect to the lens position, one can use this to determine the angle β (the angular source position with respect to the observer.)

$$\eta \approx \psi R \approx \beta D_s$$

With the source position β known, equation (9) can be used to compute the distances of closest approach in the lens plane for each of the images.

The distances of closest approach $\xi_{1,2}$ are then translated into the angular image positions $\theta_{1,2}$. The closest approach distances, the orbital radius and ψ also provide the light travel distances $D_{1,2}$ which determine the difference in angular orientation of the two beams (i.e. the angle γ). Here it is assumed that the time difference required for the beam to move from one orientation to the other is short enough that the orbital motion of the pulsar is insignificant. The pulsar does move a small distance over one pulsar rotational period and this is accounted for in computing subsequent values of the angle ψ.

Knowing the angles $\theta_{1,2}$, β and the Einstein angle allows one to compute the magnification factors $\mu_{1,2}$ for each image, while knowledge of the distances $D_{1,2}$, M, D_d and γ provide the total time delay using equation (17) between the arrival times for the two beams. This can be significant if the orientation angle γ is large or the mass of the lens is on the order of tens of solar masses.

3.1 Black Hole-pulsar Binary System

It is now assumed that the lens consists of a black hole with a mass on the order of tens of solar masses and that the source consists of a pulsar with a pulse period of one second. In the two cases presented below the orbital period of the pulsar is taken to be one hour. It is also assumed that the separation between the system and observer, D_d, is 100 parsecs. This means that the Einstein angle for the system is less than one nano-radian (i.e. a few micro-arc secs). In the tables that follow, the sign of the source and image angles assume that the pulsar is moving in a clockwise manner around the lens and that if the orientation of the source is to the left of the optical axis $\beta < 0$ and if it is to the right then $\beta > 0$, according to the geometry presented in Figure 2. One does not distinguish between the primary and secondary image locations ($\theta_{1,2}$) which would exchange roles as the source passes through superior conjunction. Instead the table below will use the notation $\theta_{\pm}$ to identify those images with a negative value of θ to those with a positive value of θ. The same notation will be used to signify the magnification of the images. The plus and minus signs are the magnifications associated with the angular image position having the same subscript. Thus μ_+ and μ_- represent the magnification factors for image positions θ_+ and θ_- respectively.

While the angular separation of the two images might be resolvable using the latest VLBI techniques, it is the changes in the pulse profiles that will produce the most noticeable measurable effects. In the discussion to follow, the orbits are assumed to be edge-on. It is easy enough to include smaller inclination angles, however in such cases, the effects of interest are suppressed significantly for small mass lenses.

In the two tables given below, the angle ψ is computed at one second intervals (i.e. the pulse period) from which β, the image angles, the magnifications and the time delays can be computed. In the time delays, the last three columns represent the sum of the differences in flat space path length plus Shapiro delay, (Δt_S), the geometric delay (δt) and the overall delay due to both (Δt).

Table 1 gives the results for a black hole lens with a mass of ten solar masses. It can be seen that the time delays in this case are dominated by the geometric effect, and that the Shapiro delay is smaller in comparison. This is due to the fact that the bending angles are relatively small. Therefore the distances of closest approach are far enough away from the lens that the gravitational field contributions are less important.

Table 1. Pulsar and $10M_\odot$ black hole lens

$T_{\rm orbit} = 1$ hr, $M_{\rm BH} = 10M_\odot$ $T_{\rm pulsar} = 1$ sec, $R = 5.06 \times 10^5 M_\odot$ $\alpha_0 = 1.5 \times 10^{-10}$								
ψ	$\beta \times 10^{-9}$	$\theta_+ \times 10^{-9}$	$\theta_- \times 10^{-9}$	μ_+	μ_-	Δt_S	δt	Δt (s)
-.200	-4.92	.0048	-4.92	9.0×10^{-7}	1.0	.014	.032	.046
-.184	-4.53	.0053	-4.53	1.3×10^{-6}	1.0	.014	.029	.043
⋮	⋮	⋮	⋮	⋮	⋮	⋮	⋮	⋮
-.004	-.098	.113	-.211	0.4	1.4	.0013	.0020	.0033
-.0005	-.013	.148	-.161	5.69	6.69	.0001	.0020	.0021
.003	.073	.195	-.122	1.64	0.64	.0009	.0021	.0030
.017	2.87	.355	-.670	1.04	0.04	.0010	.0028	.0038

Table 2. Pulsar and $50M_\odot$ black hole lens

$T_{\rm orbit} = 1$ hr, $M_{\rm BH} = 50M_\odot$ $T_{\rm pulsar} = 1$ sec, $R = 8.6 \times 10^5 M_\odot$ $\alpha_0 = 4.52 \times 10^{-10}$								
ψ	$\beta \times 10^{-9}$	$\theta_+ \times 10^{-9}$	$\theta_- \times 10^{-9}$	μ_+	μ_-	Δt_S	δt	Δt (s)
-.200	-84.1	.0024	-84.1	$< 10^{-12}$	1.0	8.56	.318	8.87
⋮	⋮	⋮	⋮	⋮	⋮	⋮	⋮	⋮
-.0041	-.168	.375	-.543	0.9	1.9	.0185	.0035	.0219
-.0005	-.002	.414	-.462	10.1	11.1	.0023	.0034	.0057
.003	.125	.518	-.393	2.35	1.35	.014	.0034	.017
.011	.492	.760	-.288	1.14	0.14	.052	.004	.056

The magnification of the the images only becomes significant when $\psi \approx 0$. As can be seen only two pulses show a total magnification increase that would exceed twice the value of the unlensed object. While there is a clear increase in magnification very close to superior conjunction, the duration of the increase does not last more than one or two pulse periods.

In Table 2 the mass of the black hole is increased to fifty solar masses. The effect of the increased mass is to increase the separation of the images as well as the time delays between the pulses. There is also an increase in the magnification factors for both beams that is higher than that for the $10M_\odot$ lens. The duration of the increased magnification is also increased.

While the time delays in Table 1 are dominated by the time required to change geometric orientations of the beams, the time delay for the larger mass lens is largely due to the Shapiro effect until the source is located close to the optical axis. As the pulsar reaches the position of superior conjunction, the magnitudes of the images increase, as expected, and the total time delay between the two beams decreases due to the fact that the differences in the angular positions of the two images on either side of the lens is minimized.

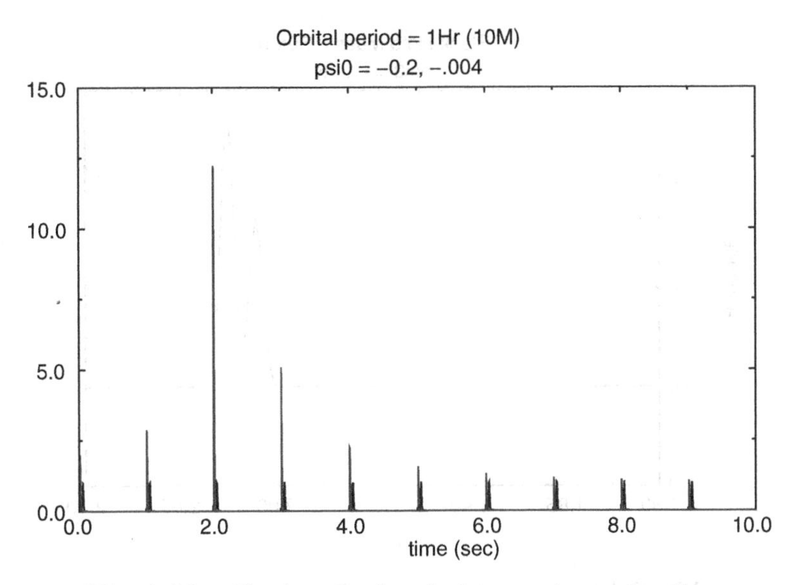

Fig. 4. Magnification of pulses during superior conjunction

When $\psi = 0$ the difference between the path lengths and Shapiro delays vanish due to an exact cancellation and the total delay is due to the time required for the beams to move through the angle γ. For black holes of intermediate mass (on the order of twenty to thirty solar masses) the Shapiro delay and the geometric delays are of comparable length.

In both cases it can be seen that when the magnification factors are large, the time delays are short and a doubling of the magnification is possible if the peak of the pulses is of sufficient duration.

In order to visualize what this effect might be on the pulse profiles, a time series graph (Figure 4) of the pulses observed at the Earth are presented where the black hole lens mass is 10 $M_{\odot}$. The pulses shown are those that occur as the pulsar passes through superior conjunction. The pulses themselves are modeled by simple gaussian pulses with a half-width of 0.01 sec without any noise. The total magnitude and the differences in magnitude are plotted next to each other. The most noticeable effect is the intensity enhancement that occurs close to superior conjunction, which if not obscured by intervening material would act as an obvious signature of gravitational lensing. However unlike the standard microlensing scenario where the total magnitude of the unresolved images is simply the sum of the magnitudes of the two images, if the time delay between the two images is significant enough there will be no overlap of the pulse profiles and the total intensity enhancement is decreased.

Such a situation is shown in Figure 5 where the time delay between the two pulses passing on either side of a 50 $M_{\odot}$ black hole is large enough that the pulse profile is more complicated than that expected from the sum of two gaussian pulses where the peaks occur at nearly the same time. Thus if the

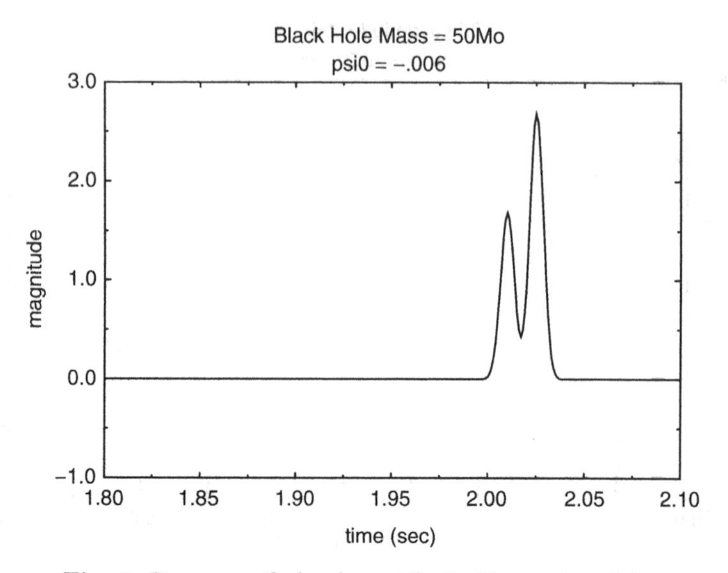

Fig. 5. Two magnified pulses with significant time delay

pulses are of short duration compared to the total delay, the magnifications do not simply add and more complicated situations arise.

Of course pulsar pulses are noisy and one can model the pulse profiles by adding a background of noise to the signal. However the intensity enhancement is such that it should provide a significant signal appearing well above the noise associated with the individual pulses. For example with a 50 $M_{\odot}$ black hole lens, adding a 15% random noise to the individual pulses will not significantly affect the intensity enhancement if the time delays between the two pulses is small as is shown in Figure 6.

3.2 Application to the Double Pulsar: PSR J0737-3039

The analysis developed above assumed that the inclination angle of the binary system was $i \approx \pi/2$ to within one or two degrees at the most. If this is not the case then the magnification of the secondary image which undergoes the most light bending is too small to be observed. If the magnification enhancements can be observed during a number of eclipses, then the intensity measurements could be used to determine the inclination angle since the maximum possible intensity should be a measurable quantity. The recent discovery of the double pulsar PSR J0737-3039 with an inclination angle close to 90° has stirred much interest in the relativity community since it should be able to act as a test of general relativity (and alternative gravity theories) with an unprecedented accuracy. The system consists of two massive compact stars ($M_A = 1.4 M_{\odot}$,

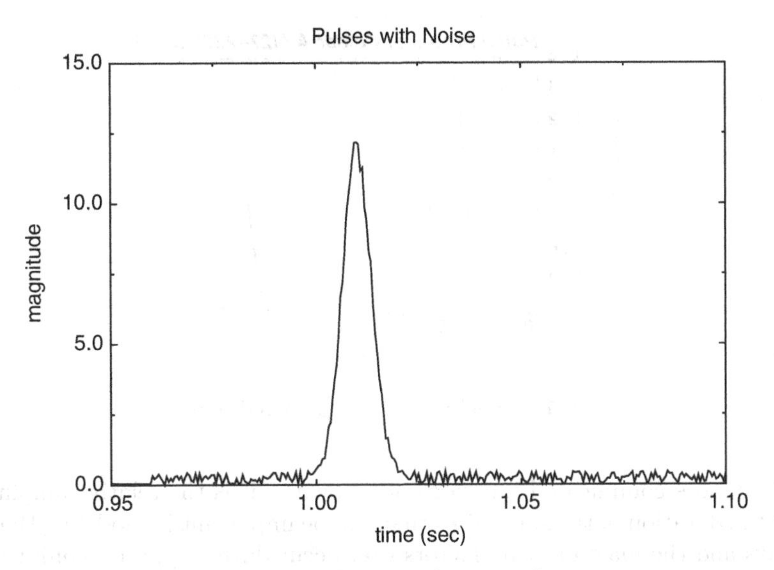

Fig. 6. Magnification of noisy pulses

Table 3. Physical Parameters for the Pulsar System: PSR J0737-3039A/B

Parameter	A	B
Pulse Period (ms)	22.70	2773
Mass ($M_\odot$)	1.338	1.249
Orbital Period (hrs)	2.453	
Distance to PSR J0737 (pc)	500-600	
Distance between lens and source (km)	4000±2000	
Inclination angle	90.29°±0.14°	
Orbital Eccentricity	0.08778	

$M_B = 1.3M_\odot$) in a nearly circular orbit ($e = 0.088$). Both stars are pulsars that can be seen from Earth. Some of the important physical characteristics of this system are shown in Table 3.

Model of Lensing in J0737-3039

The parameters given in Table 3 along with a model of the integrated pulse shapes for the pulsar sources provides the necessary information required to describe the light curve one might expect to observe during the eclipse of Pulsar A by Pulsar B. The averaged pulse profile of Pulsar A consists of two unequal pulses occurring during its 0.0227 second spin period. The averaged pulses are computed from the true light curves measured in the frequency range 427 to 2200 MHz. [8] In this analysis the light curve is approximated by two skewed gaussian pulses as shown in Figure 7. While the true light curves

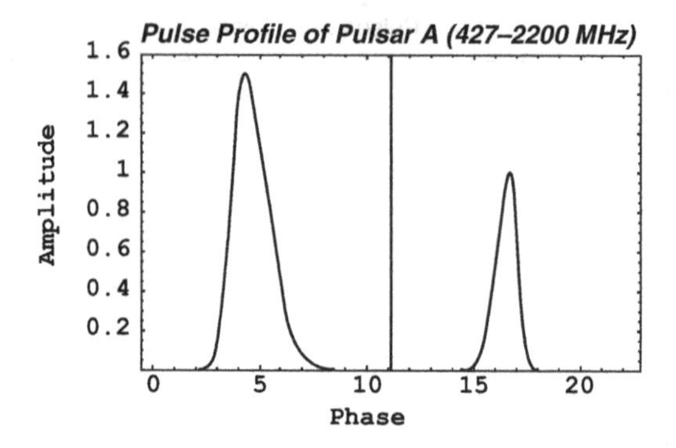

Fig. 7. Model for the Pulsar A pulse profile

will be more complicated functions of frequency, it is the relative amplitudes and the duration of the sub-pulses that will be important in modeling the time delays and the magnification factors that occur during superior conjunction.

During the eclipse of Pulsar A by Pulsar B, the observations indicate that there is an attenuation of the light curve by the atmosphere of Pulsar B. [2,5] The attenuation is not symmetric with respect to the phase. The eclipse ingress is longer in duration than egress. The depth of the attenuation is also deeper after conjunction than it is before. In addition there is also a frequency dependent behaviour and these observations have provided a challenge to theoreticians to create a model of the atmosphere of Pulsar B along with its magnetic field structure. However in order to create a model for the gravitational lensing effects we follow the method used by Kaspi, et al [2] who fit the orbit-averaged light curve attenuation with two functions of the form:

$$A(\phi) = \frac{1}{e^{(\phi-\phi_0)/\sigma} + 1}$$

where ϕ is the orbital phase defined such that at conjunction $\phi = 0$. The two free parameters ϕ_0 and width σ have been computed for both ingress and egress and for our model we use:

$$\text{Ingress}: \quad \phi_0 = -1.34 \times 10^{-3}, \quad \sigma = 0.49 \times 10^{-3}$$

$$\text{Egress}: \quad \phi_0 = 1.70 \times 10^{-3}, \quad \sigma = 0.13 \times 10^{-3}$$

The functional form for the attenuation factor is plotted in Figure 8.

Using the physical parameters presented in Table 3 and applying the method discussed in the analysis of black hole-pulsar binaries one can compute the expected pulse profiles and magnifications subject to the light curve attenuation. The resulting pulse profiles are shown in Figure 9 where it can be seen that the lensed pulses from Pulsar A can be magnified significantly even in the presence of attenuation if the inclination angle is very close to 90°. If on the other hand the inclination angle is less than 90° the enhancement for a lens

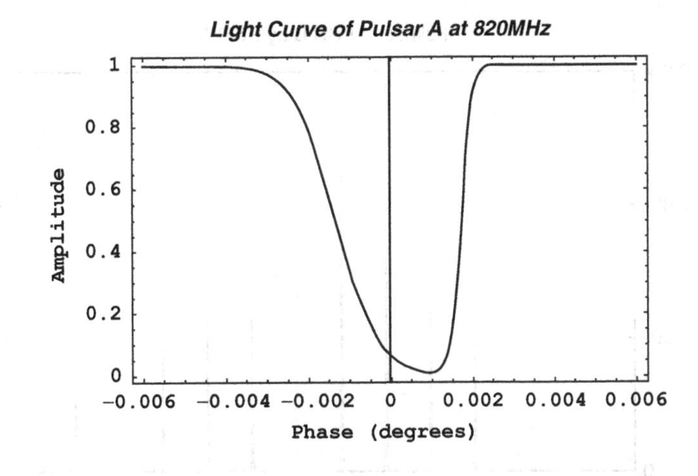

Fig. 8. The Attenuation curve for the eclipse of A by atmosphere of B

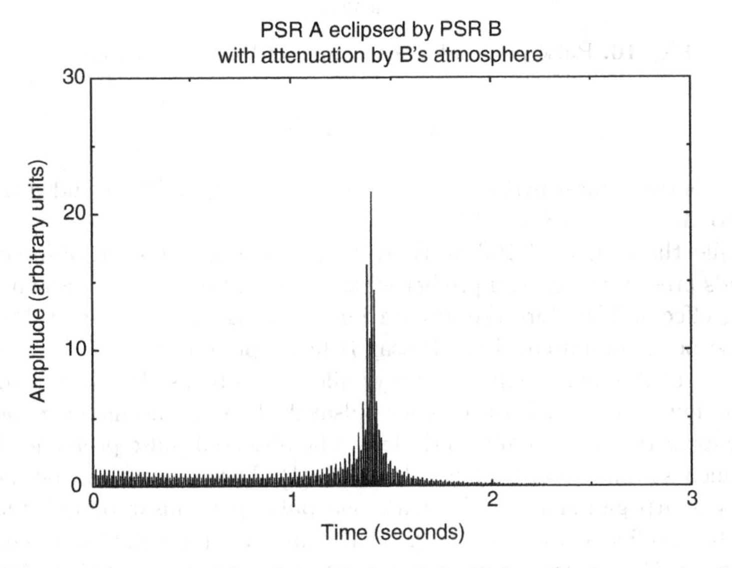

Fig. 9. Gravitational lensing of Pulsar A by Pulsar B where $\sin i = 1$

of just larger than one solar mass is not very significant beyond $\Delta i \approx 0.15^\circ$. In this case the magnification factor of the primary beam is reduced to unity and any significant attenuation of the signals close to superior conjunction would mask the gravitational lensing effects. Since Pulsar A has such a short rotational period, one can expect a very large enhancement of the magnification during conjunction since the pulsar source is now very close to being located at $\psi = 0$ when the emission of the beam occurs. The maximum value of $\Delta\psi$ away from superior conjunction can easily be estimated by:

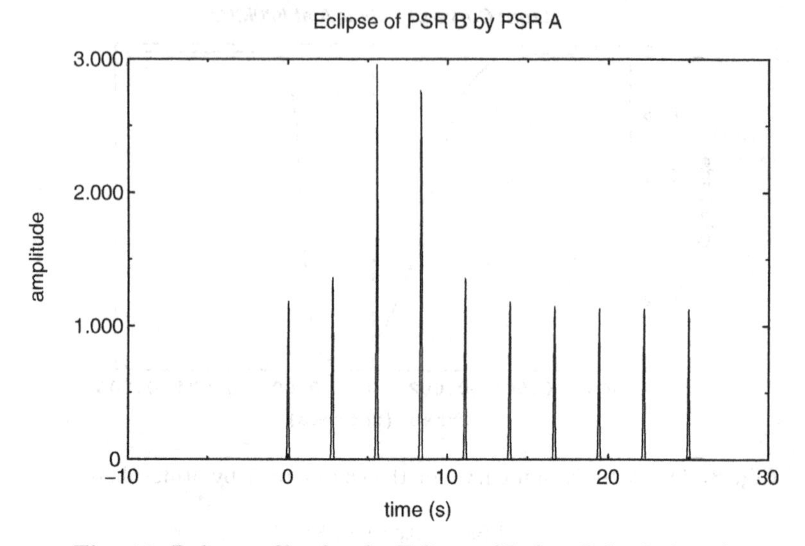

Fig. 10. Pulse profiles for the Eclipse of Pulsar B by Pulsar A

$$\Delta\psi_{\text{max}} = 2\pi \frac{T_p}{\tau}$$

where τ is the orbital period of the system. Using $T_p \approx 22$ ms and $\tau \approx 2.5$ hr leads to $\Delta\psi_{\text{max}} = 1.58 \times 10^{-5}$.

While the eclipse of Pulsar B by Pulsar A has not been observed, the methods presented here can predict what one might expect from gravitational lensing effects. Therefore the gravitational lensing of Pulsar B by Pulsar A has also been computed. Since Pulsar B has a period of about two seconds, the phase of the pulses will not always allow it to be as close to the superior conjunction position as is the case for pulsar A. The enhancements in intensity can be expected to be significantly less. The observed pulse profile for Pulsar B is much simpler than that for Pulsar A [1]. These pulses can be modeled by very sharp gaussian profiles with one pulse per pulsar period. One can compute the effects of Pulsar A's gravitational field on the light pulses emitted from Pulsar B and plot them as shown in Figure 10. As expected the intensity enhancements are found to be smaller than those for the observed eclipse. Here however both noise and/or attenuation, if present, have been ignored. As in the case of the eclipse of Pulsar A by Pulsar B the increase in magnification may be sufficient so as to render the signal observable during conjunction, even if the remainder of the light curve is of low intensity.

4 Discussion

Compact binary systems where at least one of the members is an observable pulsar might provide an opportunity to measure gravitational lensing effects, particularly when the system has a large inclination angle. The effects are most dramatic for lenses with large masses. Stellar black holes with masses greater than 20 $M_\odot$ can be expected to exist in relatively large numbers and one might be able to use gravitational lensing by such objects to search for binary systems composed of such objects.

On the other hand black holes with even larger masses are expected to lie in the centers of globular clusters and, of course, supermassive black holes lying in galactic centres might be capable of producing even more extreme enhancements of individual sources in nearby galaxies. It would be interesting to determine what values of the distances and masses provide the most promising situations that might be observable.

For the double pulsar system, the lower neutron star mass of the lens makes it much more difficult to observe gravitational lensing. Even small deviations away from $\sin i = 1$ are sufficient to lower the magnification and time delays significantly. If the anomalous enhancements seen in the eclipsing light curves are due to gravitational lensing then this places some stringent constraints on the orbital inclination angle. These limits are inconsistent with the values obtained from a direct measurement of the Shapiro delay. This conclusion agrees with that of Lai and Rafikov [3] whose analysis is based in standard microlensing formulae.

Returning to the general theory of gravitational lensing in compact binary systems, we are currently extending these techniques to include rotational and electromagnetic effects on light propagation in more general spacetimes. Extreme rotation such as that which occurs in Kerr black holes or the rotation of extended objects with multipole structure may provide enhancements in some configurations that do not occur in the non-rotating case. In addition there are a number of different rotating spacetimes for material objects which means that the differences between neutron star rotation and black hole rotation on the vacuum propagation of photons might be significant enough to determine the difference in systems where the masses of the lenses are close to the limiting mass for neutron stars.

Both exact and approximate solutions for black holes immersed in magnetic fields exist, and while they, introduce additional gravitational lensing effects at second order (i.e. those proportional to M^2/r_0^2), improvements in observational techniques may in fact provide a means for using such effects to make independent measurements of such second-order predictions. In some cases second order measurements may be capable of ruling out alternative theories of gravity that agree with general relativity at the first order.

Hopefully more detailed all sky surveys will provide us with more candidate short-period compact binary systems. These can then act as tests of gravitational theories and/or provide confirmation of the correctness of general relativity. That some of these systems might provide a means for observing gravitational lensing directly offers an exciting prospect for future observations.

5 Acknowledgements

This work was supported by an NSERC (Canada) Discovery Grant.

References

1. Burgay, M.; Possenti, A.; Manchester, R.N.; Kramer, M.; McLaughlin, M.A.; Lorimer, D.R.; Stairs, I.H.; Joshi, B.C.; Lyne, A.G.; Camilo, F.; D'Amico, N.; Freire, P.C.C.; Sarkissian, J.M.; Hotan, A.W.; Hobbs, G.B.; Ap. J., **624**, L113, (2005).
2. Kaspi, V. M., Ransom, S. M., Backer, D. C., Ramachandran, R., Demorest, P., Arons, J., & Spitkovsky, A.; Ap. J., **613**, L137, (2004).
3. Lai, D., and Rafikov, R., Ah. J., **621**, L41, (2005).
4. Lyne, A.G., Burgay, M., Kramer, M., Possenti, A., Manchester, R.N., Camilo, F., McLaughlin, M.A., Lorimer, D.R., N. D'Amico, N., Joshi, B.C., Reynolds,J., and Freire, P.C.C.; Science, **303**, 1153, (2004).
5. McLaughlin, M. A.; Kramer, M.; Lyne, A. G.; Lorimer, D. R.; Stairs, I. H.; Possenti, A.; Manchester, R. N.; Freire, P. C. C.; Joshi, B. C., Ap. J., **613**, L57, (2004).
6. Narayan, R., and Bartelmann, M., "Lectures on Gravitational Lensing" in *Formation of Structure in the Universe* (edited by Dekel, A., and Ostriker, J.P.), Cambridge Univ. Press, Cambridge (1999).
7. Paczyński, B.; Ann. Rev. Astron. Astrophys., **34**, 419, (1996).
8. Ransom, S., Demorest, P., Kaspi, V., Ramachandran, R., & Backer, D., ASP Conf. Ser., **328**, 73, (2005).
9. Schneider, P., Ehlers, J., & Falco, E.E., *Gravitational Lenses*, Springer Verlag, New York, (1992).

Part II

Accreting Neutron Star Binaries

Accreting Neutron Stars in Low-Mass X-Ray Binary Systems

Frederick K. Lamb[1,2,3] and Stratos Boutloukos[1,4]

[1] Center for Theoretical Astrophysics and Department of Physics, University of Illinois at Urbana-Champaign, 1110 W Green, 61801, Urbana, IL, USA
[2] Also, Department of Astronomy
[3] fkl@uiuc.edu
[4] stratos@uiuc.edu

Using the Rossi X-ray Timing Explorer (*RossiXTE*), astronomers have discovered that disk-accreting neutron stars with weak magnetic fields produce three distinct types of high-frequency X-ray oscillations. These oscillations are powered by release of the binding energy of matter falling into the strong gravitational field of the star or by the sudden nuclear burning of matter that has accumulated in the outermost layers of the star. The frequencies of the oscillations reflect the orbital frequencies of gas deep in the gravitational field of the star and/or the spin frequency of the star. These oscillations can therefore be used to explore fundamental physics, such as strong-field gravity and the properties of matter under extreme conditions, and important astrophysical questions, such as the formation and evolution of millisecond pulsars. Observations using *RossiXTE* have shown that some two dozen in low-mass X-ray binary systems have the spin rates and magnetic fields required to become millisecond radio-emitting pulsars when accretion ceases, but that few have spin rates above about 600 Hz. The properties of these stars show that the paucity of spin rates greater than 600 Hz is due in part to the magnetic braking component of the accretion torque and to the limited amount of angular momentum that can be accreted in such systems. Further study will show whether braking by gravitational radiation is also a factor. Analysis of the kilohertz oscillations has provided the first evidence for the existence of the innermost stable circular orbit around dense relativistic stars that is predicted by strong-field general relativity. It has also greatly narrowed the possible descriptions of ultradense matter.

1 Introduction

Neutron stars and black holes are important cosmic laboratories for studying fundamental questions in physics and astronomy, especially the properties of dense matter and strong gravitational fields. The neutron stars in low-mass

E.F. Milone et al. (eds.), Short-Period Binary Stars: Observations, Analyses, and Results, 87–109.

X-ray binary systems (*LMXB*s) have proved to be particularly valuable systems for investigating the innermost parts of accretion disks, gas dynamics and radiation transport in strong radiation and gravitational fields, and the properties of dense matter, because the magnetic fields of many of these stars are relatively weak but not negligible. Magnetic fields of this size are weak enough to allow at least a fraction of the accreting gas to remain in orbit as it moves into the strong gravitational and radiation fields of these stars, but strong enough to produce anisotropic X-ray emission, potentially allowing the spin rates of these stars to be determined. The discovery and study using *RossiXTE* of oscillations in the X-ray emission of the accreting neutron stars in *LMXBs*, with frequencies comparable to the dynamical frequencies near these stars, has provided an important new tool for studying the strong gravitational and magnetic fields, spin rates, masses, and radii of these stars (for previous reviews, see [46, 50, 51, 96, 97]).

Periodic accretion-powered X-ray oscillations have been detected at the spin frequencies of seven neutron stars with millisecond spin periods, establishing that these stars have dynamically important magnetic fields. In this review, a pulsar is considered a millisecond pulsar (*MSP*) if its spin period P_s is <10 ms (spin frequency $\nu_{\rm spin} > 100$ Hz). The channeling of the accretion flow required to produce these oscillations implies that the stellar magnetic fields are greater than $\sim 10^7$ G (see [69]), while the nearly sinusoidal waveforms of these oscillations and their relatively low amplitudes indicate that the fields are less than $\sim 10^{10}$ G [78]. The spin frequencies of these accretion-powered *MSPs* range from 185 Hz to 598 Hz (see Table 1).

Nearly periodic nuclear-powered X-ray oscillations (see Fig. 1) have been detected during the thermonuclear bursts of 17 accreting neutron stars in *LMXB*s, including 2 of the 7 known accretion-powered *MSP*s (Table 1). The existence of the thermonuclear bursts indicates that these stars' magnetic fields are less than $\sim 10^{10}$ G [38, 54], while the spectra of the persistent X-ray emission [79] and the temporal properties of the burst oscillations (see [18, 20]) indicate field strengths greater than $\sim 10^7$ G. The spin frequencies of these nuclear-powered pulsars range from 45 Hz up to 1122 Hz.

Measurements of the frequencies, phases, and waveforms of the accretion- and nuclear-powered oscillations in SAX J1808.4−3658 (see Fig. 1 and [20]) and XTE J1814−338 [89] have shown that, except during the first seconds of some bursts, the nuclear-powered oscillations have very nearly the same frequency, phase, and waveform as the accretion-powered oscillations, establishing beyond any doubt (1) that these stars have magnetic fields strong enough to channel the accretion flow and enforce corotation of the gas at the surface of the star that has been heated by thermonuclear bursts and (2) that their nuclear- and accretion-powered X-ray oscillations are both produced by spin modulation of the X-ray flux from the stellar surface. The burst oscillations of some other stars are also very stable [88], but many show frequency drifts

Table 1. Accretion- and Nuclear-Powered Millisecond Pulsars[a]

$\nu_{\rm spin}$	(Hz)[b]	Object	Reference
1122	NK	XTE J1739−285	[43]
619	NK	4U 1608−52	[29]
611	N	GS 1826−238	[93]
601	NK	SAX J1750.8−2900	[40]
598	A	IGR J00291+5934	[61]
589	N	X 1743−29	[86]
581	NK	4U 1636−53	[92, 105, 109]
567	N	MXB 1659−298	[102]
549	NK	Aql X-1	[108]
530	N	A 1744−361	[11]
524	NK	KS 1731−260	[83]
435	A	XTE J1751−305	[62]
410	N	SAX J1748.9−2021	[41]
401	ANK	SAX J1808.4−3658	[18, 103]
377	A	HETE J1900.1−2455	[71]
363	NK	4U 1728−34	[91]
330	NK	4U 1702−429	[59]
314	AN	XTE J1814−338	[60]
270	N	4U 1916−05	[25]
191	AK	XTE J1807.4−294	[58, 101]
185	A	XTE J0929−314	[24]
45	N	EXO 0748−676	[99]

[a]Defined in this review as pulsars with spin periods $P_s < 10$ ms. EXO 0748-676 is not a millisecond pulsar according to this definition. After this paper went to press, Krimm et al. (arXiv: 0709.1693) reported discovery of 182 Hz accretion-powered oscillations from the X-ray transient SWIFT J1756.9-2508, Gavriil et al. (arXiv:0708.0829) and Altamirano et al. (arXiv:0708.1316) reported discovery of intermittent 442 Hz accretion-powered X-ray oscillations from SAX J1748.9-2021 in the globular cluster NGC 6440, and Casella et al. (arXiv:0708.1110) reported discovery of intermittent accretion-powered 550 Hz X-ray oscillations from Aql X-1; Markwardt et al. (ATel#1068) reported burst oscillations from IGR J17191-2821 with a maximum frequency of 294 Hz whereas Klein-Wolt et al. (ATel#1075) reported a pair of khz *QPO*s with a separation of 330 Hz.

[b]Spin frequency inferred from periodic or nearly periodic X-ray oscillations. A: accretion-powered millisecond pulsar. N: nuclear-powered millisecond pulsar. K: kilohertz *QPO* source. See text for details.

and phase jitter [73,75,91,92]. These results confirm that burst and persistent oscillations both reveal directly the spin frequency of the star. Several mechanisms for producing rotating emission patterns during X-ray bursts have been proposed (see, e.g., [8,9,26,72,85,92]), but which mechanisms are involved in which stars is not yet fully understood.

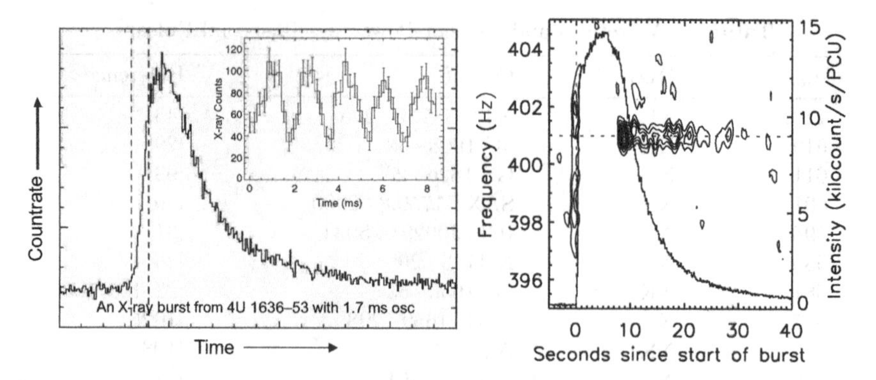

Fig. 1. *Left*: X-ray burst and millisecond burst oscillations seen in 4U 1636−53. The main panel displays the X-ray countrate in a succession of 2-second time intervals, showing the rapid rise and approximately exponential decay of the burst. The inset panel shows the strong ∼580 Hz X-ray countrate oscillations observed during the time interval bounded by the vertical dashed lines in the main panel. From T. Strohmayer, personal communication; see also [92]. *Right*: An X-ray burst with millisecond X-ray countrate oscillations observed in SAX J1808.4−3658 on 18 October 2002. The dark curve and the scale at the right show the X-ray countrate as a function of time during the burst. The contours show the dynamic power spectrum of the X-ray countrate on the scale at the left. Note the rapid increase in the oscillation frequency at the beginning of the burst, the disappearance of the oscillation at the peak of the burst, and its reappearance about 5 s later. The horizontal dashed line shows the frequency of the neutron star's spin inferred from its accretion-powered brightness oscillations. From [20].

Kilohertz quasi-periodic oscillations (*QPO*s) have now been detected in some two dozen accreting neutron stars (see [46]), including 9 of the 17 known nuclear-powered X-ray pulsars and 2 of the 7 known accretion-powered *MSP*s (Table 1). The frequencies of the kilohertz *QPO*s detected so far range from ∼1300 Hz in 4U 0614+09 [98] down to ∼10 Hz in Cir X-1 [17]. If, as expected, the frequencies of the highest-frequency kilohertz *QPO*s reflect the orbital frequencies of gas in the disk near the neutron star [69, 97], then in most kilohertz *QPO* sources gas is orbiting close to the surface of the star. The spin frequencies $\nu_{\rm spin}$ of the neutron stars in the kilohertz *QPO* sources are inferred from the periodic accretion- and nuclear-powered X-ray oscillations of these stars. In the systems in which kilohertz *QPO*s and periodic X-ray oscillations have both been detected with high confidence ($\geq 4\sigma$), $\nu_{\rm spin}$ ranges from 191 Hz to 619 Hz.

In many kilohertz *QPO* sources, the separation $\Delta\nu_{\rm QPO}$ of the frequencies ν_u and ν_ℓ of the upper and lower kilohertz *QPO*s remains constant to within a few tens of Hz, even as ν_u and ν_ℓ vary by as much as a factor of 5. $\Delta\nu_{\rm QPO}$ is approximately equal to $\nu_{\rm spin}$ or $\nu_{\rm spin}/2$ in all stars in which these frequencies have been measured. In the accretion-powered *MSP*s XTE J1807.4−294 [55] and SAX J1808.4−3658 [20], no variations of $\Delta\nu_{\rm QPO}$ with time have so far been detected. In XTE J1807.4−294, no difference between $\Delta\nu_{\rm QPO}$ and $\nu_{\rm spin}$ has been detected; in SAX J1808.4−3658, no difference between $\Delta\nu_{\rm QPO}$ and $\nu_{\rm spin}/2$ has been detected (see Fig. 2 and [20,46,55,96,101,104]). These results demonstrate conclusively that at least some of the neutron stars that produce kilohertz *QPO*s have dynamically important magnetic fields and that the spin of the star plays a central role in generating the kilohertz *QPO* pair. Consequently, $\Delta\nu_{\rm QPO}$ can be used to estimate, to within a factor of two, the otherwise unknown spin frequency of a star that produces a pair of kilohertz *QPO*s.

The kilohertz *QPO* pairs recently discovered in Cir X-1 [17] extend substantially the range of known kilohertz *QPO* behavior. In Cir X-1, values of ν_u and ν_ℓ as small, respectively, as 230 Hz and 50 Hz have been observed simultaneously. These frequencies are more than 100 Hz lower than in any other kilohertz *QPO* system. Unlike the kilohertz *QPO* pairs so far observed in other neutron stars, in Cir X-1 $\Delta\nu_{\rm QPO}$ has been observed to increase with increasing ν_u: as ν_u increased from ∼230 Hz to ∼500 Hz, $\Delta\nu_{\rm QPO}$ increased from 173 Hz to 340 Hz. The relative frequency separations $\Delta\nu_{\rm QPO}/\nu_u$ in Cir X-1 are ∼55%–75%, larger than the relative frequency separations ∼20%–60% observed in other kilohertz *QPO* systems. $\Delta\nu_{\rm QPO}$ has been seen to vary by ∼100 Hz in GX 5−1, which also has relatively low kilohertz *QPO* frequencies, but with no clear dependence on ν_u. If, as is generally thought, the frequencies of the kilohertz *QPO*s reflect the frequencies of orbits in the disk, kilohertz *QPO*s with such low frequencies would require the involvement of orbits ∼50 km from the star, which is a challenge for existing kilohertz *QPO* models. Accretion- and nuclear-powered X-ray oscillations have not yet been detected in Cir X-1 and hence its spin frequency has not yet been measured directly. Further study of Cir X-1 and the relatively extreme properties of its kilohertz *QPO*s is likely to advance our understanding of the physical mechanisms that generate the kilohertz *QPO*s in all systems.

The first 11 spins of accretion-powered X-ray *MSP*s that were measured were consistent with a flat distribution that ends at 760 Hz [20], but were also consistent with a distribution that decreases more gradually with increasing frequency [66]; the spins of 21 accretion- and nuclear-powered X-ray *MSP*s are now known. The proportion of accretion- and nuclear-powered *MSP*s with frequencies higher than 500 Hz is greater than the proportion of known rotation-powered *MSP*s with such high frequencies, probably because there is no bias against detecting X-ray *MSP*s with high frequencies, whereas detection of rotation-powered radio *MSP*s with high spin frequencies is still difficult [20]. The recent discovery of a 1122 Hz *MSP* [43] supports this argument, which is not inconsistent with the recent discovery of a 716 Hz rotation-powered radio *MSP* [31].

These discoveries have established that many neutron stars in *LMXB*s have magnetic fields and spin rates similar to those of the rotation-powered *MSP*s. The similarity of these neutron stars to rotation-powered *MSP*s strongly supports the hypothesis [2, 80] that they are the progenitors of the

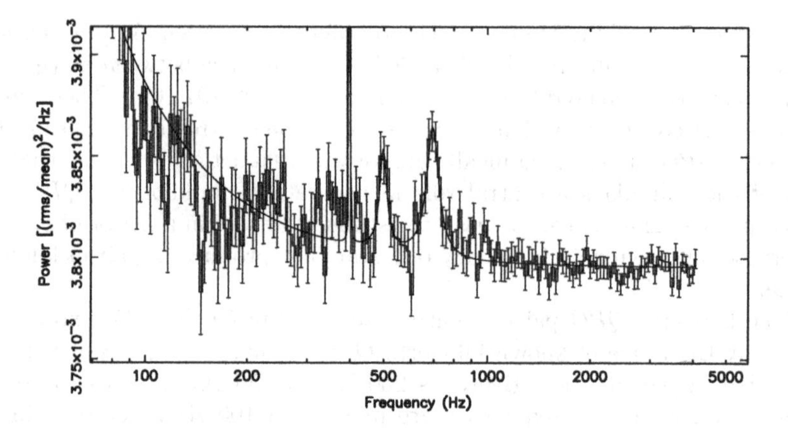

Fig. 2. Power density spectrum of the variations in the X-ray countrate from the accretion-powered *MSP* SAX J1808.4−3658 seen on 18 October 2002. The peaks correspond to the 401 Hz periodic oscillations ("pulsations") at the star's spin frequency, the lower kilohertz *QPO* at 499 ± 4 Hz, and the upper kilohertz *QPO* at 694 ± 4 Hz. In this pulsar, the separation between the two kilohertz *QPO*s is half the spin frequency. Two kilohertz *QPO*s have also been seen in the accreting millisecond X-ray pulsar XTE J1807.4−294, which has a spin frequency of 191 Hz [57, 101]. In this pulsar, the separation between the two kilohertz *QPO*s is consistent with the spin frequency. These results demonstrate that the star's spin plays a central role in the generation of kilohertz *QPO* pairs. From [104].

rotation-powered *MSP*s. After being spun down by rotation-powered emission, the neutron stars in these systems are thought to be spun up to millisecond periods by accretion of matter from their binary companions, eventually becoming nuclear- and accretion-powered *MSP*s and then, when accretion ends, rotation-powered *MSP*s.

In §2 we discuss in more detail the production of accretion- and rotation-powered *MSP*s by spin-up of accreting weak-field neutron stars in *LMXB*s, following [52], and in §3 we describe several mechanisms that may explain the nuclear-powered X-ray oscillations produced at the stellar spin frequency by such stars. In §§4 and 5 we discuss, respectively, possible mechanisms for generating the kilohertz *QPO* pairs, following [49], and how the kilohertz *QPO*s can be used as tools to explore dense matter and strong gravity, following [50, 51, 69].

2 Production of Millisecond Pulsars

Neutron stars in *LMXB*s are accreting gas from a Keplerian disk fed by a low-mass companion star. The star's magnetic field and accretion rate are thought to be the most important factors that determine the accretion flow pattern near it and the spectral and temporal characteristics of its X-ray emission

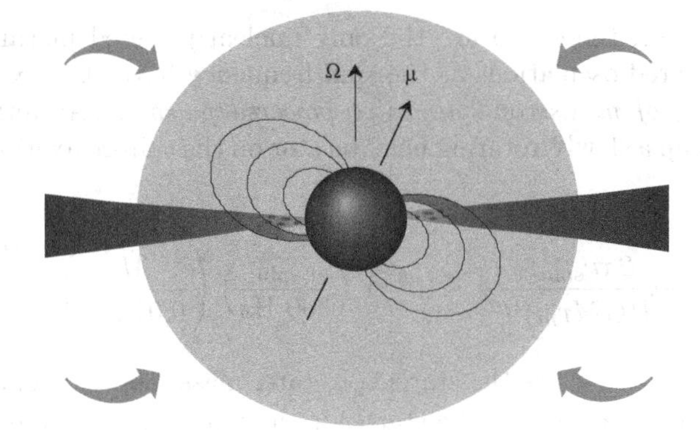

Fig. 3. Side view of a weak-field neutron star accreting from a disk, showing the complex flow pattern expected. Some accreting gas couples strongly to the magnetic field and is channeled toward the magnetic poles, but a substantial fraction couples only weakly and drifts inward in nearly circular orbits as it transfers its angular momentum to the star via the stellar magnetic field. From [69].

(see [69]). The accretion rates of these stars vary with time and can range from the Eddington critical rate $\dot{M}_E$ to less than $10^{-4}\dot{M}_E$. Their magnetic fields are thought to range from 10^{11} G down to 10^7 G or possibly less, based on their X-ray spectra [79], the occurrence of thermonuclear X-ray bursts [38], and their high-frequency X-ray variability [69,96]. Magnetic fields at the upper end of this range are strong enough to terminate the Keplerian disk well above the stellar surface, even for accretion rates $\sim\dot{M}_E$, whereas magnetic fields at the lower end of this range affect the flow only close to the star, even for accretion rates as low as $\sim10^{-4}\dot{M}_E$.

For intermediate field strengths and accretion rates, some of the accreting gas is expected to couple to the star's magnetic field well above the stellar surface. The star's magnetic field channels part of the flow toward its magnetic poles, and this flow heats the outer layers of the star unevenly. The remainder of the accreting gas is expected to remain in a geometrically thin Keplerian flow that penetrates close to the stellar surface, as shown in Figure 3. The gas that remains in orbit close to the star is thought to be responsible for generating the kilohertz *QPO*s (see [46, 48, 49]). When thermonuclear X-ray bursts occur, they also heat the outer layers of the star unevenly. Whether due to accretion or to nuclear burning, the uneven heating of the outer layers produces a broad pattern of X-ray emission that rotates with the star, making both the accretion-powered and nuclear-powered X-ray emission of the star appear to oscillate at the spin frequency. The stability of the nuclear-powered oscillations show that the heated region is strongly coupled to the rotation of the star, probably via the star's magnetic field. The phase locking of the nuclear- and accretion-powered oscillations and the strong similarity of the two waveforms in SAX J1808.4−3658 and XTE J1814−338 indicate that the stellar magnetic field is playing a dominant role, at least in these pulsars. However, these two are the only nuclear-powered pulsars in which accretion-powered oscillations at the spin frequency have also been detected.

Production of millisecond accretion-powered pulsars.—Accretion from a disk will spin up a slowly-rotating neutron star on the spin-relaxation timescale [27,28,52]

$$t_{\rm spin} \equiv \frac{2\pi\nu_{\rm spin} I}{[\dot{M}(GMr_m)^{1/2}]} \sim 10^8\,{\rm yr}\left(\frac{\nu_{\rm spin}}{300\ {\rm Hz}}\right)\left(\frac{\dot{M}}{0.01\dot{M}_E}\right)^{-1+\alpha/3}, \tag{1}$$

where $\nu_{\rm spin}$, M, and I are the star's spin rate, mass, and moment of inertia, $\dot{M}$ is the accretion rate onto the star (not the mass transfer rate), r_m is the angular momentum coupling radius, α is 0.23 if the inner disk is radiation-pressure-dominated (*RPD*) or 0.38 if it is gas-pressure-dominated (*GPD*), and in the last expression on the right the weak dependence of $t_{\rm spin}$ on M, I, and the star's magnetic field has been neglected.

The current spin rates of neutron stars in *LMXB*s depend on the average accretion torque acting on them over a time $\sim t_{\rm spin}$. Determining this average torque is complicated by the fact that the accretion rates and magnetic fields of these stars vary with time by large factors and that the accretion torque can decrease as well as increase the spin rate. Mass transfer in the neutron-star–white-dwarf binary systems is thought to be stable, with a rate that diminishes secularly with time.

While a few neutron stars in *LMXB*s accrete steadily at rates $\sim\dot{M}_E$, most accrete at rates $\sim 10^{-2}$–$10^{-3}\dot{M}_E$ or even less [30,45,69,95] and many accrete only episodically [81,95]. Important examples are the known accretion-powered *MSP*s in *LMXB*s, which have outbursts every few years during which their accretion rates rise to $\sim 10^{-2}\dot{M}_E$ for a few weeks before falling again to less than $\sim 10^{-4}\dot{M}_E$ [20,89]. Also, there is strong evidence that the external magnetic fields of neutron stars in *LMXB*s decrease by factors $\sim 10^2$–10^3 during their accretion phase, perhaps on timescales as short as hundreds of years (see [7,82]).

If a star's magnetic field and accretion rate are constant and no other torques are important, accretion will spin it up on a timescale $\sim t_{\rm spin}$ to its equilibrium spin frequency $\nu_{\rm eq}$. This frequency depends on M, the strength and structure of the star's magnetic field, the thermal structure of the disk at r_m, and $\dot{M}$ [27,28,100]. If a star's magnetic field and accretion rate change on timescales longer than $t_{\rm spin}$, the spin frequency will approach $\nu_{\rm eq}$ and track it as it changes. If instead $\dot{M}$ varies on timescales shorter than $t_{\rm spin}$, the spin rate will fluctuate about the appropriately time-averaged value of $\nu_{\rm eq}$ (see [23]). Thus $\nu_{\rm eq}$ and its dependence on B and $\dot{M}$ provide a framework for analyzing the evolution of the spins and magnetic fields of neutron stars in *LMXB*s.

Figure 4 shows $\nu_{\rm eq}$ for five accretion rates and dipole magnetic fields B_d, assumed given by $3.2 \times 10^{19}(P\dot{P})^{1/2}$ G and ranging from 10^7 G to 10^{11} G. The lines are actually bands, due to systematic uncertainties in the models.

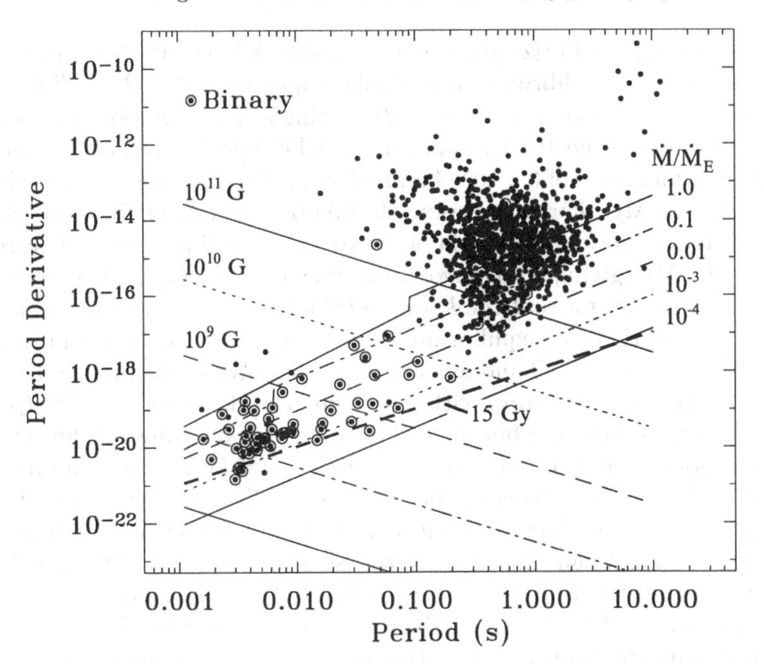

Fig. 4. Spin-evolution diagram. Lines sloping downward to the right show the P-$\dot{P}$ relation for magnetic dipole braking by a field with the strength indicated. Lines sloping upward to the right show the equilibrium spin period of a neutron star with the accretion rate indicated by the labels and a dipole field of the strength indicated by the downward-sloping lines. The dashed line sloping upward to the right shows where stars with a spin-down time equal to 15 Gy would lie. Data points are known rotation-powered pulsars; those of pulsars in binary systems are encircled. From [52]; data from [35].

The lines for $\dot{M} = \dot{M}_E$ and $\dot{M} = 0.1\dot{M}_E$ have jumps where the structure of the disk at the angular momentum coupling radius r_m changes from *RPD* (lower left) to *GPD* (upper right); in reality the transition is smooth. For $\dot{M}$ less than $\sim 0.01\dot{M}_E$, the disk is *GPD* at r_m even if the star's magnetic field is less than $\sim 3 \times 10^7$ G. Not shown are the effects of the stellar surface and the innermost stable circular orbit [52], which affect the spin evolution at spin periods less than ~ 1 ms.

The spin rates of the known *MSP*s in *LMXB*s (see Table 1) are consistent with spin-up by accretion. The existence of only a single candidate with a spin rate greater than 620 Hz could be because (1) these stars have reached accretion spin equilibrium and $\nu_{\rm eq}$ is less than 620 Hz for their appropriately (torque-weighted) time-averaged accretion rates, (2) they are still spinning up but the spin-up timescales for their current accretion rates are longer than the times they have been accreting at these rates, or (3) an additional braking torque is acting on them.

For example, the 45 Hz spin rate of the pulsar EXO 0748−676 corresponds to accretion spin equilibrium for a dipole magnetic field of 2×10^9 G and a time-averaged accretion rate of $10^{-2}\dot{M}_E$, giving a spin-evolution time scale of ~20 Myr, whereas the 191 Hz spin rate of XTE J1807.4−294 corresponds to equilibrium for a field of 3×10^8 G and $\dot{M} \approx 10^{-2}\dot{M}_E$, giving a spin-evolution time scale ~80 Myr. For comparison, the 600 Hz spin rate of 4U 1608−52 corresponds to equilibrium for a field of 3×10^7 G and a time-averaged accretion rate of $10^{-3}\dot{M}_E$, giving a spin-evolution time scale ~2 Gyr. These examples show that the spin rates of the known *MSP*s in *LMXB*s are consistent with spin-up to accretion spin equilibrium if they have magnetic fields in the range $\sim 3 \times 10^7$ G to $\sim 2 \times 10^9$ G and average accretion rates in the range $\sim 10^{-3}\dot{M}_E$ to $\sim 10^{-2}\dot{M}_E$, but that stars with accretion rates less than $\sim 10^{-3}\dot{M}_E$ may not be in spin equilibrium but instead spinning up on a timescale longer than their accretion phases. In particular, the number of *MSP*s with spin frequencies >620 Hz [20] may be small because the equilibrium spin rates of these stars are <620 Hz or because their spin-up timescales are longer than their accretion phases. As an example, the timescale to reach the 1122 Hz spin frequency reported for XTE J1739−285 [43] is 400 Myr for a long-term average accretion rate of $10^{-2}\dot{M}_E$ and 3 Gyr for an accretion rate of $10^{-3}\dot{M}_E$. The ranges of magnetic fields and accretion rates required are consistent with the other observed properties of neutron stars in *LMXB*s [20, 69, 78].

If their magnetic fields are weak enough, it is possible that the spin rates of some neutron stars in *LMXB*s are affected by gravitational radiation torques. Based on the limited information then available, some authors [13, 94] speculated that neutron stars in *LMXB*s have negligible magnetic fields and spin frequencies in a narrow range, with many within 20% of 300 Hz. Such a distribution would be difficult to explain by accretion torques and was taken as evidence that gravitational radiation plays an important role. We now know (see § 2) that most if not all neutron stars in *LMXB*s have dynamically important magnetic fields, that the observed spins of neutron stars in *LMXB*s are distributed roughly uniformly from <200 Hz to >600 Hz, and that production of gravitational radiation by uneven heating of the crust or excitation of r-waves is not as easy as was originally thought [56, 94]. At present there is no unambiguous evidence that the spin rates of neutron stars in *LMXB*s are affected by gravitational radiation.

Production of millisecond rotation-powered pulsars.—Soon after rotation-powered radio-emitting *MSP*s were discovered, it was proposed that they have been spun up to millisecond periods by steady accretion in *LMXB*s at rates $\sim\dot{M}_E$ (see [12]), with the implicit assumption that accretion then ends suddenly; otherwise the stars would track $\nu_{\rm eq}$ to low spin rates as the accretion phase ends. This simplified picture is sometimes still used (see, e.g., [3]), but—as noted above—most neutron stars in *LMXB*s accrete at rates $\ll\dot{M}_E$, many accrete only episodically, and the accretion rates of others dwindle as their binary systems evolve. The real situation is therefore more complex.

The initial spins of rotation-powered *MSP*s recycled in *LMXB*s are the spins of their progenitors when they stopped accreting. These spins depend sensitively on the magnetic fields and the appropriately averaged accretion rates of the progenitors when accretion ends. Comparison of the equilibrium spin-period curves for a range of accretion rates with the P–$\dot{P}$ distribution of known rotation-powered *MSP*s (Fig. 4) suggests three important conclusions:

(1) The hypothesis that the accretion torque vanishes at a spin frequency close to the calculated $\nu_{\rm eq}$ predicts that *MSP*s should not be found above the spin-equilibrium line for $\dot{M} = \dot{M}_E$, because this is a bounding case. The observed P–$\dot{P}$ distribution is consistent with this requirement for the *RPD* model of the inner disk that was used for $\dot{M}$ greater that $\sim 0.1\dot{M}_E$, except for two pulsars recently discovered in globular clusters: B1821−24 and B1820−30A [34]. Either the intrinsic $\dot{P}$'s of these pulsars are lower than shown in Fig. 4, or the *RPD* model of the inner disk does not accurately describe the accretion flow that spun up these stars.

(2) The accretion spin-equilibrium hypothesis predicts that *MSP*s should be rare or absent below the spin-equilibrium line for $\dot{M} = 10^{-4}\dot{M}_E$, because stars accreting at such low rates generally will not achieve millisecond spin periods during their accretion phase. The observed P–$\dot{P}$ distribution is consistent with this prediction.

(3) The *MSP*s near the 15 Gyr spin-down line were produced *in situ* by final accretion rates less than $\sim 3\times 10^{-3}\dot{M}_E$ rather than by spin-up to shorter periods by accretion at rates greater than $\sim 3\times 10^{-3}\dot{M}_E$ followed by magnetic braking, because braking would take too long. This result accords with the expectation (see above) that most neutron stars in *LMXB*s accrete at rates $\ll \dot{M}_E$ toward the end of their accretion phase.

3 Nuclear-Powered X-ray Oscillations

Accretion of matter onto the surface of a neutron star produces a fluid ocean on top of the solid crust (see [84]). Depending on the accretion rate and the initial composition of the material, the conditions needed to ignite hydrogen and helium can be reached. Ignition of the accreted matter will generally occur at a particular place on the surface of the star. For low to moderate accretion rates, burning is unstable and produces type-I X-ray bursts (see [85]). There are several important timescales in this problem, including the time required for burning to spread over the surface of the star, the time required for heat to reach the photosphere, and the timescale on which the heated matter cools by emission of radiation. The time required for burning to spread is expected to depend on the latitude(s) of the ignition point(s), because of the variation with latitude of the Coriolis force, which affects the thickness of the burning front and hence the speed at which it advances. A recent simulation [84] finds that the spreading time is shorter if the ignition point is nearer the equator, because there the burning front is less steep and propagates faster. The time required for burning to spread around the star is expected to be less than a second [10], much smaller than the ∼10–30 s observed durations of the bursts in X-rays, which may reflect the time required for heat from the burning layers to reach the photosphere. If so, nuclear burning is probably over, or almost over, by the time the burst becomes visible in X-rays to a distant observer.

Useful information about the burst evolution can be obtained from the nearly coherent X-ray oscillations seen during portions of some bursts (see [85]). The discovery of burst oscillations with very nearly the same frequencies as the spin rates of two *MSP*s [20,89], as well as the observed stability of these oscillation frequencies [88], eliminated any doubt that burst oscillations are generated by the spin of the star. However, several important questions are not yet fully resolved. In most bursters, the oscillation frequencies vary slightly, especially during the burst rise, but in burst tails the oscillation frequencies often approach an asymptotic value [87] that remains the same to high precision over long times for a given source [90]. Determining what produces these oscillations and what causes the differences in their behavior from burst to burst and star to star is important for understanding the physics of the bursts.

The most widely discussed picture for type-I bursts assumes a hotter region on the surface of the star that has been heated from below and rotates with the star. The increase in the oscillation frequency observed near the beginning of some bursts has been attributed to the collapse of the stellar atmosphere that would occur as it cools after the end of nuclear burning [21,86]. In this model, the rotation rate of the outer envelope and photosphere increases as the outer layers collapse at approximately constant angular momentum. This model is believed to capture an important aspect of the actual rotational behavior of the envelope during the rise of a burst, even though the observed frequency changes are larger than those predicted by the model by factors ∼2–3 [22] and it is not clear how uniform rotation of the gas in the envelope can be maintained in the presence of Coriolis, magnetic, and viscous forces. During the ∼0.1 s rise of bursts, oscillations with relative amplitudes as high as 75% are observed [92] with frequencies that differ from the stellar spin frequency by up to ∼1% [25]. This model probably does not provide a good description of the oscillations during the tails of bursts, when the temperature of the stellar surface is expected to be relatively uniform. During the burst tails, stable X-ray oscillations with amplitudes as large as 15% are observed for up to 30 s, with frequencies that are consistent, within the errors, with the stellar spin frequency [20, 76, 89]. The amplitudes of the oscillations during the rise of bursts appear to be anticorrelated with the X-ray flux, whereas no such relation is apparent during the tails of bursts [76]. These differences suggest that different mechanisms are responsible for the oscillations near the beginning and in the tails of bursts.

Excitation by bursts of r-waves and other non-radial modes (see, e.g., [63]) in the surface layers of neutron stars has been proposed [32,33] as a possible mechanism for producing observable X-ray oscillations during bursts. This idea has been explored further [53, 77], still without distinguishing between the oscillations observed near the beginnings of bursts and those observed in the tails of bursts. As noted above, these have significantly different characteristics. An important challenge for oscillation mode models is to explain the relatively large frequency variations observed during bursts. It has been suggested that these variations can be explained by changes in the character of the oscillations during the burst (e.g., from an r-mode to an interface mode [32]). Other challenges for models that invoke oscillations in the surface layers of the neutron star are to explain what physics singles out one or a few modes from among the very large number that could in principle be excited and how these modes can produce the X-ray oscillations with the relatively large amplitudes and high coherence observed. Further work is needed to resolve these questions.

4 Accretion-Powered Kilohertz QPOs

The properties of the kilohertz *QPO* pairs provide strong hints about the mechanisms that generate them (for a more complete discussion, see [49]):

1. It appears very likely that the frequency of one of the two kilohertz *QPO*s reflects the orbital frequency of gas in the inner disk. The frequencies of the kilohertz *QPO*s are similar to those of orbital motion near neutron stars. They also vary by hundreds of Hertz on time scales as short as minutes (see, e.g., [65, 96, 101]). Such large, rapid variations are possible if they are related to orbital motion at a radius that varies [46].

2. The star's spin is somehow involved in producing the frequency separation of the two kilohertz *QPO*s in a pair. This involvement is clear in XTE J1807.4−294, where $\Delta\nu_{\rm QPO} \approx \nu_{\rm spin}$, and in SAX J1808.4−3658, where $\Delta\nu_{\rm QPO} \approx \nu_{\rm spin}/2$. It is strongly indicated in the other kilohertz *QPO* sources, because in all cases where both $\Delta\nu_{\rm QPO}$ and $\nu_{\rm spin}$ have been measured, the largest value of $\Delta\nu_{\rm QPO}$ is consistent or approximately consistent with either $\nu_{\rm spin}$ or $\nu_{\rm spin}/2$ (see [46, 49, 96]).

3. A mechanism that produces a single sideband is indicated. Most mechanisms that modulate the X-ray brightness at two frequencies (such as amplitude modulation) would generate at least two strong sidebands. Although weak sidebands have been detected close to the frequency of the lower kilohertz *QPO* [36, 37], at most two strong kilohertz *QPO*s are observed in a given system [64, 96]. This suggests that the frequency of one *QPO* is the primary frequency while the other is generated by a single-sideband mechanism. Beat-frequency mechanisms naturally produce a single sideband. Because one *QPO* frequency is almost certainly an orbital frequency, the most natural mechanism would be one in which the second frequency is generated by a beat with the star's spin frequency.

4. Mechanisms for generating kilohertz *QPO* pairs like the 3:2 resonance proposed to explain the high-frequency *QPO*s observed in black hole can-

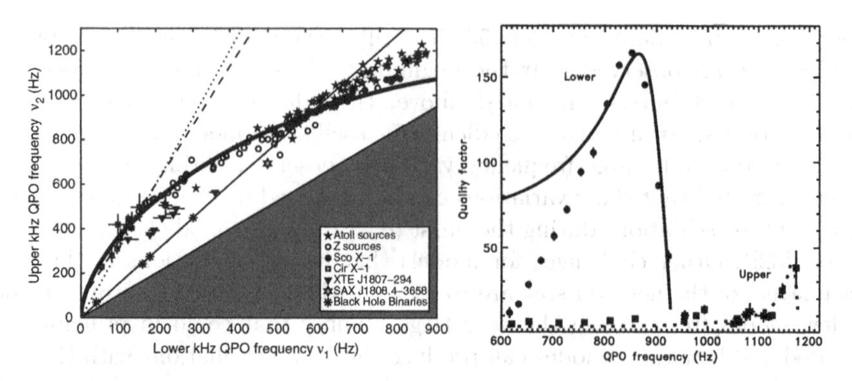

Fig. 5. *Left*: The total variation (power density times frequency) is dominated by the kilohertz *QPO*s, which in this observation are at ~600 Hz and ~900 Hz. The horizontal-branch oscillation (*HBO*) at ~50 Hz and its second harmonic at ~100 Hz are also visible, as are several broad-band noise components. From [103]. *Right*: Correlation between the upper and lower kilohertz *QPO* frequencies seen in all sources in which both have been detected. The filled stars indicate frequencies seen in Atoll sources; the empty circles indicate those seen in Z sources. The shaded boxes (bottom left) indicate the frequencies of the kilohertz *QPO* pair recently discovered in Cir X-1. The shaded stars indicate the frequencies of the *QPO* pair seen in the *MSP* SAX J1808.4−3658. A few high frequency *QPO*s from black hole systems are included; the various lines represent fits of formulas to parts of the data. From [15]. *Right*: Quality factors of the lower and upper kilohertz *QPO*s observed in 4U 1636−53 and the expected frequency-dependence of the quality factor predicted by a model of an active oscillating region approaching the *ISCO*. From [4].

didates [1] and the kilohertz *QPO*s observed in SAX J1808.4−3658 [44] are excluded as explanations for the kilohertz *QPO* pairs seen in neutron stars, because these mechanisms require a low-order resonance between the geodesic frequencies of test particles orbiting at a fixed radius, which (see, e.g., Fig. 5). As noted above, in many neutron stars the separation frequency is approximately constant, which is incompatible with a fixed frequency ratio [14, 15]. This type of mechanism also cannot explain the commensurability of $\Delta\nu_{\rm QPO}$ with $\nu_{\rm spin}$ in all neutron stars in which both frequencies have been measured [47].

5. Production of kilohertz *QPO*s by oscillating emission from a narrow annulus in the accretion disk is incompatible with their quality factors and amplitudes, for the following reason [49]: A kilohertz *QPO* peak of relative width $\delta\nu_{\rm QPO}/\nu_{\rm QPO}$ corresponds to the spread of geodesic frequencies in an annulus of relative width $\delta r/r \sim \delta\nu_{\rm QPO}/\nu_{\rm QPO}$. The emission from an annulus in the inner disk of relative width $\delta r/r$ is a fraction $\sim(\delta r/r)\, L_{\rm disk}$ of the emission from the entire disk and hence a fraction $\sim(\delta r/r)\, [L_{\rm disk}/(L_{\rm disk} + L_{\rm star})]$ of the emission from the system. Thus the relative amplitude of a *QPO* of width $\delta\nu_{\rm QPO}$ produced by oscillating emission from such an annulus is $\lesssim(\delta\nu_{\rm QPO}/\nu_{\rm QPO})\, [L_{\rm disk}/(L_{\rm disk} + L_{\rm star})]$. Some kilohertz *QPO*s have relative

widths $\delta\nu_{\rm QPO}/\nu_{\rm QPO} \lesssim 0.005$ (see [4–6, 69, 96, 97, 101]) and the accretion luminosity of a neutron star is typically ~5 times the accretion luminosity of the entire disk [67]. Consequently, even if the emission from the annulus were 100% modulated at the frequency of the kilohertz *QPO*, which is very unlikely, the relative amplitude of the *QPO* would be only ~$0.005 \times 1/6$ ~0.08%, much less that the 2–60 keV relative amplitudes ~15% observed in many kilohertz *QPO* sources (see, e.g., [69, 96, 97, 101]).

A recently proposed modification [49] of the original sonic-point beat-frequency model [69] potentially can explain within a single framework why the frequency separation is close to $\nu_{\rm spin}$ in some stars but close to $\nu_{\rm spin}/2$ in others. In this "sonic-point and spin-resonance" (*SPSR*) beat-frequency model, gas from perturbations orbiting at the sonic-point radius r_{sp} produces a radiation pattern rotating with a frequency ν_u close to the orbital frequency $\nu_{\rm orb}$ at r_{sp}, as in the original model, and this rotating pattern is detected as the upper kilohertz *QPO*. This mechanism for generating the upper kilohertz *QPO* is supported by the observed anticorrelation of the upper kilohertz *QPO* frequency with the normal branch oscillation flux in Sco X-1 [107] and the anticorrelation of the kilohertz *QPO* frequency with the mHz *QPO* flux in 4U 1608−52 [106].

A new ingredient in the modified model is preferential excitation by the magnetic and radiation fields rotating with the neutron star of vertical motions in the disk at the "spin-resonance" radius r_{sr} where $\nu_{\rm spin} - \nu_{\rm orb}$ is equal to the vertical epicyclic frequency ν_ψ. Preliminary numerical simulations show that the resulting vertical displacement of the gas in the disk is much greater at the resonant radius than at any other radius. In a Newtonian $1/r$ gravitational potential, $\nu_\psi(r) = \nu_{\rm orb}(r)$. Although $\nu_\psi(r)$ is not exactly equal to $\nu_{\rm orb}(r)$ in general relativity, the difference is <2 Hz at the radii of interest (where $\nu_{\rm orb}$ <300 Hz). Consequently, at the resonance radius where vertical motion is preferentially excited, $\nu_{\rm orb} \approx \nu_\psi \approx \nu_{\rm spin}/2$. At this radius, the orbital and vertical frequencies are both approximately $\nu_{\rm spin}/2$.

In the *SPSR* model, the clumps of gas orbiting the star at the sonic radius r_{sp} act as a screen, forming the radiation from the stellar surface into a pattern that rotates around the star with frequency $\nu_{\rm orb}(r_{sp})$. Interaction of this rotating radiation pattern with the gas in the disk that has been excited vertically at r_{sr} produces a second *QPO* with frequency $\nu_\ell = \nu_{\rm orb}(r_{sp}) - \nu_{\rm spin}/2$, if the gas at r_{sr} is highly clumped, or with frequency $\nu_\ell = \nu_{\rm orb}(r_{sp}) - \nu_{\rm spin}$, if the flow at r_{sr} is relatively smooth. This second *QPO* is the lower kilohertz *QPO*.

To see how the observed *QPO* frequency relations can be generated, suppose first that the distribution of the gas in the disk near the spin-resonance radius is relatively smooth. There may be a large number of small clumps or the flow may even be smooth. Each element of gas is oscillating vertically with frequency $\nu_{\rm spin}/2$, but together they form a pattern of raised fluid elements that rotates around the star with frequency $\nu_{\rm spin}$. Because a large number of fluid elements are scattering radiation to the observer at any given

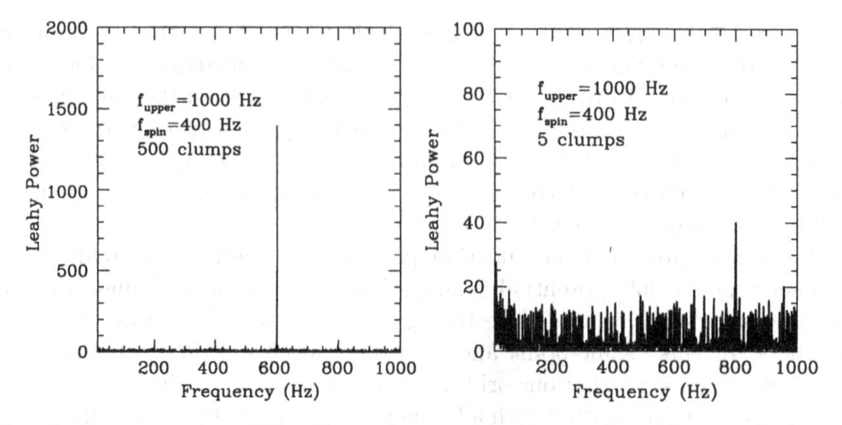

Fig. 6. Power spectra of the X-ray flux modulation produced by simulations of a disk with a large number of clumps near the spin-resonance radius (left-hand panel) and a small number of clumps (right-hand panel). The star's spin frequency is 400 Hz while the orbital frequency at the sonic radius is 1,000 Hz. These power spectra demonstrate that if the flow near the spin-resonance radius is relatively smooth, the effect of the clump pattern dominates and the dominant frequency is $\nu_{\rm orb}(r_{\rm sp}) - \nu_{\rm spin}$. If instead the flow is clumpy, the effect of individual clumps dominates and the dominant frequency is $\nu_{\rm orb}(r_{\rm sp}) - \nu_{\rm spin}/2$. This simulation did not include any signal with the orbital frequency of the gas at the sonic radius.

moment, their individual contributions blend together, so the dominant time variation has frequency $\nu_{\rm orb}(r_{sp}) - \nu_{\rm spin}$. In this case the brightness variation produced by the pattern of scattering clumps dominates the brightness variation produced by the individual clumps. The left-hand panel of Fig. 6 shows the power spectrum of the flux variation generated in a simulation in which 500 randomly-positioned clumps scatter the radiation pattern coming from the sonic radius. The peak at $\nu_{\rm orb}(r_{sp}) - \nu_{\rm spin}$ is clearly dominant.

Suppose instead that the gas in the disk near the spin-resonance radius is highly clumped. When illuminated, each clump orbiting at r_{sr} scatters radiation in all directions. In effect, each clump redirects the radiation propagating outward from the sonic radius in the modest solid angle that it subtends (as seen from the sonic radius) into all directions. From the point of view of a distant observer, each individual clump looks like a light bulb that is blinking on and off with a frequency equal to $\nu_{\rm orb}(r_{sp}) - \nu_{\rm orb}(r_{sr}) \approx \nu_{\rm orb}(r_{sp}) - \nu_{\rm spin}/2$. If there are only a modest number of clumps at r_{sr}, the scattering from the individual clumps dominates the time variation of the X-ray flux. The right-hand panel of Fig. 6 shows the power spectrum of the flux variation generated in a simulation in which five randomly-positioned clumps scatter the radiation pattern coming from the sonic radius. The peak at $\nu_{\rm orb}(r_{\rm sp}) - \nu_{\rm spin}/2$ is clearly dominant. Because the radiation is scattered in all directions, an observer does not have to be close to the disk plane to see the X-ray flux modulation.

Magnetic forces may cause the gas in the accretion disk to become more clumped as it approaches the neutron star [48, 49, 69]. Consequently, the parameters that may be most important in determining whether the flow at the spin resonance radius r_{sr} is clumpy or smooth are the star's spin frequency and magnetic field. For a given stellar magnetic field, the flow is likely to be more clumpy if the star is spinning rapidly and r_{sr} is therefore close to the star. For a given spin rate, the flow is likely to be more clumpy if the star's magnetic field is stronger.

The four sources with $\nu_{\rm spin} < 400$ Hz and measurable frequency separations have $\Delta\nu_{\rm QPO} \approx \nu_{\rm spin}$ whereas the five sources with $\nu_{\rm spin} > 400$ Hz have $\Delta\nu_{\rm QPO} \approx \nu_{\rm spin}/2$ (see [74]). With such a small sample, one cannot make any definite statements, but the apparent trend is consistent with the sonic-point and spin-resonance beat-frequency model. These trends suggest that if kilohertz *QPO*s are detected in the recently-discovered 185 Hz and 314 Hz accretion-powered X -ray pulsars XTE J0929−314 [24] and XTE J1814−338 [89], their frequency separations should be approximately equal to their respective spin frequencies. The 435 Hz spin frequency of XTE J1751−305 [62] is high enough that $\Delta\nu_{\rm QPO}$ could be either approximately 435 Hz or approximately 217 Hz; *QPO*s at both frequencies might even be detectable.

Finally, we note that there is no known reason why the mechanism for producing a lower kilohertz *QPO* proposed in the original sonic-point beat-frequency model would not operate. Apparently this mechanism does not produce a strong *QPO* in the fast rotators, but it might produce a weak *QPO* in these sources. If it operates in the slow rotators, it would produce a *QPO* near $\nu_{\rm orb}(r_{sp}) - \nu_{\rm spin}$ that might appear as a sideband to the lower kilohertz *QPO*.

The sonic-point and spin-resonance beat-frequency model appears qualitatively consistent with the basic properties of the kilohertz *QPO*s, but whether it can explain their detailed properties and the wide range of frequencies seen in different systems, such as Circinus X-1, remains to be determined.

5 Kilohertz QPOs as Tools

As explained in the previous section, despite uncertainty about the precise physical mechanisms responsible for generating the kilohertz *QPO* pairs seen in neutron star systems, there is good evidence that the upper kilohertz *QPO* is produced by orbital motion of gas in the strong gravitational field near the star. Making only this minimal assumption, the kilohertz *QPO*s can be used as tools to obtain important constraints on the masses and radii of the neutron stars in *LMXB*s and explore the properties of ultradense matter and strong gravitational fields (see [50, 51, 69]).

For example, the left panel of Fig. 7 shows how to construct constraints on the mass and radius of a nonrotating neutron star, given ν_u^*, the highest orbital frequency observed in the source. $R_{\rm orb}$ must be greater than the stellar

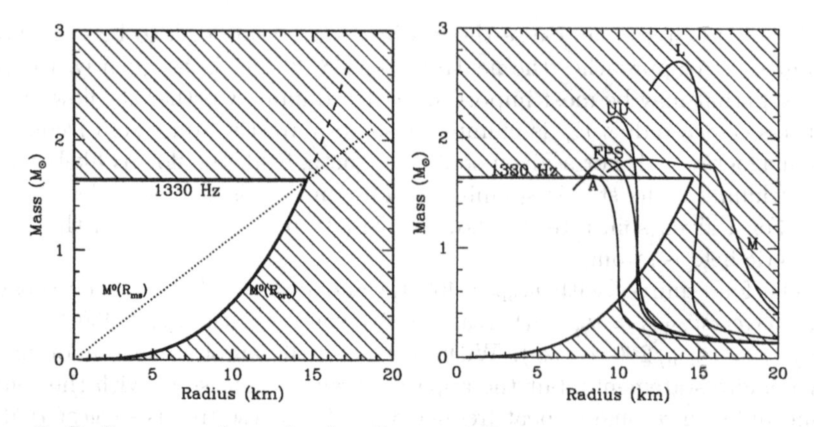

Fig. 7. Radius-mass plane showing the constraints on neutron star masses and radii and the equation of state of neutron-star matter that can be derived from the frequency of the upper kilohertz *QPO*, here 1330 Hz, which is thought to be the orbital frequency of gas accreting onto the star. *Left panel*: The dashed curved line shows the relation between the mass of the star and the radius $R_{\rm orb}$ of the orbit for a nonrotating star, which is an upper bound on the radius of a nonrotating star. The diagonal dotted line shows the relation between the mass of the star and the radius $R_{\rm ms}$ of the marginally stable orbit, which must be larger than $R_{\rm orb}$ in order for the gas to make the hundreds of orbits around the star indicated by the coherence of the kilohertz *QPO* waveform. Consequently the mass and radius of the star must correspond to a point inside the unshaded "slice of pie". If the *QPO* frequency is shown to be that of the marginally stable orbit, then $R_{\rm orb} = R_{\rm ms}$ and the mass of the star is determined precisely. *Right panel*: Curves of mass-radius relations for nonrotating stars constructed using several proposed neutron-star matter equations of state (*EOS*), showing that a 1330 Hz *QPO* is just inconsistent with *EOS* M. The higher the observed *QPO* frequency, the tighter the constraints. After [69].

radius, so the star's representative point must lie to the left of the (dashed) cubic curve $M^0(R_{\rm orb})$ that relates the star's mass to the radius of orbits with frequency ν_u^*. The high coherence of the oscillations constrains $R_{\rm orb}$ to be greater than $R_{\rm ms}$, the radius of the innermost stable orbit, which means that the radius of the actual orbit must lie on the $M^0(R_{\rm orb})$ curve below its intersection with the (dotted) straight line $M^0(R_{\rm ms})$ that relates the star's mass to $R_{\rm ms}$. These requirements constrain the star's representative point to lie in the unhatched, pie-slice shaped region enclosed by the solid line. The allowed region shown is for $\nu_u^* = 1330$ Hz, the highest value of ν_u observed in 4U 0614+09 [98], which is also the highest value so far observed in any source.

The right panel of Fig. 7 shows how this allowed region compares with the mass-radius relations given by five representative equations of state (for a description of these *EOS* and references to the literature, see [68]). If 4U 0614+09 were not spinning, EOS L and M would both be excluded. However, 4U 0614+09 is undoubtedly spinning (the frequency separation $\Delta\nu_{\rm QPO}$ between its two kilohertz *QPO*s varies from 240 Hz to 355 Hz [98]). If its spin frequency is high, *EOS* M may be allowed, but *EOS* L is excluded for any spin rate.

Assuming that the upper kilohertz *QPO* at ν_u is produced by orbital motion of gas near the neutron star, its behavior can be used to investigate orbits in the region of strongly curved spacetime near the star. For example, it may be possible to establish the existence of an innermost stable circular orbit (*ISCO*) around some neutron stars in *LMXB*s (see [39, 51, 69]. This would be an important step forward in determining the properties of strong gravitational fields and dense matter, because it would be the first confirmation of a prediction of general relativity in the strong-field regime.

The sonic-point model of the kilohertz *QPO*s predicts several signatures of the *ISCO* [50, 68]. As an example, it predicts that the frequencies of both kilohertz *QPO*s will increase with the increasing accretion luminosity until the sonic radius—which moves inward as the mass flux through the inner disk increases—reaches the *ISCO*, at which point the frequencies of the kilohertz *QPO*s will become approximately independent of the accretion luminosity. Behavior similar to this has been observed [16, 42, 110], but important issues, such as the robustness of the predicted relation between *QPO* frequency and $\dot{M}$, need further work.

The sonic-point model also predicts a steep drop in the coherence of the kilohertz *QPO*s as the orbits involved approach the *ISCO* [50, 69, 70]. Abrupt drops have been observed in the quality factors of the kilohertz *QPO*s in several atoll sources, consistent with models of their expected behavior as the orbit involved approaches the *ISCO* [4, 5].

If either of these behaviors can be shown to be caused by an *ISCO*, it will be a major advance in establishing the properties of strong-field gravity.

6 Acknowledgements

We thank D. Chakrabarty, C.J. Cook, J.M. Cook, M. van der Klis, M.C. Miller, and J. Swank for helpful discussions. This research was supported in part by NASA grant NAG5-12030, NSF grant AST0098399, and funds of the Fortner Endowed Chair at Illinois.

References

1. M.A. Abramowicz, W. Kluzniak: A&A **374**, L19 (2001)
2. M.A. Alpar, A.F. Cheng, M.A. Ruderman, J. Shaham: Nature **300**, 178 (1982)
3. Z. Arzoumanian, J.M. Cordes, I. Wasserman: ApJ **520**, 696 (1999)
4. D. Barret, J.-F. Olive, M.C. Miller: MNRAS, **370**, 1140 (2006)
5. D. Barret, J.-F. Olive, M.C. Miller: MNRAS, **376**, 1139 (2007)
6. M. Berger, M. van der Klis, J. van Paradijs, et al.: ApJ **469**, L13 (1996)

7. D. Bhattacharya, G. Srinivasan: In *X-Ray Binaries* ed. by W.H.G. Lewin, J. van Paradijs & E.P.J van den Heuvel (Cambridge, Cambridge University Press 1995) 495
8. S. Bhattacharyya, T.E. Strohmayer: ApJ **634**, 157 (2005)
9. S. Bhattacharyya, T.E. Strohmayer: ApJ **636**, 121 (2006)
10. S. Bhattacharyya, T.E. Strohmayer: ApJ **642**, L161 (2006)
11. S. Bhattacharyya, T.E. Strohmayer, C.B. Markwardt, J.H. Swank: ApJ **639**, L31 (2006)
12. D. Bhattacharya, E.P.J. van den Heuvel: Phys. Rep., **203**, 1 (1991)
13. L. Bildsten: ApJ **501**, L89 (1998)
14. T. Belloni, M. Méndez, J. Homan: A&A **437**, 209 (2005)
15. T. Belloni, M. Méndez, J. Homan: MNRAS, **376**, 1133 (2007)
16. P.F. Bloser, J.E. Grindlay, P. Kaaret, W. Zhang, A.P. Smale, D. Barret: ApJ **542**, 1000 (2000)
17. S. Boutloukos, M. van der Klis, D. Altamirano, M. Klein-Wolt, R. Wijnands, P.G. Jonker, R. Fender: ApJ **653**, 1435 (2006)
18. D. Chakrabarty: In *Binary Radio Pulsars* ed. by F.A Rasio and I.H. Stairs (ASP Conf. Series, Vol. 328 2005), 279
19. D. Chakrabarty, E.H. Morgan: Nature **394**, 346 (1998)
20. D. Chakrabarty, E.H. Morgan, M.P. Muno, D.K. Galloway, R. Wijnands, M. van der Klis, C.B. Markwardt: Nature **424**, 42 (2003)
21. A. Cumming, L. Bildsten: ApJ **544**, 453 (2000)
22. A. Cumming, S.M. Morsink, L. Bildsten, J.L. Friedman, D.E. Holz: ApJ **564**, 343 (2002)
23. R.F. Elsner, P. Ghosh, F.K. Lamb: ApJ **241**, L55 (1980)
24. D.K. Galloway, D. Chakrabarty, E.H. Morgan, R.A. Remillard: ApJ **576**, L137 (2002)
25. D.K. Galloway, D. Chakrabarty, M.P. Muno, P. Savov: ApJ **549**, L85 (2001)
26. D.K. Galloway, M.P. Muno, J.M. Hartman, P. Savov, D. Psaltis, D. Chakrabarty: astro-ph/0608259 (2007)
27. P. Ghosh, F.K. Lamb: ApJ **234**, 296 (1979)
28. P. Ghosh, F.K. Lamb: In *X-Ray Binaries and Recycled Pulsars*, ed. by E.P.J. van den Heuvel & S. Rappaport (Dordrecht: Kluwer 1992), 487
29. J.M. Hartman, D. Chakrabarty, D.K. Galloway et al.: AAS HEAD Meeting No. 35, abstract 17.38 (2003)
30. G. Hasinger, M. van der Klis: A&A **225**, 79 (1989)
31. J.W.T. Hessels, S.M. Ransom, I.H. Stairs, P.C.C. Freire, V.M. Kaspi, F. Camilo: Science **311**, 1901 (2006)
32. J.S. Heyl: ApJ **600**, 939 (2004)
33. J.S. Heyl: MNRAS **361**, 504 (2005)
34. G.B. Hobbs, A.G. Lyne, M. Kramer, C.E. Martin, C.A. Jordan: MNRAS **353**, 1311 (2004)
35. G.B. Hobbs, R.N. Manchester: ATNF Pulsar Catalogue, v1.2, http://www.atnf.csiro.au/research/pulsar/psrcat/psrcat_help.html (2004)
36. P.G. Jonker, M. Méndez, M. van der Klis: ApJ **540**, L29 (2000)
37. P.G. Jonker, M. Méndez, M. van der Klis: MNRAS, **360**, 921 (2005)
38. P.C. Joss, F.K. Li: ApJ **238**, 287 (1980)
39. P. Kaaret, E.C. Ford: Science **276**, 1386 (1997)
40. P. Kaaret, J.J.M. in't Zand, J. Heise, J.A. Tomsick: ApJ **575**, 1018 (2002)

41. P. Kaaret, J.J.M. in't Zand, J. Heise, J.A. Tomsick: ApJ **598**, 481 (2003)
42. P. Kaaret, S. Piraino, P.F. Bloser et al.: ApJ **520**, L37 (1999)
43. P. Kaaret, Z. Prieskorn, J. J. M. in't Zand, S. Brandt, N. Lund, S. Mereghetti, D. Götz, E. Kuulkers, J.A. Tomsick: ApJ **657**, 97 (2007)
44. W. Kluzniak, M.A. Abramowicz, S. Kato, W.H. Lee, N. Stergioulas: ApJ **603**, L89 (2003)
45. F.K. Lamb: In *Proc. 23rd ESLAB Symp. on X-ray Astronomy*, ed. by J. Hunt & B. Battrick (ESA SP-296 1989), 215
46. F.K. Lamb: In *X-Ray Binaries and Gamma-Ray Bursts*, ed. by E.P.J. van den Heuvel, L. Kaper, E. Rol & R.A.M.J. Wijers (San Francisco: Astron. Soc. Pacific 2003), 221
47. F.K. Lamb: In *X-Ray Timing 2003: Rossi and Beyond*, ed. by P. Kaaret, F.K. Lamb, J.H. Swank (AIP Conf. Proc., Vol. 714 2004), 3
48. F.K. Lamb, M.C Miller: ApJ **554**, 1210 (2001)
49. F.K. Lamb, M.C Miller: ApJ submitted, astro-ph/0308179 (2007)
50. F.K. Lamb, M.C Miller, D. Psaltis: In *Accretion Processes in Astrophysical Systems: Some Like it Hot*, ed. by S.S Holt and T.R. Kallman (AIP Conf. Proc. No. 431 1998), 389
51. F.K. Lamb, M.C Miller, D. Psaltis: In *The Active X-ray Sky, Results from Beppo-SAX and Rossi-XTE*, ed. by L. Scarsi, H. Bradt, P. Giommi, & F. Fiori, (Nucl. Phys. B. 69 1998), 113
52. F.K. Lamb, W. Yu: In *Binary Radio Pulsars*, ed. by F.A. Rasio and I.H. Stairs (ASP Conf. Series, Vol. 328 2005), 299
53. U. Lee: ApJ **600**, 914 (2004)
54. W.H.G Lewin, J. van Paradijs, R. Taam: In *X-Ray Binaries*, ed. by W.H.G. Lewin, E.P.J. van den Heuvel & J. van Paradijs (Cambridge Univ. Press 1995), 175
55. M. Linares, M. van der Klis, D. Altamirano, C.B. Markwardt: ApJ **634**, 1250 (2005)
56. L. Lindblom, B. Owen: Phys. Rev. D, **65**, 063006 (2002)
57. C.B. Markwardt: personal communication (2003)
58. C.B. Markwardt, E. Smith, J.H. Swank: IAU Circ. 8080, 2 (2003)
59. C.B. Markwardt, T.E. Strohmayer, J.H. Swank: ApJ **512**, L125 (1999)
60. C.B. Markwardt, J.H. Swank: IAU Circ., 8144, 1 (2003)
61. C.B. Markwardt, J.H. Swank, T.E. Strohmayer: ATel **353**, 1 (2004)
62. C.B. Markwardt, J.H. Swank, T.E. Strohmayer, J.J.M. in 't Zand, F.E. Marshall: ApJ **575**, L21 (2002)
63. P.N. McDermott, R.E. Taam: ApJ **318**, 278 (1987)
64. M. Méndez, M. van der Klis: MNRAS, **318**, 938 (2000)
65. M. Méndez, M. van der Klis, E.C. Ford, R.A.D. Wijnands, J. van Paradijs: ApJ **511**, L49 (1999)
66. M.C Miller: personal communication (2005)
67. M.C. Miller, F.K. Lamb: ApJL **413**, L43 (1993)
68. M.C Miller, F.K. Lamb, G.B. Cook: ApJ **509**, 793 (1998)
69. M.C Miller, F.K. Lamb, D. Psaltis: ApJ **508**, 791 (1998)
70. M.C Miller, F.K. Lamb, D. Psaltis: In *The Active X-ray Sky, Results from Beppo-SAX and Rossi-XTE*, ed. by L. Scarsi, H. Bradt, P. Giommi, & F. Fiori (Nucl. Phys. B. 69 1998) 123
71. E. Morgan, P. Kaaret, R. Vanderspek: ATel, 523 (2005)

72. M.P. Muno: AIP conference proceedings, Vol. 714, pp 239–244 (2004)
73. M.P. Muno, D. Chakrabarty, D.K. Galloway, D. Psaltis: ApJ **580**, 1048 (2002)
74. M.P. Muno, D. Chakrabarty, D.K. Galloway, P. Savov: ApJ **553**, L157 (2001)
75. M.P. Muno, D.W. Fox, E.H. Morgan, L. Bildsten: ApJ **542**, 1016 (2000)
76. M.P. Muno, F. Özel, D. Chakrabarty: ApJ **581**, 550 (2002)
77. A.L. Piro, L. Bildsten: ApJ **629**, 438 (2005)
78. D. Psaltis, D. Chakrabarty: ApJ **521**, 332 (1999)
79. D. Psaltis, F.K. Lamb: In *Neutron Stars and Pulsars*, ed. by N. Shibazaki, N. Kawai, S. Shibata, & T. Kifune (Tokyo: Univ. Acad. Press 1998), 179
80. V. Radhakrishnan, G. Srinivasan: Curr. Sci. **51**, 1096 (1982)
81. H. Ritter, A.R. King: In *ASP Conf. Ser. Vol. 229, Evolution of Binary and Multiple Star Systems*, ed. by P. Podsiadlowski, S.A. Rappaport, A.R. King, F. D'Antona, & L. Burderi (San Francisco, Astron. Soc. Pac. 2001), 423
82. N. Shibazaki, T. Murakami, J. Shaham, K. Nomoto: Nature **342**, 656 (1989)
83. D.A. Smith, E.H. Morgan, H. Bradt: ApJ **482**, L65 (1997)
84. A. Spitkovsky, Y. Levin, G. Ushomirsky: ApJ **566**, 1018 (2002)
85. T. Strohmayer, L. Bildsten: In *Compact Stellar X-ray Sources*, ed by W.H.G. Lewin & M. van der Klis, (Cambridge University Press 2006) pp 113–156, astro-ph/0301544
86. T.E. Strohmayer, K. Jahoda, A.B. Giles, U. Lee: ApJ **486**, 355 (1997)
87. T.E. Strohmayer, C.B. Markwardt: ApJ **516**, L81 (1999)
88. T.E. Strohmayer, C.B. Markwardt: ApJ **577**, 337 (2002)
89. T.E. Strohmayer, C.B. Markwardt, J.H. Swank, J.J.M. in't Zand: ApJ **596**, L67 (2003)
90. T.E. Strohmayer, W. Zhang, J.H. Swank, I. Lapidus: ApJ **503**, L147 (1998)
91. T.E. Strohmayer, W. Zhang, J.H. Swank, A. Smale, L. Titarchuk, C. Day, U. Lee: ApJ **469**, L9 (1996)
92. T.E. Strohmayer, W. Zhang, J.H. Swank, N.E. White, I. Lapidus: ApJ **498**, L135 (1998)
93. T.W.J. Thompson, R.E. Rothschild, J.A. Tomsick, H.L. Marshall: ApJ **634**, 1261 (2005)
94. G. Ushomirsky, C. Cutler, L. Bildsten: MNRAS, **319**, 902 (2000)
95. E.P.J. van den Heuvel: In *X-Ray Binaries and Recycled Pulsars*, ed. by E.P.J. van den Heuvel & S.A. Rappaport (Dordrecht: Kluwer 1992), 233
96. M. van der Klis: ARA&A **38**, 717 (2000)
97. M. van der Klis: In *Compact Stellar X-ray Sources*, W.H.G. Lewin and M. van der Klis, (Cambridge University Press 2006), p. 39
98. S. van Straeten, E.C. Ford, M. van der Klis, M. Méndez, P. Kaaret: ApJ **540**, 1049 (2000)
99. A.R. Villarreal, T.E. Strohmayer: ApJ **614**, L121 (2004)
100. N. White, L. Stella: MNRAS **231**, 325 (1987)
101. R.A.D. Wijnands: In *Pulsars New Research*, in press (New York: Nova Science Publishers 2007), astro-ph/0501264
102. R. Wijnands, T. Strohmayer, L.M. Franco: ApJ **549**, L71 (2001)
103. R. Wijnands, M. van der Klis: Nature **394**, 344 (1998)
104. R.A.D. Wijnands, M. van der Klis, J. Homan, D. Chakrabarty, C.B. Markwardt, E.H. Morgan: Nature **424**, 44 (2003)
105. R.A.D. Wijnands, M. van der Klis, J. van Paradijs, W.H.G. Lewin, F.K. Lamb, B. Vaughan, E. Kuulkers: ApJ **479**, L14 (1997)
106. W. Yu, M. van der Klis: ApJ **567**, 67 (2002)

107. W. Yu, M. van der Klis, P.G. Jonker: ApJ **559**, L29 (2001)
108. W. Zhang, K. Jahoda, R.L. Kelley, T.E. Strohmayer, J.H. Swank, S.N. Zhang: ApJ **495**, L9 (1998)
109. W. Zhang, I. Lapidus, J.H. Swank, N.E. White, L. Titarchuk: IAU Circ. 6541, 1 (1997)
110. W. Zhang, A.P. Smale, T.E. Strohmayer, J.H. Swank: ApJ **500**, L171 (1998)

Observations and Modeling of Accretion Flows in X-ray Binaries

D.A. Leahy

University of Calgary, Calgary, Alberta, Canada, T2N 1N4
leahy@ucalgary.ca

1 Introduction

This article describes what we have learned from observations and from modeling of the observations about several aspects of accretion in X-ray binaries. In particular, what we have learned about the geometry of accretion disk structure, accretion columns, and stellar wind structure will be described. The main discussion will be limited to just two X-ray binaries, Hercules X-1 and GX 301-2, which represent low mass X-ray binary systems and high mass X-ray binary systems, respectively. Both of these systems have orbits determined from X-ray pulse timing studies which make possible the detailed modeling and interpretation of the X-ray data. Although the discussion here is only about two systems, the methods and also the qualitative part of the results can be applied to other similar X-ray binary systems.

2 Hercules X-1: a Low Mass X-ray Binary

2.1 Overview

Her X-1/HZ Her is a low-mass accretion-powered pulsar system exhibiting a great variety of phenomena. This eclipsing system contains a 1.24 second period pulsar in a 1.7 day circular orbit around its optical companion HZ Her.

This binary system also displays a longer 35-day cycle that was first discovered as a repeating pattern of High and Low X-ray flux states ([8]). A Main High state, lasting about ten days, and Short High state, lasting about five days, are each followed by a ten day long Low state, to comprise the full 35-day cycle (reviewed in [45]). X-ray pulsations are visible during the High states but cease during the Low states. Other properties of the 35-day cycle include: pre-eclipse X-ray absorption dips that repeat at nearly the 35-day and 1.7 day beat period ([4, 8, 17, 21, 45]); optical pulsations that occur at certain 35-day and orbital phases ([35]); systematic 35-day variations in the orbital

E.F. Milone et al. (eds.), Short-Period Binary Stars: Observations, Analyses, and Results, 111–133.

optical light curve ([5, 7]); and High state evolution of the X-ray pulse profile ([6, 44, 46]). The complex pulse profile has been modeled successfully using an accretion column model with gravitational light-bending ([31]) and subsequently the accretion column model was used to constrain mass and radius of the neutron star ([32]).

Long term X-ray monitoring of Her X-1 was conducted in the early 1970's using *UHURU* which observed several Main High states and one Short High state ([8, 10]) in the 2–6 keV range. Many of the characteristic features of the 35-day lightcurve such as the High and Low states, the High state eclipses, absorption dips and the tendency of the Main High states to turn on near orbital phases 0.2 and 0.7 were discovered with these observations. Higher resolution short term monitoring has been conducted with *HEAO*, Tenma, *EXOSAT* and Ginga ([18, 26, 36, 37, 44, 49]) which has confirmed many of the X-ray phenomena first reported with the *UHURU* observations, added significant details and discovered several new phenomena such as: eclipses during the Low state ([40]); an extended Low state in the mid-1980's ([36]); pre-eclipse dips caused by photoelectric absorption ([17, 20]); varible eclipse durations with 35-day phase due to changing extended source size at Her X-1 ([19]); and evidence for sub-synchronous rotation of HZ Her ([22]).

Continued monitoring of Her X-1 has been carried out with the *RXTE/ ASM* ([45]) along with modeling of the resulting 35-day light-curve ([27, 30]) to strongly constrain the shape and nature of the accretion disk . Observations with the *Extreme Ultraviolet Explorer* (*EUVE*) have also led to very useful information on the binary system. Reflection of extreme UV photons off of the companion HZ Her during low state and off of the inner edge of the accretion disk during Short high state was discovered by [25] and modeled in detail by [29].

2.2 The Accretion Disk 1: 35-day X-ray Flux Cycle

A twisted tilted accretion disk is required to be present in Her X-1 to explain the 35-day pulse evolution ([46]). The disk consists of a continuous series of tilted rings of increasing radius, with the warp due to an increase of twist (angle of line of nodes) with radius. The physical mechanism which maintains the tilt and twist is radiation reaction force from illumination by the central X-ray source ([55]).

The disk model is illustrated in Fig. 1 from the point of view of the observer for 35-day phase 0.06. The observer is taken to be at 5° above the orbital plane, and the disk shape is rotating counterclockwise as time increases. For illustration purposes, the ratio of inner edge radius to outer edge radius is taken to be much larger than the actual value (~0.0002). The neutron star and central part of the disk are fully visible at phase 0.06. At other 35-day phases the disk is at a different rotation phase about the orbital axis (which is a vertical line through the center of the disk in Fig. 1). The outer and/or

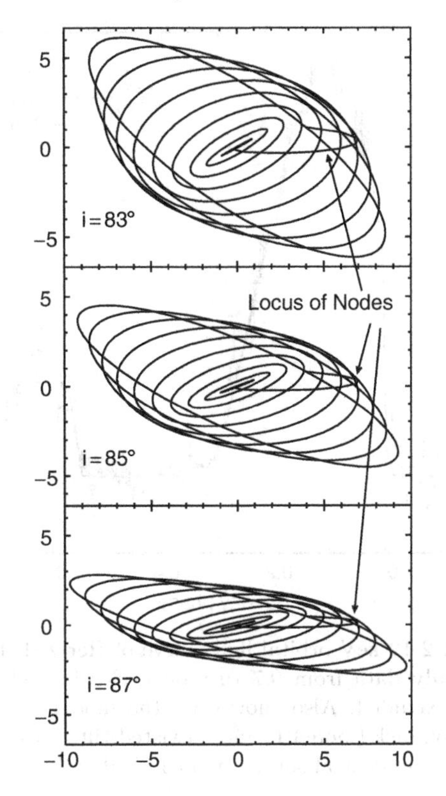

Fig. 1. Illustration of twisted-tilted accretion disk shape from the viewpoint of observer, for system inclinations of 83°, 85° and 87°: circular rings are drawn with linearly increasing radius from 10% to 100% of the disk outer edge radius.

inner edges of the disk will progressively cover and uncover the line of sight to the neutron star as 35-day phase progresses.

The All-Sky-Monitor (*ASM*) on *RXTE* is described by Levine et al. ([33]) and consists of three scanning shadow cameras (*SSC*'s), each with a field of view of 6° by 90° *FWHM*. The *SSC*'s are rotated in a sequence of "dwells" with an exposure typically of 90 seconds, so that the most of the sky can be covered in one day. The dwell data are also averaged for each day to yield a daily-average. The *RXTE*/*ASM* dwell data and daily-average data were obtained from the ASM web site.

Fig. 2 shows the shape of the 35-day light curve of Her X-1, which has been best determined using *RXTE*/*ASM* observations ([27,45]). Eclipse times have been removed from the data here in order to emphasize the 35-day variability. Less precise measurements of the 35-day cycle of Main High – Low – Short High – Low were the original motivation for postulating the existence of the twisted-tilted disk. Also shown in Fig. 2 are some model fits to the light curve (discussed in detail in [27,30]).

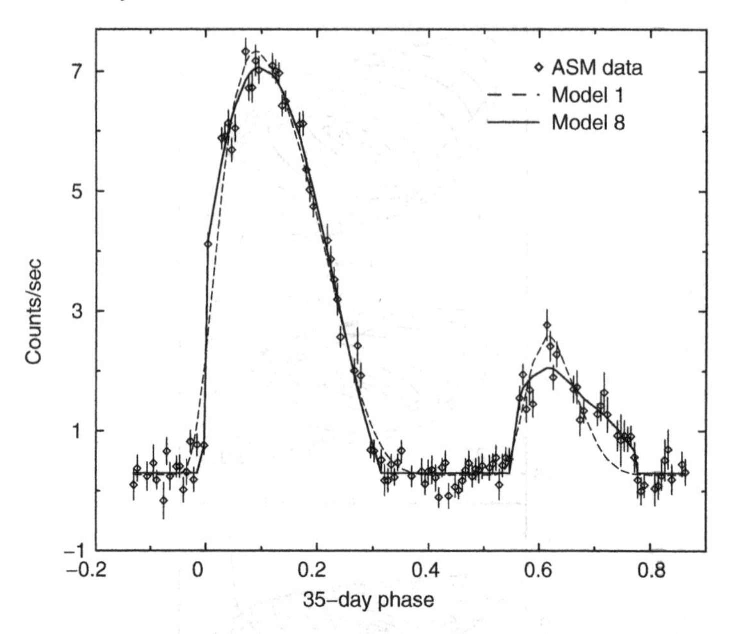

Fig. 2. RXTE/ASM 2-12 keV orbital light-curve of Her X-1, folded at the 35 day long term period. Only data from 0.7 turn-on cycles in included and data from Her X-1 eclipses is excluded. Also shown are the model light curve fits using a twisted-tilted accretion disk (model 1) and a twisted-tilted accretion disk plus inner illuminated disk edge and central source (model 8)- details are given in the text.

One can determine geometric parameters of the disk by calculating the light curve which results from a compact or extended emission source which is eclipsed by a precessing disk. The simplist model is a disk with sharp edges and a point source of X-rays from the neutron star. In this case the photoelectric absorption by the disk edge sets in completely as soon as the disk edge crosses the line-of-sight to the neutron star, yielding a light curve inconsistent with observation. One expects a smoothly decreasing column density at the disk edge and a reasonable model is an exponential atmosphere at the disk edge, characterized by a scale height. The Main High and Short High turn-ons are determined by uncovering of the source by the outer disk edge, and the turn-offs are determined by covering of the source by the inner disk edge. Since the atmosphere is expected to have different properties at the inner disk radius and outer disk radius, it is natural to have two different disk scale heights, one for inner disk edge and one for outer disk edge. The curve labelled model 1 in Fig. 2 is a disk model with different atmospheric scale heights for the inner edge and outer edge and is the simplest model that can give the overall shape of the light curve, although the model does not fit the shape in detail.

In order to accurately reproduce the observed light curve, one can introduce more realistic models of the disk, both inner and outer edges, and of the source region. As discovered by [46], the pulse shape evolution of Her X-1 gives compelling evidence that the neutron star emission region is extended (pencil beam plus reversed fan beam configuration) compared to the sharpness of the inner disk edge. Thus there is strong physical justification for the more complex modeling. Here is briefly outlined the approach taken by [27]. The next more physical disk model has an outer disk edge with a two-layer exponential atmosphere and inner edge with a single atmosphere. The basic idea is that the outer disk has a cold thin atmosphere which is responsible for the sharp turn-on of Main High, and additionally a more extended, likely hot, atmosphere which is responsible for the gradual increase of Main High after turn-on. This hot atmosphere has been observed by a number of studies (e.g. [18]) as extended scattering material in the Her X-1/HZ Her system. The Main High turn-on (uncovering of the neutron star by the outer disk edge) for this model is illustrated in Fig. 3 (the disk atmospheres and source size are not to scale). This is basically a zoomed-in view of Fig. 1 around 35-day phase 0 (recall the disk is rotating clockwise in the view in Fig. 1). As can be

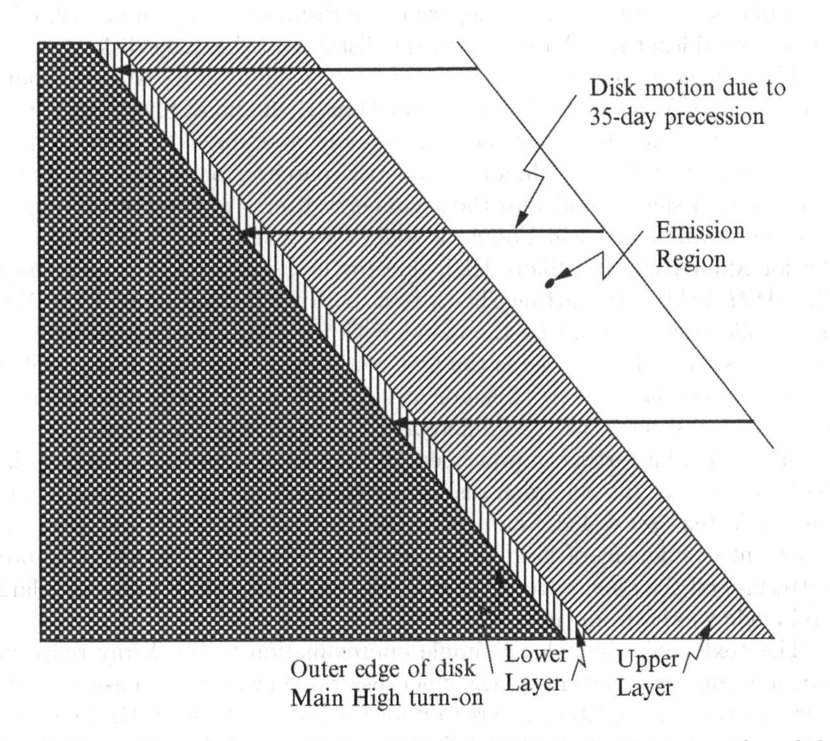

Fig. 3. Illustration of the occultation of the neutron star by the outer edge of the accretion disk: the less-dense outer disk layer results in a slower decrease in intensity, followed by a rapid decrease due to the higher density lower disk layer.

seen from Fig. 1, the orientation of the disk outer edge as it crosses the line-of sight is opposite (inverted up-down) for Short High as for Main High. Thus the equivalent of Fig. 3 for Short High has the disk inverted up-down. The physical size of the emission region is of the order of the neutron star radius, much smaller than the scaleheight of the thin layer at the outer disk edge, which is $\simeq$0.007 radian in angular size, or 10^4 km. The two-layer outer disk model is a much better fit than the single layer model: it has a lower χ^2 and fits the Main High turn-on much better. The scaleheight of the thin layer is equal to better than 1σ (determined by $\delta\chi^2$) of the value determined by the Ginga data (above), 0.0069 radian, so it was fixed at this value for subsequent fitting. The scaleheight of the thick layer was 0.7 radian, much larger than for the thin layer.

The outer disk radius for Her X-1 is $R_o \simeq 2 \times 10^{11}$cm ([4]), whereas the inner disk radius is $R_i \sim 30R_n \sim 4 \times 10^7$cm ([46]) with R_n the neutron star radius, about a factor of 5000 smaller. Since the disk (and its edges) move at a constant angular rate (360° in 35 days) the quantities which determine the shape of the light curve are the angular sizes of the disk edges and of the emission region. The emission from the accretion columns , of size $R_c \sim 2R_n$, is a point source (in angular size) from the distance of the outer disk edge, but is a significant angular size from the distance of the inner disk edge.

Thus the next complexity introduced into the model is an extended source at the neutron star. Since the turn-off of Main High and of Short High states are both due to occultation of the source by the inner disk edge, and the turn-offs is sharper for Main High, an asymmetric source shape at the neutron star is required. A simple model for the asymmetric source is an elliptical uniform disk. This is illustrated in Fig. 4, as well as the cause of the different turn-offs for Main High and Short High due to the asymmetrical source shape. The *ASM* light curve fitting shows that both Main High and Short High are significantly better fit by the model with the addition of the asymmetric extended source. This model is the first model so far which can fit the Short High state reasonably well. This indicates that the asymmetry of the emission region with respect to the two different disk crossings is important.

Motivation for further improvement to the model is based on the fact that the inner edge of the accretion disk is a significant source of emission below a few keV (e.g. see [24, 25] and references therein). There should also be a significant component of hard x-rays which is reflected (primarily Compton scattering) off the inner edge of the disk: the inner edge of the disk has a large solid angle viewed from the neutron star.

The next model includes a simple approximation to the X-ray reflection emission from the inner side of the inner disk ring but returns to using a point source at the neutron star. The reason for the latter is to see if the inner disk edge reflection can produce the difference between the Main High and Short High light curves. The ring is assumed to have a uniform brightness (this is partly for simplicity, but might be expected since the rotation of the pulsar beam over the disk edge is rapid compared to the data sample time). Self-

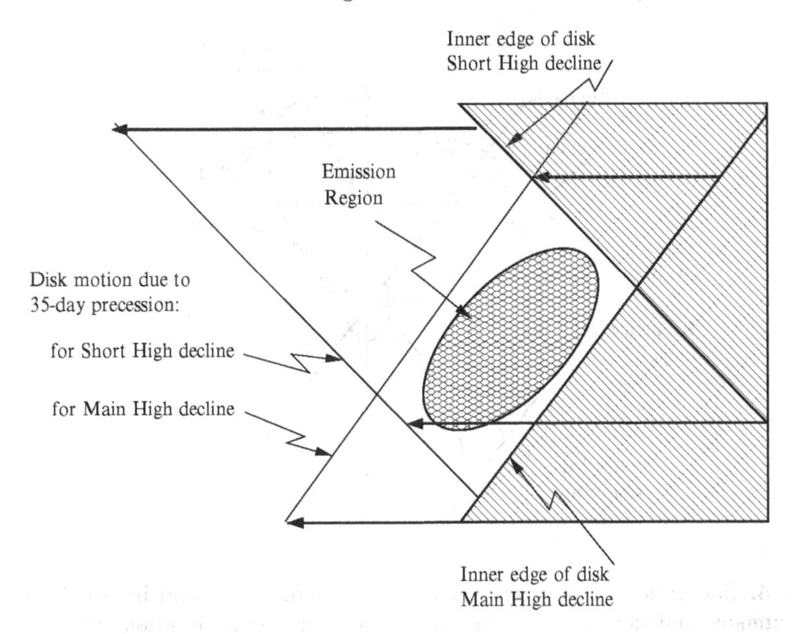

Fig. 4. Illustration of the occultation of a spatially extended source at the neutron star: the source is averaged over a rotation period and represented by an ellipse. The inner disk edge has a different sense of tilt for the Main High occultation and for the Short High occultation, as shown. This results in a more extended decline during Short High than during Main High, as observed.

occultation of the ring is calculated in the model as well as occultation by the accretion disk which extends from the ring outward to the outer edge of the disk. The ring geometry is illustrated in Fig. 5 for the observer viewing the ring at 35-day phase near peak of Main High. For the light curve fitting, the observer inclination to the orbital plane was at first varied giving best-fit models for inclinations near 85°. Then the inclination was fixed at 85° for subsequent fits. The free parameters of the ring are the ring tilt, θ_i, the ring half-angle, θ_r, and the ring precession phase at 35-day phase 0, p_r (defined as 0 for the ring axis pointing most closely at the observer). The visible ring area for Main High is larger than for Short High due to the observer location above the orbital plane: the ring is viewed as more open. The simplest ring model can have no central source but only the ring emission, however for Her X-1 it is known that the central source is visible during Main High. The central source can be modeled as a point source, but the light curve fitting is improved when the central source is extended.

The central source is taken to be circular and described by two parameters: the total intensity, c_c, and angular radius, α_1 (the radius divided by the ring radius). The maximum part of the central source that is visible during Short High is seen when the ring axis is pointing away from the observer. This final model is labelled Model 8 in Fig. 2 and well reproduces the features of

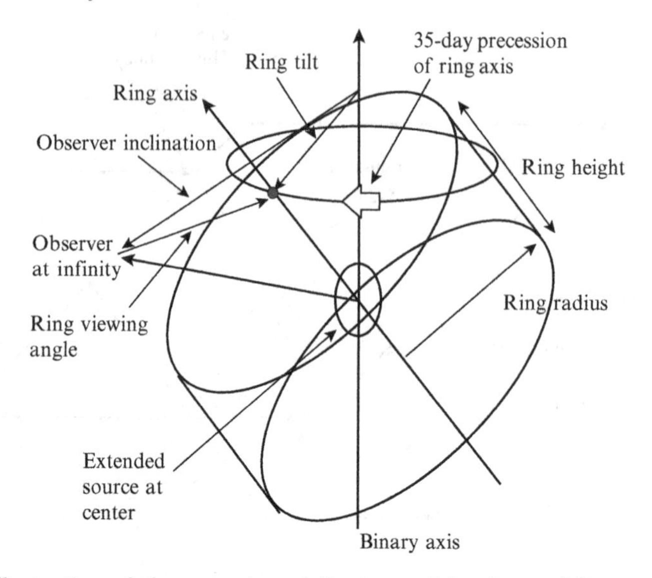

Fig. 5. Illustration of the geometry of the inner disk edge and its occultation of an extended emission region at the neutron star due to disk rotation. The observed emission is from the unobscured part of the central extended source and from the visible part of the inner edge of the disk, which is illuminated by the central source-details are given in the text.

the *RXTE/ASM* observed light curve. The outer disk edge properties are as follows: Main High turn-on occurs at 35-day phase -0.002, defined by position at which the thin layer has optical depth unity. Since the fit gives the thick layer to have an optical depth of 1.30 at this position, the total optical depth at this phase is 2.30. Short High turn-on occurs at 35-day phase interval 0.548 after Main High turn-on. Refering to Fig. 1, it can be seen that the interval between Main High and Short High turn-ons is determined by the tilt of the outer ring, for a fixed observer inclination: a smaller tilt results in a longer interval between turn-ons. Spherical geometry is used to calculate the ring tilt from the turn-on difference: for a system co-inclination 5°, the outer ring tilt is $\theta_o = 30.2°$. The thin layer effective scale height is 0.0069 radians, and the thick layer effective scale height is 0.47 radians, both measured in the orbital plane. These translate to physical scaleheights perpendicular to the disk of $1.4 \times 10^9 sin(\theta_o)cm = 6.9 \times 10^8 cm$ and $9.4 \times 10^{10} sin(\theta_o)cm = 4.7 \times 10^{10} cm$, resp. This assumes the outer disk ring is a planar circle, but if the outer disk ring is distorted then the mean ring tilt, θ_o, and the local tilt of the disk where it crosses the line-of-sight to the neutron star can be different.

Main High decline due to coverage of the central source by the inner ring starts at 35-day phase -0.067, and is complete at phase 0.309. Short High decline due to coverage by the inner ring starts at 35-day phase 0.602 and is complete at phase 0.676. Not including outer disk occultation, the inner ring is visible from the top for 35-day phase -.210 to .322, and from the bottom for 35-day phase .322 to .790. However it is covered by the outer disk edge prior to phase -.002 for Main High and prior to phase .546 for Short High.

In summary, Main High involves the following sequence of events. At 35 day phase -0.002, the outer disk edge starts to uncover the source, which appears like a point source compared to the large physical size of the outer disk structure. The source consists of an extended region inside a tilted thick ring (Fig. 5). The central region has a varying brightness due to opening and closing of the view through the ring. The emission from the inside surface of the ring varies due to changing occultation by the rest of the ring. The ring emission and the central source increase and decline at slightly different times.

The Short High sequence is similar, with a few important differences. The maximum emission from the inside of the tilted ring is not much different than for Main High (see Fig. 5). This is due to the relatively high inclination (40.8^o) of the ring compared to the observer co-inclination (5°), so that most of the inside of the ring is visible whether the ring axis is tilted toward the observer (during Main High ring maximum) or away from the observer (during Short High ring maximum). The timing of events for Short High is as follows. The central source is simultaneously emerging from behind the outer disk double layer atmosphere and being covered up by the inner disk edge from 35-day phase 0.546 to .676. The emission from the inner side of the ring is emerging from behind the outer disk double layer atmosphere starting at 35-day phase 0.546 and is being increasingly occulted by the rest of the ring up until 35-day phase 0.790. Thus for Short High the ring emission is much more important, and dominates the emission for most of Short High.

What is the new information obtained here from the model fitting to the *ASM* light-curve? It has been shown that a two layer structure at the outer disk edge fits the turn-on of Main High state, and the scale heights of the layers are well determined if they are exponential layers. (The two layer structure could be at an intermediate disk radius if the disk twist is not monotonic in which case statements about the outer disk edge apply to the intermediate disk ring). Assuming planar disk rings, the outer ring tilt is large (30°). This is consistent with the conclusion from iron line studies ([26]) that during low state most of the iron line is likely due to reprocessing on the disk surface. To fit the details of the Short High state, a thick inner ring and an extended central source give the best results. The results of this fitting are as follows. The inner disk ring is thick (half-angle of $\simeq 40°$) and significantly tilted ($\sim 40°$). The model emission region is found to have significant extent compared to the radius of the inner disk edge, consistent with the finding that the inner disk edge radius should be small, of order $\sim 30 R_n$ ([46]).

2.3 The accretion disk 2: companion shadowing

Emission from Her X-1/HZ Her covers the optical, ultraviolet, *EUV* and X-ray regime. The hard X-rays (>1 keV) are believed to arise mostly as a result of mass accretion onto the neutron star and are modulated by the neutron star rotation and obscuration by the accretion disk (e.g. [46] and references therein), companion star and moving gas "blobs" that cause the well known absorption dips (e.g. [4]). A small reflected/reprocessed X-ray component is also present that is observable during the low state and eclipses (e.g. [3, 18]). [52] observed Her X-1 during the anomalous low state in 1999 and detected modulation which was shown to be due to reflection of X-rays off of the companion HZ Her. A major portion of the observed optical/ultraviolet emission is believed to arise from X-ray heating of HZ Her and the accretion disk. The X-ray heating causes the surface temperature of HZ Her facing the neutron star to be approximately 10,000 degrees higher than the cooler shadowed side ([2]).

Observations of the broad band optical emission of HZ Her/Her X-1 have been presented by [5] and [53], among others. The broadband optical emission exhibits a complex, systematic variation pattern over the course of the 35-day cycle in addition to the orbital modulation due to X-ray heating of HZ Her. This pattern can be explained as a consequence of disk emission and disk shadowing/occultation of the heated face of HZ Her by a tilted, counter-precessing accretion disk ([7]). The precessing disk also causes the alternating pattern of high and low X-ray intensity states by periodically blocking the neutron star from view, e.g. as modeled by Leahy (2002). [27]. The soft X-ray/extreme ultraviolet band ($\sim$0.016 – 1 keV) has been observed many times during the main or short high states, showing a blackbody spectral component with a temperature of about 0.1 keV (e.g. [48]; [38]). This has generally has been attributed to reprocessing of hard X-rays in the inner region of the accretion disk (e.g. [34], [24]).

Her X-1 has the advantage of a high galactic latitude and hence a low interstellar hydrogen column density, making extreme ultraviolet (EUV) observations feasible. Her X-1 has previously been observed in the EUV energy range ([23, 25, 41, 54]). Rochester et al. (1994) ([41]) detected Her X-1 during a declining phase of the Short High state with the Rosat Wide Field Camera. The Vrtilek et al. (1994) ([54]) observations occurred over the 35-day phase 0.14–0.245, normally associated with the peak and flux decline of an average Main High state ([45]). X-ray observations with BeppoSAX of the middle and later part of a Short High state are reported in Oosterbroek et al. (2000). [38].

During the low state of Her X-1 the inner emission region of Her X-1 is blocked from view. This inner emission region consists of the direct emission from the pulsar plus reprocessed emission at the inner edge of the disk. The geometry of the direct emission and of the reprocessed emission is discussed in [46], deduced from the 35-day evolution of the X-ray pulse profile. Further evidence that a significant part of the reprocessed emission is from the inner edge of the accretion disk is given by, e.g., [24] and [26].

Leahy & Marshall (1999) ([23]) observed Her X-1 with *EUVE* at 35-day phase 0.76-0.88 which normally covers the end of the Short High state and the low state. The 35-day phases quoted are based on nearby Main High state Turn-ons observed with the Burst and Transient Source Experiment (*BATSE*) on the Compton Gamma Ray Observatory (*CGRO*) at JD $-$ 2440000.0 $=$ 9205.14 and 9936.22 for [54] and [23], respectively, rather than the values quoted in those papers that rely on longterm phase extrapolations. The average Short High state ends at 35-day phase 0.76-0.80 ([45]). The count rate during the Leahy & Marshall (1999) ([23]) observation was $\sim$ 0.02 c/s and strongly modulated at the binary period.

The *EUV* modulation during low state can be explained by the accretion disk causing varying illumination of the companion HZ Her with orbital phase. [29] determined that most of the *EUVE* low state emission is reflected emission from the companion star HZ Her but that the accretion disk has a significant contribution. Here in Fig. 6 is shown the shadow of the accretion disk on the celestial sphere (centered on the neutron star). Three cases of system inclination are illustrated using the thin disk model. The reflection flux is calculated numerically by integrating over the surface of HZ Her. HZ Her is very close to filling its Roche lobe (e.g. [22]), thus its surface is taken to

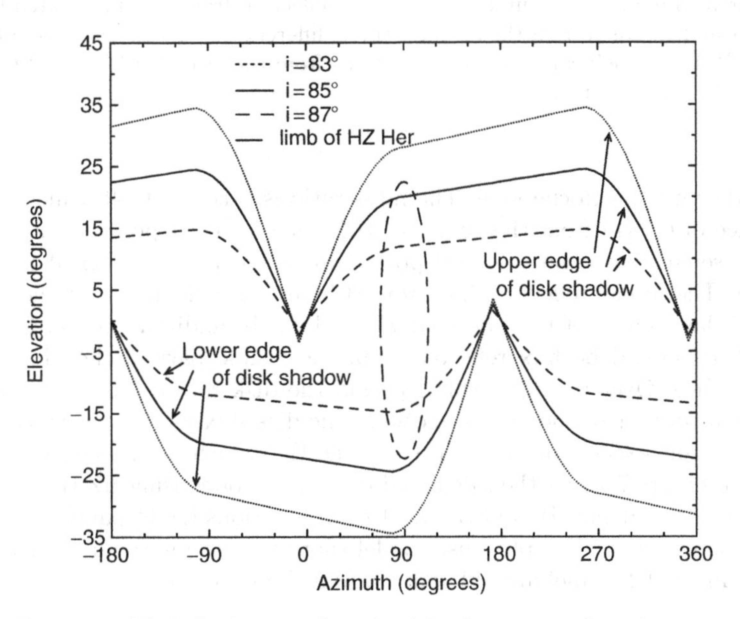

Fig. 6. Diagram of disk shadow, on the celestial sphere centred on the neutron star, for three different system inclinations (83°, 85° and 87°). The face of HZ Her is the circle (seen as an ellipse due to different scales for longitude and latitude): relative to the disk shadow, HZ Her moves 360° in azimuth with frequency $\nu_{orb} - \nu_{35day}$.

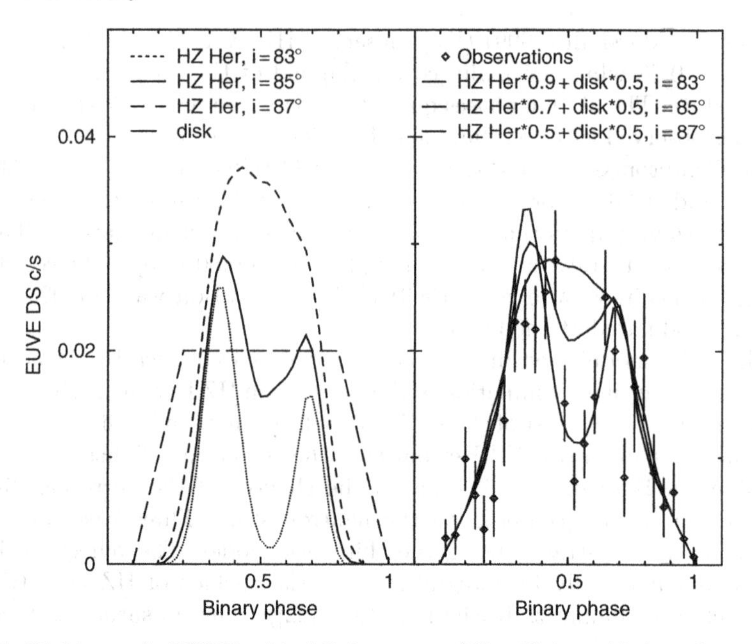

Fig. 7. Right panel: *EUVE* orbital light-curve of Her X-1, at 35-day phase 0.795 compared to model light-curve consisting of reflection from the illuminated face of HZ Her and emission from the disk, for three different system inclinations (83°, 85° and 87°). Left panel: calculated contributions from the face of HZ Her and the disk for the three different inclinations.

coincide with the Roche lobe. The mass ratio is taken as 0.59, which gives a distance of 0.554a from the center of HZ Her to the L1 point, with a is the binary separation. The side and pole radii of HZ Her are then 0.424a and 0.400a. The model reflection flux from HZ Her was calculated for the orbital and 35-day phases of the observed *EUV* data. In addition to flux from HZ Her there should be flux reflected from the visible part of the illuminated disk surface. Over a single orbital period the disk rotates only 17.6° in the direction counter to the orbital motion. The disk flux will be nearly constant over an orbit except for occultation by HZ Her which occurs around orbital phase 0. Figure 7 shows the calculated reflection model using HZ Her and disk components (left panel), and the *EUVE* observations (right panel) compared to the summed HZ Her plus disk model (note that the normalizations of the curves in the left panel are different than in the right panel).

2.4 The Accretion Disk 3: 35-day Pulse Shape Change Cycle

The nature of the large changes in pulse shape seen over the 35-day cycle in Her X-1 has been a long-standing puzzle until recently. Early models included neutron star precession and occultation of the emission region by some material in the system. The puzzle was conclusively solved in the 1990's using *GINGA* and *RXTE* data. A good overview of the observations of pulse shape changes and the most plausible explanation is in a PhD dissertation ([44]). The conclusive observations and argument, which also included proving that neutron star precession cannot work, is given in [46].

2.5 The accretion column

X-ray pulsars pulse due to a relatively constant emission region at a fixed position on the neutron star surface , which due to rotation about an axis different from the emission region location, acts like a light-house to the observer. [46] studied the sequence of appearance and disappearance of different distinctive features in the pulse shape as a function of 35-day phase. Since 35-day phase is itself caused by a precessing accretion disk, there was strong evidence for an extended emission region being occulted to a varying amount by the inner edge of the accretion disk. Using details of the observations, yields arguments in general for pencil beam emission (narrow beaming perpendicular to the neutron star surface) from the near magnetic pole, and fan beam emission (broad beaming parallel to the neutron star surface) from the far magnetic pole.

[31] (and references therein) developed a modeling code which can calculated observed pulse shapes from accretion columns. The shape of the accretion column was taken to be follow magnetic dipole field lines. If the column does not extend high above the neutron star surface, it looks approximately like a pill-box. This is illustrated by the wireframe model in Fig. 8. This figure also illustrates that the two accretion columns (one for each magnetic pole) in general are not antipodal. The calculation of observed brightness vs. rotation phase (i.e. pulse shape) necessarily must include light bending in the strong gravitational field of the neutron star. Figure 9 illustrates the light bending for a 1.4 solar mass object with different radii, as noted. The observer is far off to the left so all rays reaching the observer end up parallel to the z-axis, although they start off in different directions. The main effects of light bending are to distort the image of the surface and to magnify the flux leaving from the back half of the neutron star.

For Her X-1, the detailed pulse shape code was used to model the observed pulse shape by [31]. In addition to the pencil beam (near pole) component and reversed fan beam (far pole) component in the pulse profile, there are other components. In addition it was found necessary to have azimuthally dependent surface emissivity around the sides of both accretion columns. This latter factor is not surprising in retrospect as any realistic model of accretion from a disk has a varying accretion rate around the disk azimuth due to the changing magnetic field geometry. Even in the simplest geometry, the changing magnetic field geometry at the disk is due to a tilted dipole intersecting a

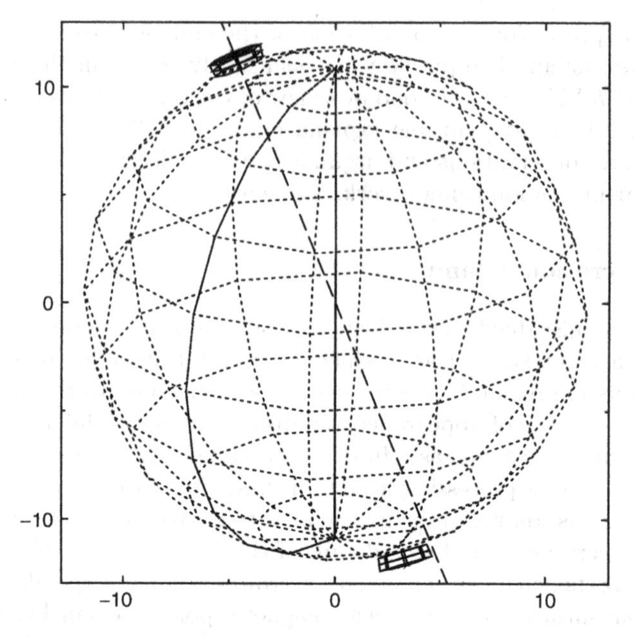

Fig. 8. Wireframe model of vertically extended accretion columns on the surface of neutron star, with offset of the axis of one column from the axis of the second column.

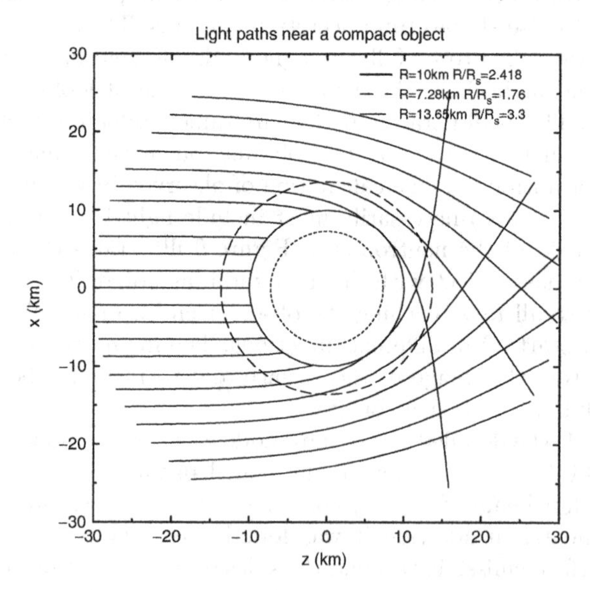

Fig. 9. Illustration of ray paths from the surface of a compact object of 1.4 $M_\odot$ to an observer at $z = -\infty$. Three different radii for the surface of the compact object are shown: 7.28, 10 and 13.65 km.

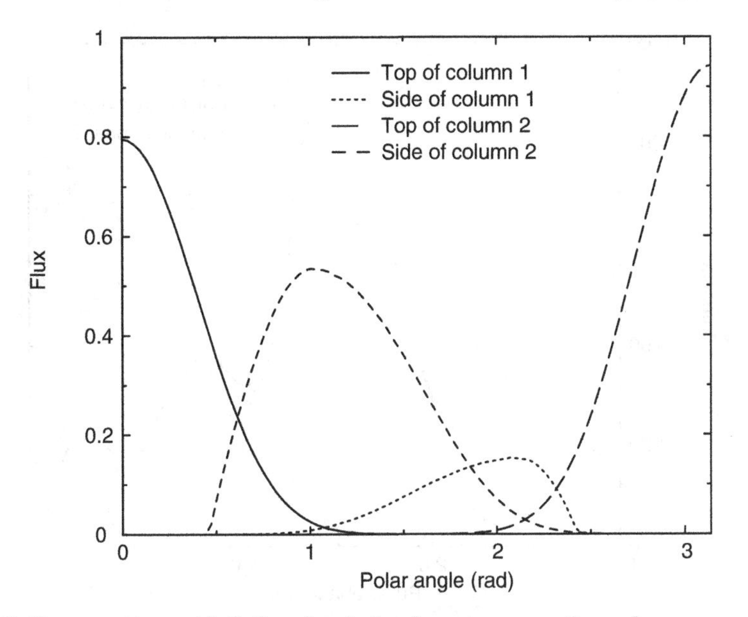

Fig. 10. Beam pattern at infinity of emission from two accretion columns, measured along a single line of fixed longitude. The azimuthal dependence of the surface emissivity of both columns results in the side of column 2 having larger flux than that of column 1.

planar accretion disk in a plane different than the magnetic equator. Figure 10 here shows a cross-section of the two-dimensional beam pattern, defined as the angular flux distribution far from the star, traced from one magnetic pole of the star to its antipodal point along a particular azimuth.

Figure 11 shows the resulting pulse shape when the beam pattern is rotating about the appropriate rotation axis. The main contributions to the observed pulse shape are: the pencil beam from near pole (highest and narrowest peak); the fan beam from far pole (two peaks flanking the highest peak) and the fan beam from the near pole (resulting in significant smoothly varying component in the pulse profile).

An improved model ([32]), mainly in the efficiency of the calculation, has allowed a fairly complete search of the complex parameter space (10 parameters to specify the emissivity, geometry of the accretion columns, and the observer's orientation with respect to the neutron star). This has refined the various parameters, but also allowed a mass and radius constraint to be placed on the neutron star. The result is quite interesting and puts the mass near 1.3 solar masses, and radius near 11 km. This constraint can be compared ([32]) to mass-radius calculations for various equations of state: a number of equations of state are ruled out. However since the mass and radius obtained are not extreme, a number of equations of state still agree with the constraint.

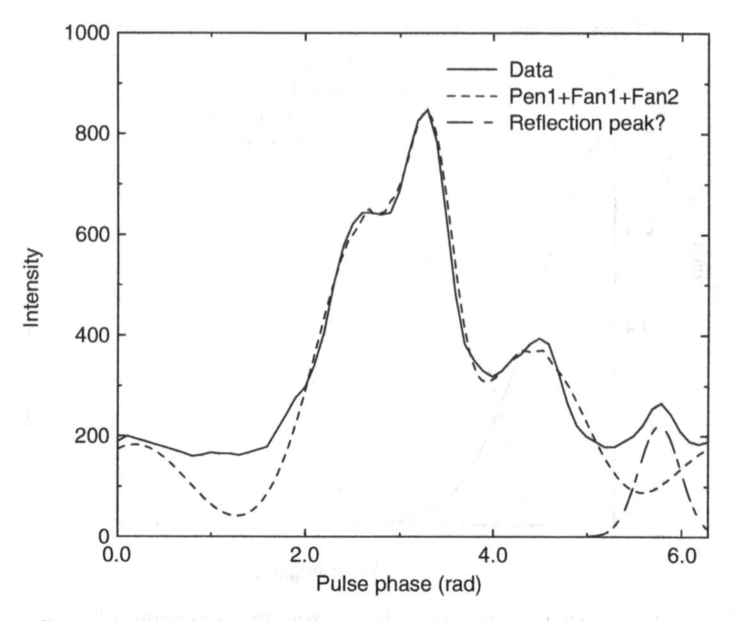

Fig. 11. Comparision of observed pulse shape of Her X-1 in the 9-14 keV band with the accretion column model including top (pencil beam) and side (fan beam) emission from two columns. The pencil beam emission from column 2 is not visible to the observer.

3 GX 301-2: A High Mass X-ray Binary

3.1 Overview

GX 301-2 (4U 1223-62) is a pulsar with a 680 s rotation period, in a 41.5 day eccentric orbit ([43]). The mass function is 31.8 $M_{\odot}$, making the minimum companion mass 35 $M_{\odot}$ for a 1.4 $M_{\odot}$ neutron star. The companion, Wray 977, has a B2 Iae spectral classification ([39]). The orbital parameters of GX 301-2, updated with the *BATSE* observations ([11]) are as follows. $P_{\rm orb} = 41.498$days, $a_x sin(i) = 368.3$ lt-s, eccentricity $e = 0.462$, longitude of periastron $\omega = 310°$, time of periastron passage T_0 =MJD 48802.79. Figure 12 illustrated the orbital geometry of GX 301-2 for an inclination of 62°.

The neutron star flares regularly in X-rays approximately 1-2 days before periastron passage, and several stellar wind accretion models have been proposed to explain the magnitude of the flares and their orbital phase dependence (e.g. [9, 11, 16]). The modeling by [16] and [9] was done using *TENMA* and *EXOSAT* observations, respectively, which cover many short data sets spaced irregularly over orbital phase. Better orbital phase coverage has been obtained by *CGRO/BATSE* ([11]), which however has much lower sensitivity than the previous studies.

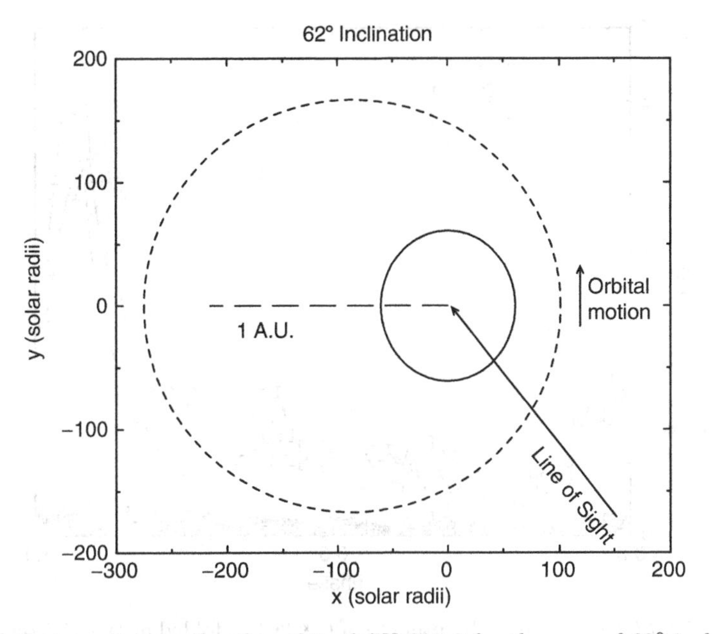

Fig. 12. System diagram for the orbit of GX 301-2 for the case of 62° inclination, indicating observer direction.

The X-ray spectrum of GX 301-2 has been studied by *TENMA* ([13–15]) and *ASCA* measurements ([42]). The latter study illustrates the complexity of the GX 301-2 spectrum. Four components are necessary: i) an absorbed power law with high column density; ii) a scattered power law with much lower column density; iii) a thermal component with temperature of 0.8 keV; iv) a set of six emission lines (including the iron line at 6.4 keV). Of the above, ii) and iv) are due to reprocessing in the stellar wind of Wray 977. Of relevance to the discussion in the next section, an strong intensity dip due to occultation by a dense gas stream was observed in GX 301-2 by [12].

3.2 Orbital X-ray Flux Cycle and Stellar Wind

A good quality orbital light curve has been obtained using the *RXTE/ASM* ([28]). The *RXTE* orbital light curve is shown in Fig. 13. The band 1 and band 2 light curves are strongly affected by X-ray absorption in the strong stellar wind of Wray 977, but band 3 reflects well the intrinsic X-ray flux from the neutron star. The light curve has large peak just prior to periastron and a secondary broad peak around orbital phase 0.5.

A spherically symmetric wind has been shown to not produce the main flux peak prior to periastron ([9, 16]). However, neither of those studies included an azimuthal component in the wind. This component is small but does shift the phase of the main flux peak. The wind velocity law is usually

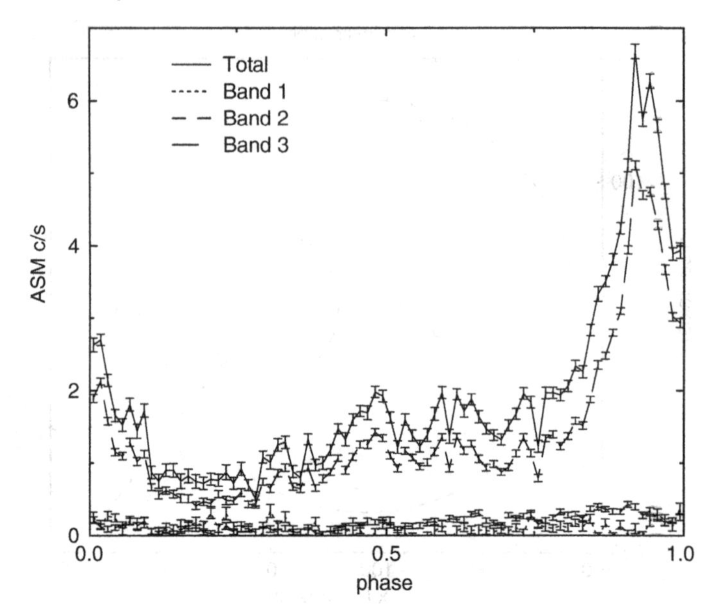

Fig. 13. *RXTE/ASM* orbital light-curve of GX 301-2, folded at the 41.5 day orbital period. Band 1 is 1.3-3 keV; band 2 is 3-5 keV; band 3 is 5-12 keV and Total is the sum of the count rates for the 3 bands.

taken to be of the form (e.g. [1]): $v_w(r) = v_o(1 - R_s/r)^{\beta} + c_s$ with $\beta = 1$ and c_s the speed of sound in the wind. There is an additional azimuthal component to the velocity. This is due to stellar rotation and is equal to the equatorial rotation velocity of the primary at $r = R_s$, with R_s the radius of the primary. The azimuthal component decreases with r due to angular momentum conservation. The accretion rate is taken to be the Bondi-Hoyle accretion rate (e.g. [16]). The wind model was fit to the *RXTE/ASM* data with free parameters v_o and normalization, using the standard case for the inclination, masses, stellar radius and effective temperature. The normalization can be related to the mass-loss rate after conversion of *ASM* counts/s to flux, conversion to luminosity given a distance to GX 301-2, and applying assumed accretion efficiency and neutron star mass. The wind parameter β has a small effect on the fits, and is fixed at 1.0 after comparing results with different β. The best-fit wind model is shown in Fig. 14: it can give a peak just prior to periastron but it cannot occur early enough, nor does the shape of the light curve fit the data.

The physical model which does describe the data adequately includes a gas stream from the L1 point of the companion star in addition to the stellar wind. The basic idea, as proposed by Stevens ([50]), is that a gas stream flows out from the point on the primary facing the neutron star. The fitting of a phenomenological stream model to the *RXTE/ASM* data is described first. Then a numerical calculation of the spiral stream geometry as a function of binary phase is described.

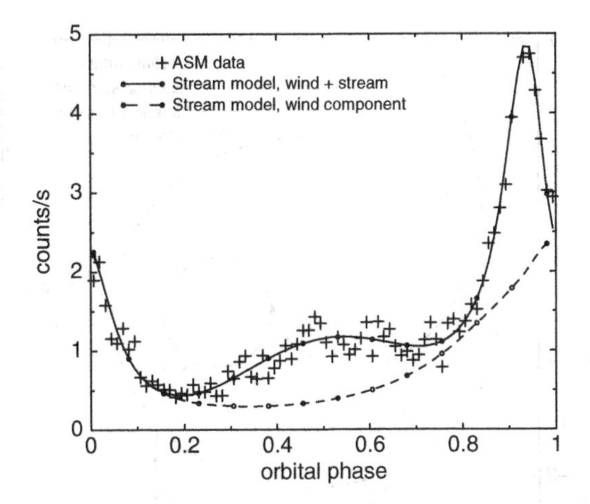

Fig. 14. Stellar wind plus archimedes-type stream light-curve compared with *RXTE/ASM* band 3 folded orbital light-curve of GX 301-2.

The stream has a geometry such that it is crossed twice by the neutron per binary orbit. This is motivated by the observations of two flux peaks at orbital phases 0.93 and 0.55, and confirmed by the physical stream calculations that follow. For the observed X-ray flux, only the stream properties where the neutron star crosses the stream are important. The model parameters are the stellar wind velocity v_o, the position angle ϕ_1, over-density α_1 and Gaussian width σ_1 of the stream for the first crossing, and ϕ_2, α_2 and σ_2 for the second crossing. The best fit stream model is shown in Fig. 14 assuming the standard case (inclination 62°, mass $50.5 M_\odot$, $T_{\rm eff} = 21000$K). It has a best-fit χ^2 of 901 for 72 degrees of freedom (dof). The main contribution to χ^2 is from the large scatter in the data compared to the smooth model curve between binary phases 0.2 and 0.8. Taking this into account, the model does produce a satisfactory explanation of the light curve.

The location of the gas stream from the primary star is calculated as follows. The stream's origin is the point on the primary along the line of centers. The stream is has the same radial velocity as the stellar wind. The azimuthal velocity is determined by conservation of angular momentum. The stellar angular velocity, ω is described by the parameter f: $\omega = f \times \omega_{\rm orb} + (1-f) \times \omega_{\rm per}$, with $\omega_{\rm orb}$ is the average orbital angular velocity ($2\pi/P_{\rm orb}$) and $\omega_{\rm per} = \omega_{\rm orb}(1+e)^{0.5}/(1-e)^{1.5} = 3.06\omega_{\rm orb}$ is the periastron angular velocity. Thus the primary is taken to be rotating at some angular velocity between $\omega_{\rm orb}$ and $\omega_{\rm per}$. The large difference in $\omega_{\rm orb}$ and $\omega_{\rm per}$ is due to the high eccentricity of the orbit. For any given orbital phase, the stream position is integrated using the radial and azimuthal velocities. The stream takes an Archimedes-type spiral pattern, with a shape which varies with orbital phase. Figure 15 shows the stream shape at the two times during the neutron star orbit where the neutron star crosses the center of the stream. The shape change is due to

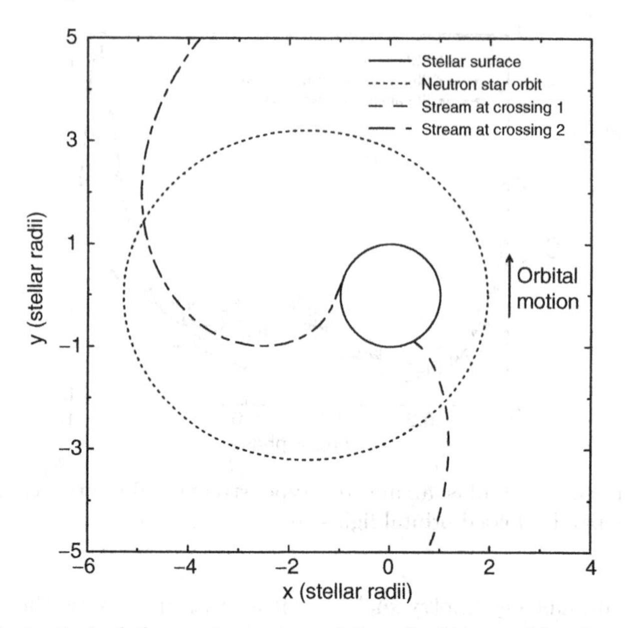

Fig. 15. Illustration of the neutron star orbit and of the archimedes-type stream from Wray 977 at two particular orbital phases (at the times of the two neutron star - stream crossings). The diagram is for 76° system inclination.

the variation in angular velocity of the point of origin of the stream on the primary, which is caused by the variable angular velocity of the neutron star.

3.3 Inclination and stellar mass constraints

The orbital inclination of GX 301-2 is not constrained by the pulse timing data alone, but by including constraints on the mass and radius of the companion star, Wray 977. This is discussed by [11], and an updated discussion is given by [28]. The exact form of the Roche potential for a synchronously rotating star was computed by [28], rather than the standard simple approximation. The surface area of the critical Roche surface (or mean radius defined by the surface area) is the correct quantity to compare to the radius derived from the luminosity, rather than other measures of the Roche lobe such as the Roche radius.

How the first good estimate of inclination for GX 301-2 ([28]) was obtained is reviewed here. The rotation rate of the primary star affects the stream angular velocity and position vs. orbital phase. Thus it affects the orbital phases where the neutron star crosses the stream. As one increases the inclination from 55° to 76°, the agreement between calculated and the fit for the first crossing (near 5.1 rad) occurs for f decreasing from 0.66 to 0.42. The agreement between calculated and the fit for the second crossing (near

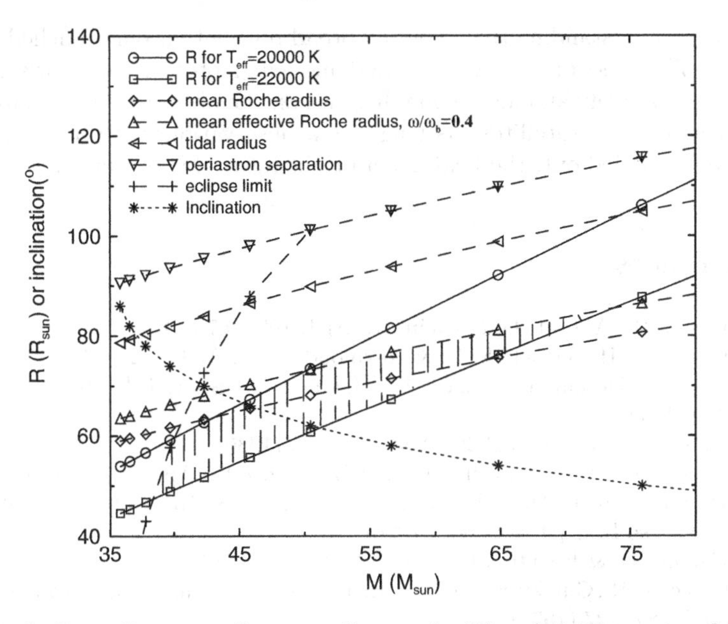

Fig. 16. Stellar radius vs. stellar mass diagram for Wray 977. The various labelled curves are discussed in the text.

3 rad) occurs for f increasing from 0.43 to 0.59. Both crossings match for $f \simeq 0.51$ at inclination of $\simeq 65°$.

Also shown here (Fig. 16) is a summary figure of the constraints on mass for the companion, Wray 977 and the system inclination. These use the mass-luminosity relation derived from Shaller, Schaerer & Maeder ([47]), so depend on $T_{\rm eff}$, which observationally is between 20000K and 22000K. For the orbit not to give eclipses of the neutron star, upper limits on inclination (lower limits on mass and radius) are: 76.8° ($38.4 M_\odot$ and $47.5 R_\odot$) for $T_{\rm eff}$ = 22000K; and 73.6° ($40.0 M_\odot$ and $59.6 R_\odot$) for $T_{\rm eff}$ = 20000K. For the companion to fit inside its Roche lobe, lower limits on inclination (upper limits on mass and radius) are: 54.3° ($64.0 M_\odot$ and $75.2 R_\odot$) for $T_{\rm eff}$ = 22000K; and 43.2° ($43.1 M_\odot$ and $63.7 R_\odot$) for $T_{\rm eff}$ = 20000K. The case with $T_{\rm eff}$ = 21000K and inclination of 62° is near the center of the allowed region, with mass of $50.5 M_\odot$ and stellar radius of $66.7 R_\odot$. This case has a distance of 4.1 kpc in order to yield the observed optical flux from Wray 977.

4 Summary

In this paper, I have reviewed a subset of a large body of work on observations and modeling of accretion flows in X-ray binaries. I have concentrated on a few systems and just a few topics for these systems. The goal has been to give some more details of the type of work that has occured in order to encourage a deeper understanding than would a broad review that only touched on these topics. The field of observations and modeling of accretion flows in X-ray binaries is a lively and active one. It relies on the observational capabilities of X-ray astronomy satellites. As long a new observations with new capabilities are being carried out, the field has a bright future for continued progress.

References

1. Castor, J., Abbott, D. & Klein, R., ApJ, 195, 157 (1975)
2. Cheng, F. H., Vrtilek, S. D. & Raymond, J. C., ApJ, 452, 825 (1995)
3. Choi, C., Dotani, T., Nagase, F., Makino, F., Deeter, J. & Min, K., ApJ, 427, 400 (1994)
4. Crosa, L. & Boynton P. E., ApJ, 235, 999 (1980)
5. Deeter, J., Crosa, L., Gerend, D. & Boynton P.E., ApJ, 206, 861 (1976)
6. Deeter, J., Scott, D.M., Boynton, P., Miyamoto, S., Kitamoto, S. Takahama, S., & Nagase F., ApJ, 502, 802 (1998)
7. Gerend, D. & Boynton, P.E., ApJ, 209, 562 (1976)
8. Giacconi, R., Gursky, H., Kellogg, E., Levinson, R., Schreier, E. & Tananbaum, H., ApJ, 184, 227 (1973)
9. Haberl, F., ApJ, 376, 245 (1991)
10. Jones, C. & Forman, W., ApJL, 209, L131 (1976)
11. Koh, D., Bildsten, L., Chakrabarty, D. Leahy, D.A. et al., ApJ, 479, 933 (1997)
12. Leahy, D.A., Nakajo, M., Matsuoka, M., et al., PASJ, 40, 197 (1988)
13. Leahy, D.A., Matsuoka, M., Kawai, N. & Makino, F., MNRAS, 236, 603 (1989)
14. Leahy, D.A., Matsuoka, M., Kawai, N. & Makino, F., MNRAS, 237, 269 (1989)
15. Leahy, D.A. & Matsuoka, M., ApJ, 355, 627 (1990)
16. Leahy, D.A, MNRAS, 250, 310 (1991)
17. Leahy, D.A., Yoshida, A. and Matsuoka, M., ApJ, 434, 341 (1994)
18. Leahy, D.A., A&A Supp., 113, 21 (1995)
19. Leahy, D.A., ApJ, 450, 339 (1995)
20. Leahy, D.A. & Yoshida, A., MNRAS, 276, 607 (1995)
21. Leahy, D.A., MNRAS, 287, 622 (1997)
22. Leahy, D. & Scott, D. M., ApJ, 503. L63 (1998)
23. Leahy, D.A. & Marshall, H., ApJ, 521, 328 (1999)
24. Leahy, D.A. 2000, MNRAS, 315, 735
25. Leahy, D., Marshall, H. & Scott, D., ApJ, 542, 446 (2000)
26. Leahy, D.A. ApJ, 547, 2001 (2001)
27. Leahy, D.A. MNRAS, 334, 847 (2002)
28. Leahy, D.A. A&A, 318, 219 (2002)
29. Leahy, D.A. MNRAS, 342, 446 (2003)
30. Leahy, D.A. AN, 325, 205 (2004)
31. Leahy, D.A. MNRAS, 348, 932 (2004)
32. Leahy, D.A. ApJ, 613, 517 (2004)
33. Levine, A., Bradt, H., Cui, W., et al., ApJ, 469, L33 (1996)
34. McCray, R., Shull, M., Boynton, P., Deeter, J., Holt, S. and White, N., ApJ, 262, 301 (1982)
35. Middleditch J., ApJ, 275, 278 (1983)

36. Ögelman H. & Trümper J., *Mem. Soc. Astron. Ital.*, 59, 169-1 (1988)
37. Ohashi, T., Inoue, H., Kawai, N., Koyama, K., Matsuoka, M., Mitani, K., Tanaka, Y., Nagase, F., Nakagawa, M. & Kondo, Y. PASJ, 36, 719 (1984)
38. Oosterbroek, T., et al., A&A, 353, 575 (2000)
39. Parkes, G., Mason, K., Murdin, P. & Culhane, J., MNRAS, 191, 547 (1980)
40. Parmar, A.N., Pietsch, W., Mckechnie, S., White, N.E., Trümper, J., Voges, W. & Barr, P., Nature, 313, 119 (1985)
41. Rochester, G., Barnes, J., Sidher, S., Sumner, T., Bewick, A., Corrigan, R. & Quenby, J., AAP, 283, 884 (1994)
42. Saraswat, P., Yoshida, A., Mihara, T. et al., ApJ, 463, 726 (1996)
43. Sato, N., Nagase, F., Kawai, N., et al., ApJ, 304, 241 (1986)
44. Scott, D.M., PhD Thesis, University of Washington (1993)
45. Scott, D.M. & Leahy, D., ApJ, 510, 974 (1999)
46. Scott, D.M., Leahy, D. & Wilson, R., ApJ, 539, 392 (2000)
47. Shaller, G., Schaerer, D. & Maeder, G., A&AS, 96, 269 (1992)
48. Shulman, S., Friedman, H., Fritz, G., Henry, R.C. & Yentis, D. J., ApJL, 199, L101 (1975)
49. Soong Y., Gruber D.E., Peterson L.E. & Rothschild R.E., ApJ, 348, 634 (1990)
50. Stevens, I.R. MNRAS, 232, 199 (1988)
51. Still, M., Quaintrell, H., Roche, P. & Reynolds, A., MNRAS, 292, 52 (1997)
52. Still, M., O'Brien, K., Horne, K., Boroson, B., et al., ApJ, 554, 332 (2001)
53. Voloshina, I., Lyutyi, V. & Sheffer, K., Sov. Astron. Lett. 16, 257 (1990)
54. Vrtilek, S., Mihara, F., Primini, A., et al., ApJ, 436, L9 (1994)
55. Wijers, R. & Pringle, J., MNRAS, 308, 207 (1999)

Part III

Cataclysmic Variable Systems

Modeling the Hot Components in Cataclysmic Variables: Info on the White Dwarf and Hot Disk from GALEX, FUSE, HST and SDSS

Paula Szkody

University of Washington, Department of Astronomy,
Box 351580, Seattle, WA 98195
szkody@astro.washington.edu

The close binary systems with active accretion called cataclysmic variables contain a white dwarf that is heated by the mass transfer from the late type secondary star. These heating effects are best visible in the *UV* with satellites like *GALEX*, *FUSE*, and *HST*, while some clues also emerge from optical spectra obtained through the *SDSS*. Modeling these light curves and spectral energy distributions determines the temperatures of the white dwarfs and of the hot spots created by the accretion. At high accretion rates, the accretion disk or column dominates the light, but at low accretion rates, the white dwarf is visible and may be directly studied. Recent results from studies of this type have determined that pulsating white dwarfs in accreting systems exist in a different instability strip than single pulsators and that highly magnetic white dwarfs still have hot spots even when the mass transfer stream has ended.

1 Introduction

Cataclysmic variables (*CV*s) consist of close binaries that contain a late main-sequence or brown-dwarf like secondary star transferring material to a primary white dwarf. The nature of the accretion and its resulting effects on the white dwarf depend on the strength of the magnetic field on the white dwarf. If the field strength is below 1 MG, the material forms an accretion disk and the accretion takes place at a boundary layer surrounding the white dwarf. At the highest fields (greater than 10 MG), the mass transfer stream goes directly to the magnetic poles of the white dwarf, forming an accretion column that emits X-ray and cyclotron radiation that heats up a small hot spot on the white dwarf surface. These systems are termed Polars. For intermediate field strengths of 1-10 MG, the material forms an outer disk ring, while the inner parts of the disk are disrupted and flow in broad accretion curtains to the magnetic poles of the faster spinning white dwarf. These types of systems are

E.F. Milone et al. (eds.), Short-Period Binary Stars: Observations, Analyses, and Results, 137–146.

called Intermediate Polars (*IP*s). Good reviews of the different types of *CV*s are available in [17] and [18].

Optical spectra have shown a wide variety of energy distributions for *CV*s. The Sloan Digital Sky Survey (*SDSS*) [31] has produced the largest range of spectra that are the least biased in terms of covering the widest range of mass transfer rates and therefore brightness range. These spectra [24] (and references therein) show some spectra that are dominated by continuum, indicating high accretion rates, while most objects have short orbital periods and low mass transfer, revealing the underlying stars. Figure 1 shows some typical examples, where SDSS J0748 has a strong blue continuum, SDSS J0131 shows the absorption lines from a white dwarf and SDSS J1553 shows TiO bands from the secondary star. To correctly model this wide array of spectral energy distribution requires four primary components: the white dwarf, the secondary star, the accretion disk or accretion column and hot spots on the disk or white dwarf. The modeling of these components can be done using light curves or spectra, but the modeling will depend on the state of the system. This paper

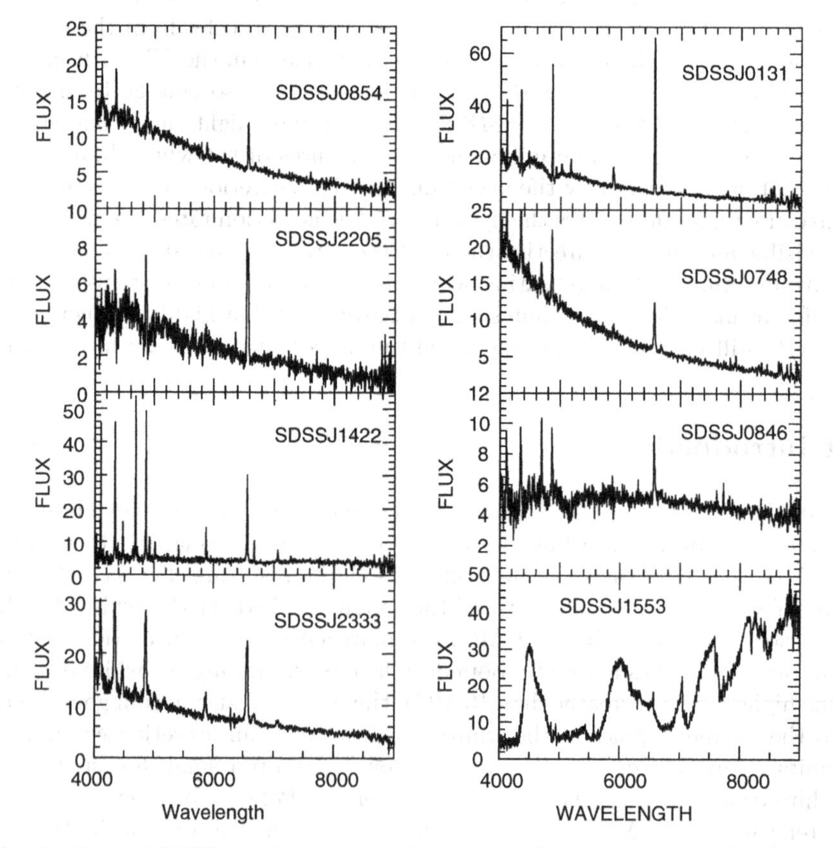

Fig. 1. Typical *SDSS* spectra showing the variety of spectral energy distributions possible in *CV*s.

will consider results on the hot components in the systems that are primarily studied through ultraviolet satellite data. The paper by Howell in this volume will review the cool components (the secondary star and outer disk).

2 Disk Systems

2.1 High Accretion Rates

The best results from modeling high accretion rate systems comes from high resolution *UV* and *EUV* spectra, as the peak flux emerges in this region of the spectrum. The models uusually invoke Hubeny TLUSTY and SYNSPEC codes [11] that produce disks for a range of accretion rates and/or white dwarfs. However, there are shortcomings to this approach. The disk models cannot reproduce the broad emission lines that are typically present in these systems (especially high excitation lines such as NV and CIV) and it is difficult to fit a broad range of wavelengths from 900-10,000Å well. The long orbital period, high accretion rate system SS Cyg [15] illustrates the problem. The *SED* can be fit equally well with a *WD* or a disk, but each presents problems. The fit to a white dwarf is with a temperature of 46,000K, but the known distance of 166 pc for SS Cyg (from an *HST* parallax measuremnt [8]) results in the white dwarf having a radius that is too large for normal mass-radius relations. The fit with a model disk produces an accretion rate that is too high for dwarf nova outbursts to happen. In addition, neither model can explain the higher observed fluxes in the *FUV* below 1000Å. Thus, the modeling of outburst disks can be problematic because the contributions of all the components are not known, other than that the disk is dominant. Another hot component is required, but its nature is not clear (accretion belt, boundary layer, disk corona). Thus, until the physics of quiescent accretion disks can be solved, it is easier to model systems of lower accretion rate where the stars are visible and the contribution of the accretion disk is smaller and can be estimated.

2.2 Low Accretion Rates

At low mass transfer rates, the white dwarf can be separated from the accretion disk or column light and a hot spot can be visible in the light curve. The best way to directly view the white dwarf is to 1) observe in the ultraviolet, where the peak energy of the white dwarf is emitted, 2) observe when the system is at quiescence or the mass transfer has turned off, and/or 3) work on systems with short orbital periods, as these have the lowest mass transfer rates. The evolution of *CV*s envisions systems starting mass transfer at orbital periods of about 6-10 hrs. The angular momentum loss from magnetic braking causes the orbits to shrink until the magnetic braking turns off near periods of 2-3 hrs, and then gravitational radiation alone drives the evolution at low mass transfer rates until the secondary becomes degenerate at about periods near 70 min and the orbit expands. The predictions from evolution models [10] are that most systems should have evolved to near the minimum period in the lifetime of the galaxy.

One of the best examples of a low accretion rate system is that of WZ Sge.This system has very infrequent outbursts (20-33 yrs) but when they do occur, they are of extreme amplitude (7 mag). The last outburst in 2001 was followed with *FUSE* [14] and *HST* [20] as the system returned to quiescence. The results revealed that the white dwarf became visible about 2 months past outburst and gradually cooled from a temperature of 25,000K to a temperature of 15,000K at 3 years past outburst [7].

An *HST* study that concentrated on the shortest orbital period systems in order to find the coolest temperature white dwarfs showed that the lowest temperature disk accreting white dwarfs are about 12,000K [21]. Since it is not possible to determine both log g and T independently (they depend on each other), the best procedure for modeling the white dwarf is to plot the change of Mass, T_{eff}, distance and optical V magnitude with log g and then find a solution that is compatible with all 4 variables. A lower limit to log g comes from the optical V magnitude, which must be less than the observed V magnitude, since there will be some contribution from the accretion disk. An upper limit comes from the white dwarf mass limit. The best results for a model match are obtained for datasets that encompass large wavelength coverage from *UV* through optical. An example of this procedure is shown in Figures 2 and 3 for the fits accomplished for SW UMa [4]. Figure 2 shows the dependence of the mass, T_{eff}, distance and V magnitude on log g. Figure 3 shows how models that are quite similar in the *UV* produce different amounts of optical flux. The correct model for the white dwarf must fall below the observed optical flux, with some room for a contribution by the accretion disk.

The clue to the amount of contribution of the white dwarf comes from the Lyα line. In the coolest white dwarfs, a broad Lyα turnover is evident and the flux at the bottom of Lyα goes to zero if the white dwarf contributes all the flux. If the accretion disk contributes to the light, emission lines are

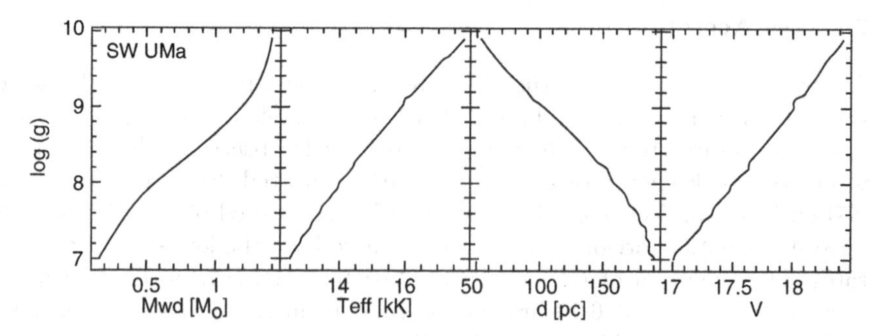

Fig. 2. The dependence of log g on mass, temperature, distance and V mag of the white dwarf. From [4], copyright 2005, AAS.

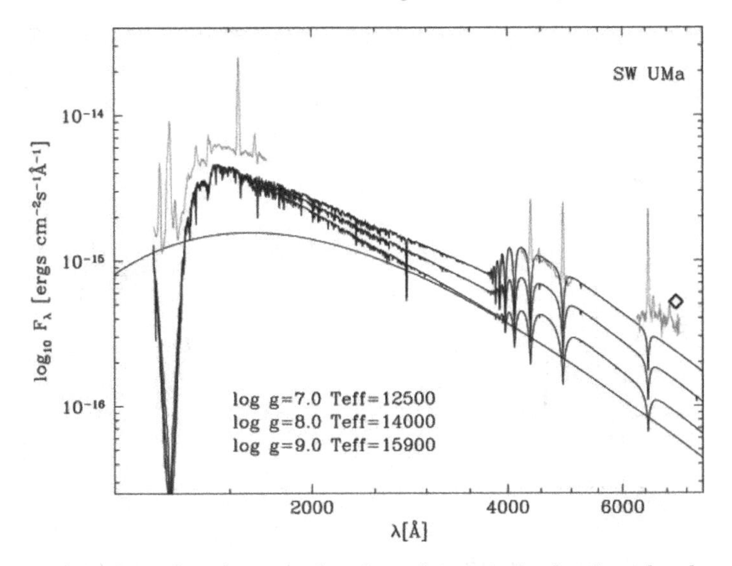

Fig. 3. Model white dwarf extrapolated to the optical, showing the dependence on log g. The diamond is the measured optical magnitude at the time of the *HST* observation. From [4], copyright 2005, AAS.

usually prominent and the continuum near Lyα does not reach zero. In this case, a disk component (either a black body or a power law) can be added to fill in the correct amount of continuum near Lyα. Figure 4 from [4] shows the result for SW UMa. To fill in the Lyα core, a disk that contributes about 20% of the light at 1400Å is needed (shown as a dotted line in the lower part of Figure 4). This then yields a white dwarf of temperature 13,900K (for log g = 8) and gives a distance of 159 pc and a V magnitude of 17.5 (Figure 2), which is below that observed. The contribution of this disk to the entire spectrum is shown in Figure 3 as well.

The rotation and composition of the white dwarf can also be obtained if the resoluton is high enough (better than about 1Å). The model lines can be broadened with different amounts of rotation until they provide a match to the absorption lines seen in the observed spectrum. The lines of CI at 1660 and SiII near 1530 are especially good to use. In all cases, the spin of the white dwarf generally appears to be low (200-400 km/s) [4,18,21]. Once rotation is determined, the composition of individual lines can be adjusted to match particular elements. Several systems show enhancements of N and Al, while C is depleted, implying CNO processed material [4,18,21].

As in the high accretion rate systems, several objects require an additional *FUV* component rather than a standard disk. Sometimes, this component is visible as an upturn in the far *UV*, especially at times of outburst, which indicates a hotter source, such as the boundary layer. At other times, this second component could be a hotter temperature on the white dwarf itself, possibly a belt heated by equatorial accretion [19]. The system GW Lib, which

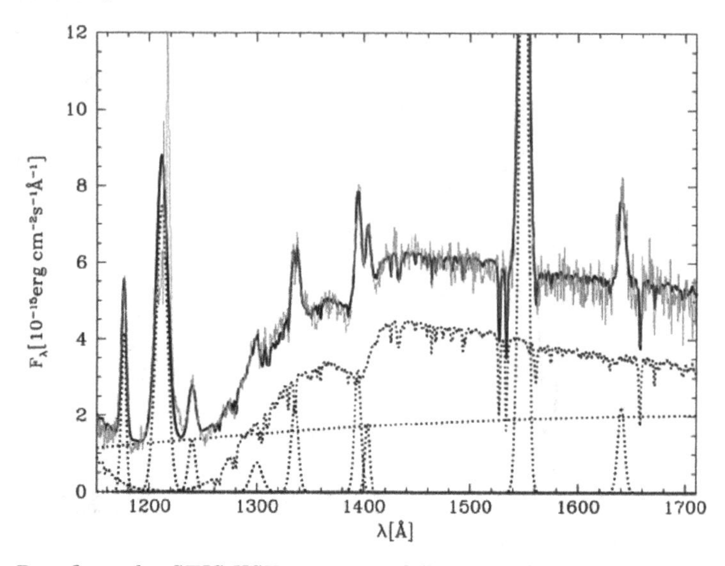

Fig. 4. Best fit to the *STIS HST* spectrum of SW UMa. Simple Gaussians are used to fill in the emission lines (no modeling). The accretion disk is modeled as a black body with enough contribution to fill in the bottom of Lyα, while the white dwarf is the remainder. The sum of all components is shown superposed on the observed fluxes. From [4], copyright 2005, AAS.

contains a pulsating white dwarf, is best fit with a two component white dwarf, with 63% of the surface at 13,300K and 37% at 17,100K [22], rather than a white dwarf plus a disk component.

The *SDSS* data show that about 15% of the lowest accretion rate disk systems show the presence of the white dwarf through broad absorption lines flanking the Balmer emission [24], such as SDSS J0131 in Figure 1. Follow-up photometric observations on these systems has led to the discovery of several more pulsating white dwarfs [5,16,28,30]. Followup low resolution UV spectroscopy obtained with the Solar Blind Channel and prism on the Advanced Camera System on *HST* [26] has shown high temperatures (15,000K) for three white dwarf pulsators in accreting *CV*s. These three, together with GW Lib, imply accreting white dwarfs have a different instability strip than for single white dwarf pulsators. However, Araujo-Betancor et al. [2] have determined that the white dwarf in the accreting pulsator HS 2331 + 3905 has a normal temperature of 10,500K. Thus, it is not clear if there is a bifurcated distribution of temperatures in accreting pulsators or if HS 2331 + 3905 is just a strange system, as it has peculiarities evidenced by other periodicities not due to pulsations.

Besides quiescence for dwarf novae, the other opportunity to view the white dwarf exists for novalike systems in the 3-4 hr period range. These objects typically have accretion rates that are too high to undergo dwarf nova outbursts and the high mass transfer rate keeps the disk in an extended state

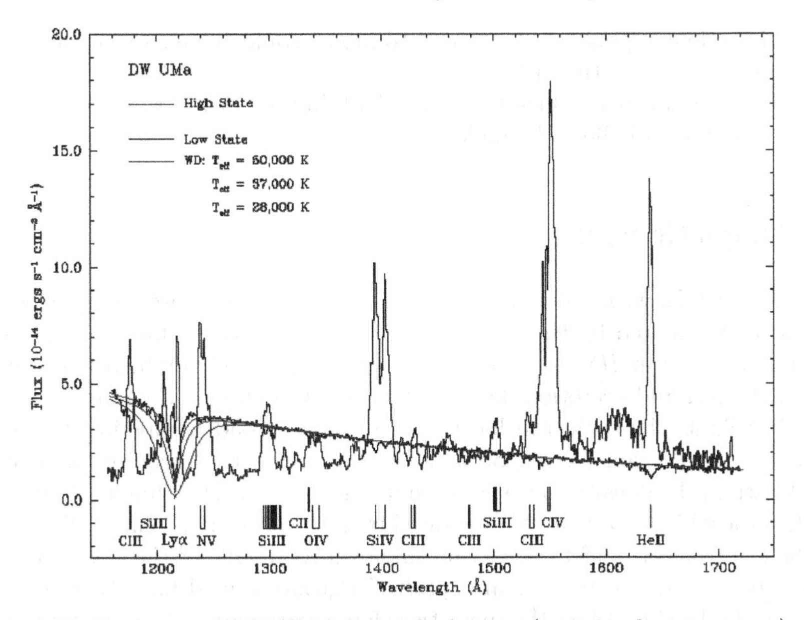

Fig. 5. *STIS* spectra of DW UMa during high state (emission line spectrum) and low state (absorption line spectrum fit with 3 different *WD* temperatures showing the 50,000K as best fit) on the same scale. Note the low state fluxes are higher than the high state fluxes shortward of 1450Å. From [13], copyright 2004, AAS.

that obscures the white dwarf. Fortunately, these objects cease mass transfer (for unknown reasons) and enter low states when the disk disappears and the underlying white dwarf can be seen and its temperature measured. An example of this case is DW UMa which has been observed with the *STIS* instrument on *HST* (resolution about 1Å) both at high accretion [13] and during a serendipitous low state [12]. The low state fluxes were actually larger than the high state (Figure 5), and showed a pure absorption spectrum that could be nicely fit with a white dwarf at a temperature of 50,000K. The interpretation of these differences is that during the high state, the disk becomes extended and blocks the white dwarf from view, so the *UV* spectrum shows only strong, broad emission lines and continuum from the disk. Other nova-like systems in this period range such as MV Lyr observed with *FUSE* [9] also show high temperature white dwarfs, indicating that the high accretion rates produce a significant heating of the white dwarf.

2.3 Summary of Disk Systems

In summary, we have learned the following from modeling disk systems:

- The white dwarf is easily visible when the accretion rate is low or off.
- The white dwarf temperatures range from 12,000K at short orbital periods (low accretion) to 50,000K at 3-4 hr periods (high accretion).

- Multiple components are needed to model some sytems (boundary layers, heated belts on the *WD*, coronae?).
- The optical gives clues to a model fit but the wide range of *UV-IR* is needed for a believable model fit.

3 Magnetic Systems

As in the *CVs* with disks, when the mass transfer rates are high, the system light is dominated by the accretion process. For Polars, this is the accretion column, while in *IPs*, it is an even more complicated combination of accretion column and accretion disk. When cyclotron emission dominates the *IR*-optical light, the cyclotron harmonics produce broad humps in the spectral energy distribution (see SDSS J1553 in Figure 1) that require modeling by field strength, density and shock temperature [29]. The simpler case occurs when the white dwarf can be separated from the accretion column. This occurs when the mass transfer ceases (for unknown reasons) and the system enters a low state, or when the system is part of a group termed Low Accretion Rate Polars (*LARPs*). When the mass transfer stream is off, the only accretion is likely via a wind from the secondary star [17], and in the case of the *LARPs*, the accretion may never have started and the systems may still be evolving toward becoming Polars. A recent study [1] of *STIS* spectra of 11 polars, 8 of which were at a low state, showed that the white dwarfs were all colder than those in disk systems. In addition, they found that the fits were improved when a hot spot was added to the white dwarf flux.

The prototype of Polars (AM Her) has been studied in detail during a low state [6]. The observed *FUSE* and *HST* ultraviolet fluxes were fit with a 19,800K white dwarf and a hot spot of T_{spot} = 34,000K that covers 12% of the white dwarf surface. The light curves constructed from the time-resolved spectra show modulations consistent with a hot spot of this type remaining on the surface even after the mass transfer has stopped. *GALEX* results on EF Eri [25] push the duration of this hot spot even further. EF Eri showed large amplitude *FUV* variations even though it was 7 years past the start of a low state. Although the spectral energy distribution could not be easily fit with a typical hot spot model on its 9500K white dwarf, there is no denying that a large and variable source of *UV* light remains long after there is no evidence of active mass transfer. While a wind could still be providing a source of material that is funnelled to the magnetic pole, the secondary in EF Eri is likely a brown dwarf and the wind rates are not expected to be high. Part of the problem of getting good fits to the white dwarfs with high magnetic fields is that there are still problems with the models [3], where the flat observed distributions in the *UV* compared to the optical do not match the models.

The *LARPS* are the lowest accretors (10^{-14}M$_\odot$ yr^{-1}) [23], (see SDSS J1553 in Figure 1), with only wind accretion likely and their white dwarfs are the coolest (5000-8000K). Thus, these systems may be the oldest white dwarfs among *CVs* and may have never been heated by active accretion.

In general, we have learned the following from modeling Polars:

- The white dwarfs with high magnetic fields are cooler (less heated by accretion) than those in disk systems.
- Hot spots remain on the white dwarf for long periods (many years) after the mass transfer stops.
- *FUV* models for magnetic white dwarfs are not yet adequate to account for the observed fluxes from *FUV* to optical wavelengths.

4 Acknowledgments

This research was supported by *NASA FUSE* grants NAG5-13656 and NNG04GC97G, *HST* grants GO-09724 and GO-10233.01-A, *GALEX* grant NNG05GG46G, and *NSF* grant AST-0607840.

References

1. S. Araujo-Betancor et al: Ap. J. **622**, 589 (2005)
2. S. Araujo-Betancor et al: A. Ap. **430**, 629 (2005)
3. B. T. Gänsicke, G. D. Schmidt, S. Jordan, P. Szkody: Ap. J. **555**, 380 (2001)
4. B. T. Gänsicke, P. Szkody, S. B. Howell, E. M. Sion: Ap. J. **629**, 451 (2005)
5. B. T. Gänsicke et al: MNRAS **365**, 969 (2006)
6. B. T. Gänsicke, K. S. Long, M. A. Barstow, I. Hubeny: Ap. J. **639**, 1039 (2006)
7. P. Godon, E. M. Sion, F. Cheng, K. S. Long, B. T. Gänsicke, P. Szkody: Ap. J. **642**, 1018 (2006)
8. T. E. Harrison, B. J. McNamara, P. Szkody, R. L. Gilliland: A. J. **120**, 2649 (2000)
9. D. W. Hoard et al: Ap. J. **604**, 346 (2004)
10. S. B. Howell, L. A. Nelson, S. Rappaport: Ap. J. **550**, 897 (2001)
11. I. Hubeny, T. Lanz: Ap. J. **439**, 875 (1995)
12. C. Knigge, K. S. Long, D. W. Hoard, P. Szkody, V. S. Dhillon: Ap. J. **539L**, 49 (2000)
13. C. Knigge et al: Ap. J. **615**, 129 (2004)
14. K. S. Long, C. S. Froning, B. T. Gänsicke, C. Knigge, E. M. Sion, P. Szkody: Ap. J. **591**, 1172 (2003)
15. K. S. Long, C. S. Froning, C. Knigge, W. P. Blair, T. R. Kallman, Y-K, Ko: Ap. J. **630**, 511 (2005)
16. A. Mukadam et al: in prep (2006)
17. G. Schmidt et al: Ap. J. **630**, 1037 (2005)
18. E. M. Sion: PASP **111**, 532 (1999)
19. E. M. Sion, F. H. Cheng, M. Huang, I. Hubeny, P. Szkody: Ap. J. **471**, L41 (1996)
20. E. M. Sion et al: Ap. J. **592**, 1137 (2003)

21. P. Szkody, B. T. Gänsicke, E. M. Sion, S. B. Howell: Ap. J. **574**, 950 (2002)
22. P. Szkody, B. T. Gänsicke, S. B. Howell, E. M. Sion: Ap. J. **575**, 79L (2002)
23. P. Szkody et al: Ap. J. **583**, 902 (2003)
24. P. Szkody et al: A. J. **131**, 973 (2006)
25. P. Szkody et al: Ap. J. **646L**, 147 (2006)
26. P. Szkody et al: Ap. J. submitted (2006)
27. B. Warner: *Cataclysmic Variable Stars*, (CUP, Cambridge 1995)
28. B. Warner, P. A. Woudt: ASP. Conf. Ser. **310**, 382 (2004)
29. D. T. Wickramasinghe, L. Ferrario: PASP **112**, 873 (2000)
30. P. A. Woudt, B. Warner: MNRAS **348**, 599 (2004)
31. D. G. York et al: A. J. **120**, 1579 (2000)

The Cool Components in Cataclysmic Variables: Recent Advances and New Puzzles

Steve B. Howell

WIYN Observatory and NOAO
howell@psi.edu

1 Introduction

The past few years have brought about a renaissance in the study of cataclysmic variables. *CV*s, as they are called, are close, interacting binary stars containing a white dwarf primary and a low-mass secondary star. These old star systems interact via mass transfer from the Roche-Lobe filling secondary to the higher gravitational potential of the primary star. The white dwarf primary has a mass of 1.4 $M_{\odot}$ or less while the secondary object ranges from $\geq$1.2 $M_{\odot}$ to as low as $\sim$0.05 $M_{\odot}$, the latter being a brown dwarf-like or He core degenerate object. The two stars orbit each other in close proximity having orbital periods from near 12 hours to less than 80 minutes. Warner (1995) provides a review of the entire class of *CV*s.

The current paradigm for the evolution of *CV*s is that they start as fairly wide binaries containing two main sequence stars. The more massive evolves and becomes the present day white dwarf while the less massive spirals closer toward the primary as the orbital period decreases with time. Gravitational radiation and magnetic braking are the key components in this long dance leading to a close interacting pair of stars. Howell et al. (2001) reviews the current paradigm for the evolution of these stars while concentrating on the initial and final fate of the secondary object. As we will see below, however, the entire process of how *CV*s initially form and evolve is now in question.

The white dwarfs in *CV*s can contain strong magnetic fields and the strength of their field is the deciding factor as to the type of *CV* they produce. If the field is weak, up to perhaps 5 MG, the material overflow from the secondary star L1 point will form an accretion disk around the white dwarf. As the field strength increases above 5 MG or so, this disk will become an annulus in which an inner hole is produced by magnetic accretion onto the

E.F. Milone et al. (eds.), Short-Period Binary Stars: Observations, Analyses, and Results, 147–159.

white dwarf magnetic poles. These two types of *CV*s are called dwarf novae and intermediate polars in turn. As the primary star magnetic field strength increases, reaching from 8-10 MG up to >200 MG, the material coming from L1, is threaded along the magnetic field lines and forced into an accretion column as it funnels its way onto the white dwarf surface at or near one or both of the magnetic poles. No accretion disk forms. These highly magnetic *CV*s, the polars or AM Herculis stars, have periods of high mass transfer through L1 (high state) and periods of low mass transfer through L1 (low state) which have long been thought to be somehow related to starspot-type activity on the secondary star. The simple view of a single gas stream connecting the two stars and a direct one-to-one cause for the variations in the mass transferred between the two stars is now also in question.

One parameter is well known for *CV*s and that is their orbital period. The orbital period of even a poorly studied *CV* is often known to high accuracy and many orbital periods are known to a few seconds or better. Using this well known property, Howell et al. (2001) presented theoretical models that related the orbital period of a given *CV* to the stellar properties of its component stars. For example, the run of secondary star mass or spectral type with orbital period was determined. Single normal main sequence stars have a well determined relationship between mass and radius and corresponding theoretical relations exist for the mass donors in *CV*s as well (see Howell, *et al.*, 2001; Howell 2004). The basic result is that the longer the orbital period of the *CV*, the earlier the spectral type and the higher the mass of the secondary star will be. For example, a *CV* with an orbital period near 6-7 hours will have a G2V-like secondary. However, these same theoretical models (also see Beuermann et al. 1998; Baraffe & Kolb 2000) make predictions about how the mass donor in a *CV* will deviate from the normal main sequence star relation. The relationship of mass and radius for the secondary star deviates from the main sequence result once the orbital period of the binary is shorter than ~5 hours (Fig. 1). The relationship for *CV* secondary stars is far from that for main sequence stars until the orbital period becomes very short, near 2.5 hours, at which time it returns to a main sequence value. Once a *CV* reaches the so-called orbital period minimum, near 75-80 minutes, the secondary star is predicted to be of low mass, near 0.06 $M_\odot$, have no further internal energy generation, and the orbital period of the binary begins to increase as mass transfer continues at extremely low levels ($\lesssim 10^{-11}$ $M_\odot$ yr^{-1}). Accordingly, binary evolution models predict that most of the present-day *CV*s in the Galaxy are of very short orbital period, contain very low mass secondaries, and are extremely faint objects. The results and predictions of these models for the evolution of *CV*s have yet to be fully confirmed or refuted by observation.

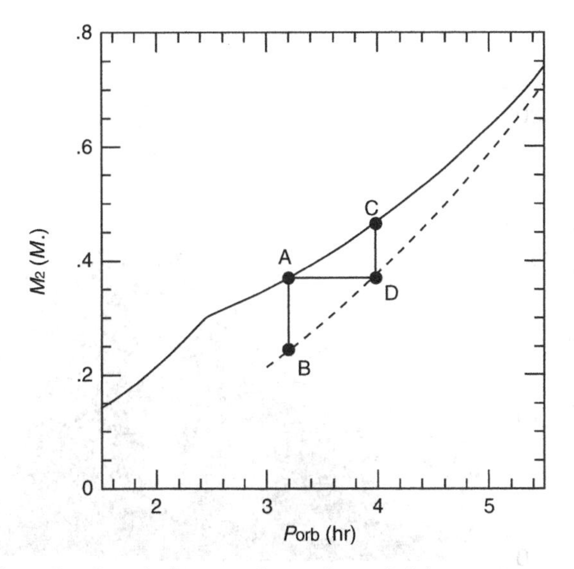

Fig. 1. For the orbital period range from about 2.5 hours up to near 5.5 hours, the theoretical mass-radius relation for *CV* secondaries (dashed line) deviates from the theoretical (and observed) mass-radius relationship for normal main sequence stars (solid line). The four labeled points show example systems at equal orbital period and for equal masses illustrating the predicted differences between the real secondary star and the main sequence assumption. From Howell 2001.

2 A Few Details from the Current Paradigm for *CV* Evolution

Two details of the current evolutionary predictions for *CV*s are of interest here; the mass and age of the secondary star at the time of contact and the necessary core mass for a star in order to produce a CO core and/or CNO burning in its core. The first prediction comes from secular evolution models of binary stars specifically aimed at the formation of *CV*s. It has been shown that nearly all secondary stars in *CV*s are believed to have started as main sequence stars of less than about 1.2 $M_\odot$ and come into contact, that is begun mass transfer to the white dwarf, after only a small amount of time on the main sequence. Their ages being only 30% or far less of their main sequence lifetime (Fig. 2). This prediction suggests that as they start the mass loss process and get less massive with time, their original core was not yet near hydrogen exhaustion and therefore far from being dominated by helium let alone heavier CNO products.

The second prediction of interest is from basic stellar evolution such as that presented by Iben (1985). A star, even if it fully evolves on the main sequence

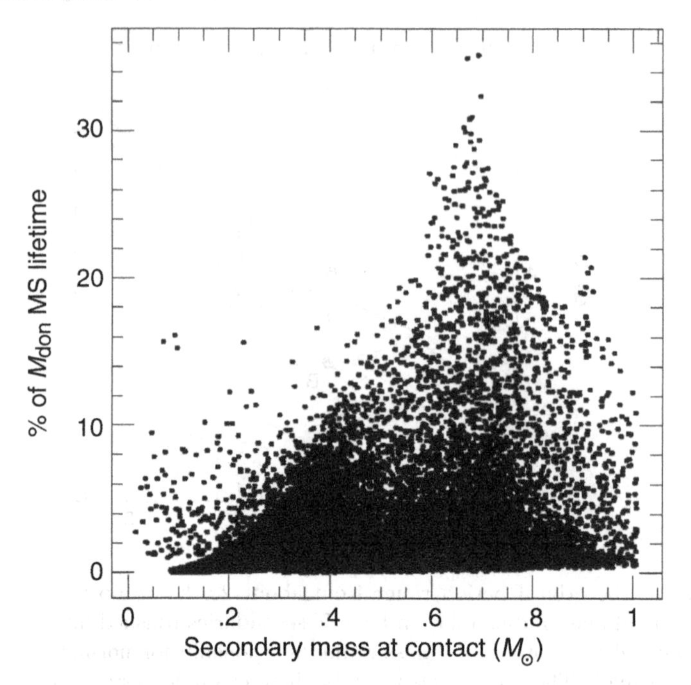

Fig. 2. Theoretically predicted percentage of the main sequence lifetime a *CV* secondary star undergoes prior to contact. Once contact occurs, the secondary loses mass and further core evolution is suppressed. From Howell 2001.

and reaches its terminal age, will only produce a CO core if its initial mass is $\geq$2 $M_{\odot}$. A single star will not undergo significant CNO burning until later in its life (post-TAMS) and only if it has a massive enough core, one which was formed in an initial main sequence star of $\geq$8 $M_{\odot}$.

3 Recent Infrared Spectroscopic Observations of Cataclysmic Variables

*CV*s contain three main emitting components, the white dwarf, the secondary star, and the accretion disk (in the weak magnetic systems) or accretion column/region (in the polars). In the optical part of the electromagnetic spectrum, the accretion disk or column is often the dominant light source. Thus, studies of the component stars in a *CV* are limited to *UV* or shorter wavelengths (for the white dwarf) and the *IR* for the secondary. Even these non-optical bands are not perfect as the accretion disk/column in many dwarf novae/polars can dominate the light even in the *UV* or *IR* bands. However, for longer orbital period *CV*s (which have intrinsically brighter secondary stars), for the very short period *CV*s (which can have low surface brightness, optically thin accretion disks), and for the polars in low accretion states, *IR* spectroscopy has a shot at making a direct detection of the secondary star.

The K-band (2.0-2.5 microns) is about the best choice for *CV* secondary star detection as any disk/column flux will be weaker than in the *J* and *H* bands and the secondary star contribution is often near its maximum. Additionally, *IR* spectrographs on large telescopes are available today with spectral resolutions of 2000-6000 or higher. The *K* band has a number of very good, late type star spectral diagnostic features such as the atomic transitions due to Na I, Ca I, hydrogen and helium lines, and molecular bands due to steam (water) and CO (see Fig. 3; Harrison, *et al.* 2004; Dhillon et al. 2000).

Recent observations with good *IR* spectrographs, 4-m class and larger telescopes, and moderate to high spectral resolution has allowed the collection of spectra for a few tens of *CV*s (*e.g.*, Harrison, *et al.* 2005a and refs therein). The observations to date are mostly a single, phase-summed spectrum but a smaller set of *CV*s have phase resolved spectroscopy available (*e.g.*, Howell, *et al.* 2006a).

Observers have been attempting to determine the spectral type and other properties of the secondary star in *CV*s for decades. There are four main ways in which one can get at the secondary star spectral type. 1) Under the

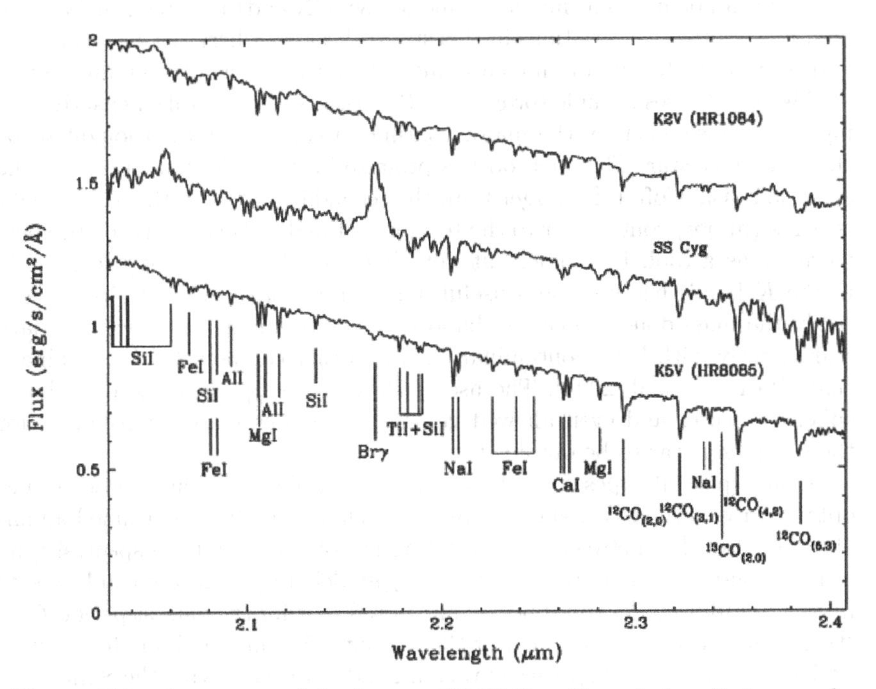

Fig. 3. *K* band spectrum of the famous *CV* SS Cyg. Normal stars that are a close match to the secondary in SS Cyg are shown and a number of *K* band absorption features are identified. The H and He emission lines (not labeled) are from the accretion disk. From Harrison et al. 2004.

assumption that the secondary star must fill its Roche Lobe in order to have mass transfer, the radius of the Roche Lobe can be taken to approximate the radius of the secondary star. A simple calculation presented in Howell *et al.* (2001) (with P_{orb} in hours and f is the "bloating factor" equal to 1 for normal stars)

$$R_2 \simeq 0.10 f^{-0.65} P_{orb}^{1.1} \quad (1)$$

can determine the size of the secondary Roche Lobe from only the orbital period of the binary essentially independently of the stellar component masses. The above equation is valid for *CV*s with non-degenerate secondary stars. A similar relation exists for the theoretically predicted post orbital period minimum systems that contain degenerate secondaries (Howell 2004). Taking the Roche Lobe size equal to that of a normal main sequence star of the same radius, one has a guess at the spectral type and other properties of the secondary. This method, as we have seen, is often poor at getting the correct answer. 2) At times, a near-*IR* (~7000-9500Å) spectrum is used to try to set limits on the secondary star or used to directly estimate its type based on perceived bumps in the spectrum assigned to specific TiO bandheads. The quality of a single spectrum and/or the total phase covered by the time averaged spectrum calls this method into question. The "bumps" observed can be misidentifications, due to low S/N and/or terrestrial features, or simply caused by variations in this spectral region where the accretion disk and secondary star are often competing for control of the continuum light.

Method 3) uses a single (or co-added) optical spectrum taken exactly during mid-eclipse whereby the major continuum contribution is thought to be the secondary star. This method has promise in principle, but in reality, the accretion disk is often far larger than the secondary star and the former light is still a (major) contributor to the total flux. Finally, 4) is the use of *IR* spectroscopy as a tool. For some, but not all *CV*s, *IR* spectroscopy (especially in the *K* band) has become a useful diagnostic tool to directly observe and study the mass donor object. *K* band spectroscopy may view the secondary star directly with little contamination from other components in the binary (*e.g.*, Harrison et al. 2004). The use of *IR* spectroscopy has revealed photospheric features and even allowed velocity (and thus mass) information for the secondary star to be obtained.

Using spectral types for *CV* secondaries available in the literature and obtained by methods 1-3 above, Smith and Dhillon (1998) determined an analytic relationship between the orbital period of a *CV* and the spectral type of its secondary star. Howell *et al.* (2001) modeled the run of secondary star mass and radius with orbital period and noted that for the longest period *CV*s (those with orbital periods greater then about 5.5 hours) and for the shortest period systems (<2.5 hr), theory was in good agreement with the Smith and Dhillon result. However recent *K* band spectroscopy, arguably the most accurate of the four methods listed above, disagrees with the determined relation and in fact provides a further conundrum. Essentially all the secondary stars observed, independent of orbital period, appear to be K class stars in their apparent spectral type. For example, Harrison, *et al.* (2004) found that 12 out of 14 secondary stars did not fit the Smith and Dhillon relation and all 12 stars were of apparent spectral type K2 to K7. Furthermore, three ultra-short period *CV*s, with very small mass donor Roche Lobes (Howell 2005) appear to (impossibly) contain K type secondary stars.

Recent work on non-interacting double-lined spectroscopic binaries (*e.g.*, Sing, *et al.*, 2004) provide a clear indication that their *K* spectral type secondaries have masses of early M stars. The inconsistent masses and main sequence spectral type lead to a confused picture. What are the true secondary stars in *CV*s and do their spectral types have any relation at all to their mass and therefore their true nature?

An even deeper mystery is the fact that *K*-band spectroscopy of non-magnetic and magnetic *CV* secondary stars show large differences. The non-magnetic *CV* secondaries show odd abundances and present clear evidence that they are carbon poor (Harrison et al. 2004). Some of these same stars show normal oxygen and increased nitrogen in their ultraviolet spectra. All these signs point to some form of CNO processing occurring in the secondary star (see Bonnet-Bidaud & Mouchet 2004; Howell 2005). At the same time, the magnetic *CV* secondary stars appear to be completely normal (Harrison et al. 2005b). Our present ideas of *CV* evolution can not account for the differences observed in the secondary stars in magnetic *vs.* non-magnetic *CV*s or the apparent CNO processed material.

Can we formulate some evolution path that allows non-magnetic secondary stars to show CNO processing as well as be too large and too luminous for their mass (compared with normal main sequence stars)? The answer appears to be yes and, in fact, an evolutionary path has been available in the literature for many years. It is the path assumed to be taken by post Algol binaries. Stepping back for a moment, we can ask if there are any binaries known to contain a present day secondary star that must have had a more massive past? Two well-known examples are the star systems Sirius and Procyon. Procyon is a binary containing, at the present epoch, a 1.74 $M_\odot$ F5V primary and a 0.63 $M_\odot$ white dwarf secondary. The two stars orbit each other every 40 years with a separation of $\sim$16 a.u. Sirius is a similar system containing a 2.3 $M_\odot$ A1V primary and a 1.1 $M_\odot$ white dwarf secondary. The orbital period in this case is 50 years with a separation of $\sim$20 a.u. Both of these binaries provide irrefutable proof that in the past, the present day white dwarf was the more massive of the pair. These systems are too far separated for the former primary to have filled its Roche lobe and had episodes of mass transfer. However, if a binary pair is close enough, mass transfer episodes can cause interesting outcomes.

Algol binaries are such a case. Algols are a class of non-degenerate interacting binaries consisting of a main sequence star and a mass losing evolved star. The present-day lower mass evolved object used to be the more massive of the pair. This observational fact is the reason for the famous "Algol

Paradox" (Kopal 1978). If the two components are earlier than G0, the system is referred to as a "hot" Algol while if one component is later then G0, the system is a "cool" Algol. A comparison of the space density of Algols, their orbital period distribution, and the masses of the component stars, especially in the cool Algols, all lead one to speculate that Algols may be the progenitor of the non-magnetic *CV*s. The present day evolved star in an Algol is of low mass (albeit large in radius), contains a partially degenerate He-rich core and has some CNO processed material in its atmosphere. The present-day primary star will evolve into a white dwarf and a pre-non-magnetic *CV* will be born. Eggleton & Kiseleva-Eggleton (2002) present a theoretical view of this evolution (see their §§4.2, 4.3) and list a number of Algol and post-Algol binaries that may soon evolve into bone-fide *CV*s. The binary stars FF Aqr, AY Cet, and V651 Mon are interesting examples.

4 Irradiation or Stellar Activity

Observers have made the case for years that the secondary stars in close (interacting) binaries are highly irradiated by their hot white dwarf companion. While this seems a reasonable occurrence, the actual evidence is mostly circumstantial. What are the likely signs of irradiation? Chromospheric emission lines? Photospheric temperature changes? Will these observables be modulated with the orbit with stronger signatures on or near the white dwarf facing hemisphere of the secondary star?

The few attempts at the calculation of the expected emission line signatures in an irradiated spectrum (Barman et al. 2002;2004) have provided generally comparable results, including emission lines of hydrogen and metals, for pre-*CV*s with hot (T_{WD}~30,000K or more) primaries in short period orbits. The spectral continuum shapes expected for the heated side of the secondary, the forest of metal emission lines, and the changes in line flux expected with orbital phase are not consistent, however, with observations of the majority of *CV*s that generally contain cooler primary stars.

Authors have argued their observations both ways: Secondary star irradiation during a dwarf nova outburst could be correct (Schmidt, *et al.* 2005); Apparent photospheric temperature changes with orbital phase might explain the observations (Howell, *et al.* 2000); and stellar activity seems a likely scenario (Rottler et al. 2002; Kafka et al. 2006; Howell et al 2006b).

While each of the authors above makes a case for the cause of emission lines present on or near the secondary star, they all agree that *UV* photons are needed to produce irradiation effects. In the cases of low mass X-ray binaries, black hole systems, *CV*s in outburst (including *CN*), or close white dwarfs of 30,000K or so, the primary is very hot indeed and irradiation is easily observed and believed. However, for *CV*s and Pre-*CV*s the case is less clear as the white

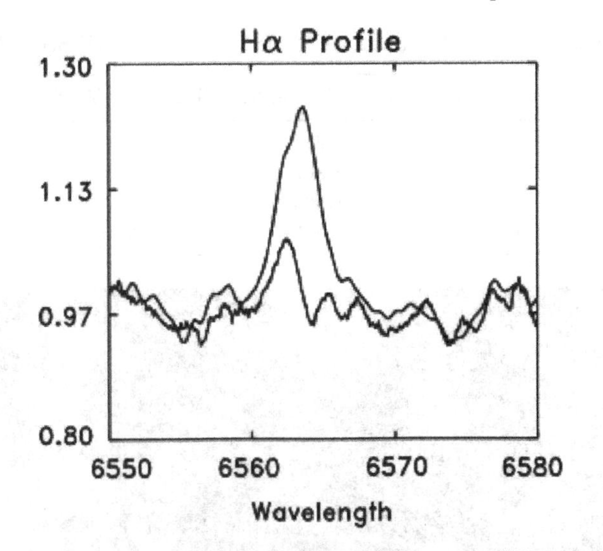

Fig. 4. Hα emission line profiles from the secondary star in V471 Tau for the same orbital phase from 1985 (top curve) and 1990 (bottom weaker emission line). From Rottler et al. 2002.

dwarfs are not always hot. Rottler et al. (2002) examined about a decade of spectroscopic observations of the Pre-*CV* V471 Tau as well as other similar systems. Pre-*CV* systems contain hot white dwarfs (hot compared with typical *CV* white dwarfs) and a close non-interacting late-type secondary. While the paradigm was that V471 Tau is the quintessential irradiated system, the fact is, it is not. Observations over time have shown that the Hα emission has weakened (Fig. 4) and recently went away completely, yet the system has not changed, including the white dwarf temperature (i.e., *UV* flux output). The conclusion is that the Hα emission is not due to irradiation but stellar activity and the fact that it was concentrated on the white dwarf facing side of the secondary fooled us all. The white dwarf facing hemisphere concentration of active regions in close binaries is well known in RS *CV*n systems and others and is discussed for *CV*s in Howell, *et al.* (2006b).

The fractional area of the sky subtended by the secondary star in a *CV*, as seen from the white dwarf primary, is given by $f = R_2^2/(4a^2)$ and is ≈0.012 for a typical short-period *CV*. Using the (*UV*) flux limit derived in the Rottler et al. study as that required to produce irradiation of the secondary star ($\geq 10^{10}$ ergs/cm^{-2}/sec/ster), and calculating the fractional area available to a typical *CV* over at range of orbital period (*i.e.*, R_2 is set to the Roche Lobe radius for each period considered), we find that the fluxes available at the secondary star are generally far too low to produce irradiation of the secondary star. Table 1 presents the results of the calculations for the incident flux on the secondary star for two example white dwarf temperatures.

Table 1. Incident Flux (ergs $cm^{-2} sec^{-1} ster^{-1}$) at *CV* Secondary Star

P_{orb} (hr)	T_{WD}=10,000K	T_{WD}=20,000K
6	2e7	1e9
4	3.2e7	1.6e9
2	1.3e8	6.3e9
1	5e8	2.6e10

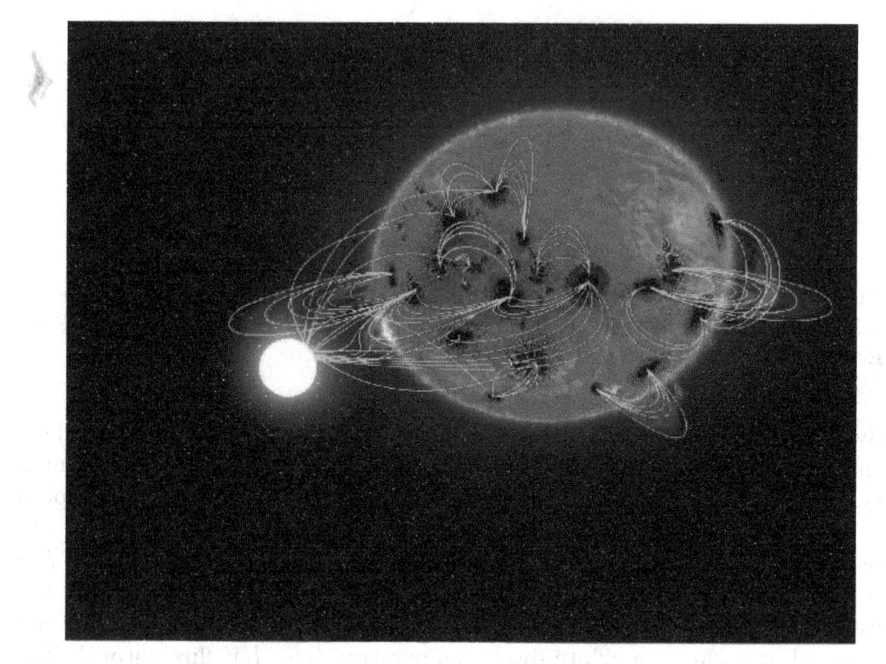

Fig. 5. Artist view of hyperactivity on the low mass secondary star in a polar during low mass accretion. Based on observations presented in Howell, *et al.* (2006b) and Kafka, *et al.* (2006).

Note: *CV* white dwarfs are rarely as hot as 20,000K, especially in short period systems, and photospheric temperatures of 10,000K or less produce essentially no *UV* photons to speak of.

For magnetic *CV*s (*i.e.*, polars) Kafka, *et al.* (2006) and Howell, *et al.* (2006b) have argued for a model whereby the strong magnetic field of the nearby white dwarf drives electrical currents in the interior of the low mass secondary star; some of the secondaries being sub-stellar. These electric currents produce a (pseudo)dynamo in the secondary and cause high levels of stellar activity on its surface. An artist conception of a low-state polar showing hyper-activity on the low mass secondary is shown in Figure 5.

5 Molecules and Dust Disks

The past decade has seen an explosion in infrared observations of *CV*s. Large telescopes, such as Gemini, *VLT*, and Keck, and the Spitzer Space Telescope provide good access to *IR* windows and have dramatically changed our view of the nature of the secondary stars in *CV*s.

Howell, *et al.* (2004) obtained a moderate resolution Keck *K* band spectrum of WZ Sge revealing, for the first time, molecular emission due to CO and H_2. The location of the molecular emission is still not completely resolved, but the outer cool disk (T=3000K or less), invisible in the optical, is the likely culprit. To date, no other *CV* has been shown to contain molecular emission and there is some evidence that the level of molecular emission in WZ Sge is modulated on the superoutburst cycle. The presence of CO emission provides a yet to be modeled, highly effective cooling mechanism for the accretion disk. Perhaps such emission is the cause of the long inter-outburst time periods (decades) between superoutbursts in the class of *CV*s known as *TOAD*s.

Spitzer observations of short period polars surprised us all with the fact that they showed excess emission in the 3-8 micron region over that expected from the secondary star alone. Simple models presented in Howell, *et al.* (2006c) with follow-up detailed modeling in Brinkworth, *et al.* (2007) assign the excess emission to a large, cool (T$\sim$700K) circumbinary disk of material (see Figure 6). The physical properties of these tenuous circumbinary disks

Fig. 6. Artist conception of a short orbital period magnetic *CV* showing the (low state) highly magnetic white dwarf, the brown dwarf-like secondary star, and the circumbinary dust disk. Based on the observations presented in Howell, *et al.* 2006.

are not known but the disks do not appear to be nearly massive enough to be those predicted to exist and affect binary evolution (Dubus, *et al.* 2004). Are the disks isothermal? Do they contain large or small dust grains? What is their origin? The answers to these questions are likely to change our present view of the entire evolutionary process of binary stars.

6 Conclusions

Infrared observations of *CV*s, especially recent spectroscopic work using large telescopes, have provided us new puzzles to solve. These data have shown us that a number of our long held beliefs about how these interacting binaries form and evolve must be questioned and reexamined in this new (*IR*) light.

We have seen that spectral type of the secondary star in a *CV* by itself is essentially meaningless to use as a proxy of the true properties of the star including its mass. The current idea of how *CV*s form and evolve can not account for a number of observational properties, most notably the difference in spectral properties of the secondary stars in non-magnetic and magnetic systems. Phase-resolved *IR* spectroscopy can provide velocity measurements and thus mass estimates for the secondary stars. This line of research holds promise to help resolve the currently unknown spectral-type-mass-radius ambiguity for the secondary stars. Irradiation of the secondary star may not be as prevalent and important as previously thought and studies of the emission lines from chromospherically to active secondary stars will likely open a new field of *CV* research. It seems likely that some or all of the stellar activity in the low mass companions, including those of sub-stellar mass, may be induced by the close presence of the high field white dwarf. Molecular and dust emission in *CV*s is an infant field of study. Further *IR* observations are required in this new area to determine if all *CV*s have molecular emission and what is the site of the formation. Circumbinary dust disks require far-*IR* observations and much is yet to be learned about them. Are they specific to short period *CV*s and/or polars or do all interacting binaries have them?.

7 Acknowledgments

I'd like to thank my collaborators C. Brinkworth, T. Harrison, D. Hoard, K. Honeycutt, S. Kafka, P. Szkody, and F. Walter for their help and work on many of the projects discussed above. Figures 5 and 6 were produced by P. Marenfeld & S. Howell and provided by National Optical Astronomy Observatory/Association of Universities for Research in Astronomy/National Science Foundation.

References

1. Baraffe, I. & Kolb, U., MNRAS, **318**, 354 (2000)
2. Barman, T. S., ASPC, **261**, 49 (2002)
3. Barman, T. S., ApJ, **614**, 338 (2004)
4. Beuermann, K., et al., A&A **339**, 518 (1998)
5. Bonnet-Bidaud, J.-M. & Mouchet, M., ASPC, **315**, 149 (2004)
6. Brinkworth, C., et al., ApJ, in press, (2007)
7. Dhillon, V., et al., MNRAS, **314**, 826 (2000)
8. Dubus, G., et al., MNRAS, **349**, 869 (2004)
9. Eggleton, P. & Kiseleva-Eggleton, L., ApJ, **575**, 461 (2002)
10. Harrison T. E., et al., AJ, **127**, 3493 (2004)
11. Harrison T. E., et al., AJ, **129**, 2400 (2005a)
12. Harrison T. E., et al., ApJ, **623**, L123 (2005b)
13. Howell, S. B., et al., ApJ, **530**, 904 (2000)
14. Howell, S. B., PASJ, **53**, 625 (2001)
15. Howell, S. B., et al., ApJ, **550**, 897 (2001)
16. Howell, S. B., et al., ApJ, **602**, L49 (2004)
17. Howell, S. B., ASPC, **315**, 353 (2004)
18. Howell, S. B., ASPC, **330**, 67 (2005)
19. Howell, S. B., et al., AJ, **131**, 2216 (2006a)
20. Howell, S. B., et al., ApJ, **652**, 709 (2006b)
21. Howell, S. B., et al., ApJ, **646**, L65 (2006c)
22. Iben, I., QJRAS, **26**, 1 (1985)
23. Kafka, S., et al., AJ, **131**, 2673 (2006)
24. Kopal, Z., *Dynamics of Close Binary Systems* (D. Reidel Publishing Company, Dordrecht, Holland, 1978)
25. Rottler, L., et al., A&A **392**, 535 (2002)
26. Schmidt, G., et al., ApJ, **630**, 1037 (2005)
27. Smith, D. & Dhillon, V., MNRAS, **301**, 767 (1998)
28. Sing, D., et al., AJ, **127**, 2936 (2004)
29. Steeghs, D., et al., ApJ, **562**, L145 (2001)
30. Warner, B., *Cataclysmic Variable Stars* (Cambridge Astrophysics Series, Cambridge, 1995)

References

Models for Dynamically Stable Cataclysmic Variable Systems

Albert P. Linnell

Department of Astronomy, University of Washington,
Box 351580, Seattle, Washington,USA, 98195-1580
linnell@astro.washington.edu

1 Introduction

Spectra provide the most detailed information about Cataclysmic Variables (CVs), and synthetic spectra potentially represent the best avenue to study the characteristics of these fascinating objects. Warner [74] (sect. 2.6.1) lists theoretical flux distribution studies. Also see the discussion by la Dous [12] and references therein. Early representation of accretion disk continua by a sum of black body curves [58,69] was supplanted by model stellar atmosphere spectra applied to finite width annuli of similar $T_{\rm eff}$ and log g [34, 35, 45]. The synthetic spectra studies by Wade [71] and Shaviv & Wehrse [60] may be mentioned in particular.

Wade [72] examined the fits of synthetic accretion disk spectra, based both on black body radiation curves and model stellar atmosphere spectra, to a sample of nine novalike CVs. He found appreciable discrepancies between the models and observations, and concluded in particular that model stellar atmosphere spectra do not reflect the physics of accretion disks.

la Dous [12] calculated line synthetic spectra based on a star-like model for accretion disk annuli. The accretion disk temperature profile agreed with the standard model (see below). Shaviv & Wehrse [61] developed a model for accretion disk annuli with detailed vertical structure, and with $T_{\rm eff}$ in agreement with the standard model (see below). They calculated continuum spectra, and applied the theory to RW Sex and IX Vel. They found reasonably accurate fits. As they state, the theoretical models for the optically thick case have clearly visible Balmer jumps, while many CVs show little or no Balmer jump. la Dous [13] discussed the issue of Balmer jumps in dwarf novae during outburst and concluded that star-like accretion disk spectra cannot account for the observations. She suggests that a much flatter average vertical temperature profile is needed in annuli.

Hubeny [25] (hereafter H90) provides a detailed theoretical model for accretion disk annuli which we discuss below. This model has been the basis of several studies. Wade & Hubeny [73] have used the model to provide

E.F. Milone et al. (eds.), Short-Period Binary Stars: Observations, Analyses, and Results, 161–176.

a large database of mid- and far-ultraviolet model spectra for *CV* accretion disks. Huang et al. [23] modeled VW Hyi in superoutburst and included a separately-calculated white dwarf (*WD*) synthetic spectrum to produce a complete system synthetic spectrum.

Eclipse mapping (see [21] for a review) shows that there can be departures from the standard model temperature profile (e.g., [57]), particularly for systems in quiescence.

To the best of our knowledge, the literature does not previously describe a general purpose program to calculate synthetic spectra and synthetic light curves of *CV* systems, based on detailed models of accretion disk annuli, including the contributions of the *WD*, the secondary star, and the accretion disk rim, and tailored to specific systems. Such a program should provide an option to calculate an accretion disk model that is assumed to be quasi-stationary but for which the radial temperature profile does not follow the standard model. This paper describes a program suite with those characteristics.

2 CV Accretion Disks

Frank, King, & Raine [14] discuss the theory of accretion disks. They show that the effective temperature of an accretion disk, in hydrostatic equilibrium and timewise invariant, at a distance R from the accretion disk center, is given by the equation

$$T_{\rm eff}(R) = \left\{\frac{3GM\dot{M}}{8\pi R^3\sigma}\left[1 - \left(\frac{R_*}{R}\right)^{1/2}\right]\right\}^{1/4}, \quad (1)$$

where M is the mass of the *WD*, $\dot{M}$ is the mass transfer rate, and R_* is the radius of the *WD*. G is the gravitation constant and σ is the Stefan-Boltzmann constant. This is the standard model, and is our representation of the accretion disk in dynamically stable cataclysmic variable systems. Such systems also are known as nova-like (*NL*) systems and are characterized by large mass transfer rates. Inclusion of heating from tidal torques and impact by the mass transfer stream introduces additional terms [8, 38]. (Also see [66]).

Dwarf nova outbursts occur at smaller mass transfer rates than for *NL* systems. Osaki [53] reviews the categories of *CV* systems and their relation to the limit cycle instability model. See Cannizzo [9] for an excellent discussion of dwarf novae and the limit cycle instability model. On the assumption that the accretion disk is in quasi-equilibrium, the model described in this paper can be applied as a crude model to the high state or low state of a dwarf nova.

The accretion disk is heated by viscous shear. It is believed that the physical basis for viscosity is the magneto-rotational instability (*MRI*) [6]. Lack of knowledge of the physical cause of viscosity in the original study [63] of accretion disks led to the parameterization

$$\nu = \alpha c_s H, \tag{2}$$

where ν is kinematic viscosity, c_s is the velocity of sound, H is the thickness of the accretion disk at distance R from the center of the accretion disk, and α is a dimensionless parameter [63].

A different viscosity prescription uses a Reynolds number. Pringle [54] discusses both prescriptions and shows that, for *CV* accretion disks, the α prescription requires values in the range 0.1–1.0, and the Reynolds number prescription should use values in the approximate range 10^2–10^3.

An important point is that (1) has no explicit dependence on the viscosity parameter. Details are in [14].

As described in Linnell et al. [42], we model a *CV* accretion disk as a nested series of cylindrical annuli, using program `BINSYN` [40]. The theory of stationary annulus structure in H90, based in part on an earlier development in [37], provides the model for the individual annuli, with additional development in [29, 30] and [31, 32]. Computer implementation of the model is in `TLUSTY` (v.200) [24,28]. Calculation of a model for a single annulus in many respects parallels calculation of a model stellar atmosphere, but with important differences: The structure of a model stellar atmosphere is largely controlled by the incident flux from the stellar interior, while the flux at the central plane of the accretion disk is zero. Generally there are no thermal sources in a stellar atmosphere, while the emergent radiation from an annulus results from viscous dissipation within the annulus.

An annulus model calculation in `TLUSTY` progresses through a sequence of stages of increasing complexity. The first stage is calculation of a gray model. This is followed by an *LTE* model. Next is a first approximation *NLTE* model in which all lines are assumed to be in detailed radiative balance. The final model is an *NLTE* model which treats all lines explicitly. Additional details are in the `TLUSTY` Users Manual, avilable from the `TLUSTY` web site (`nova.astro.umd.edu`). Our experience is that models for inner annuli typically converge but convergence becomes more and more fragile for larger radius annuli until only a gray model can be calculated. Depending on the mass transfer rate, even the gray model calculation may fail for the largest radius annuli.

The calculation of annulus synthetic spectra from the `TLUSTY` models uses program SYNSPEC (v.48) [25, 27]. An annulus synthetic line spectrum includes physical line-broadening effects, such as natural, van der Waals, Stark, or from turbulence. By default, `SYNSPEC` adopts solar composition, but a composition modified species by species is possible. Generally, `SYNSPEC` will produce a line spectrum, requiring a line list with appropriate parameters for individual lines. Line lists are available covering the separate wavelength ranges *EUV*, *FUV*, *UV* (800–3000Å), optical (3000–10,000Å), and *IR*. An input control specifies the wavelength resolution to use in calculating the spectra. A useful option in `SYNSPEC` is to calculate a continuum spectrum, but also including H lines. This option permits calculation of a synthetic spectrum covering, e.g., the range 800–10,000Å, providing comparison with *FUSE*, *HST*, and optical spectra of a given object with a single synthetic system spectrum.

Based on extensive tests, we find that synthetic spectrum morphology progresses smoothly from smaller to larger annulus radii, even though the models that can be calculated transition from *LTE* to gray models. In the comparisons with observational data that have been performed, continuum fits using gray models have been adequate. The largest discrepancy has been the inability of optically thick annuli to produce emission lines. A later section discusses this topic.

H90 shows that the temperature at level m within an equilibrium annulus connects to the annulus effective temperature by

$$T^4(m) = \frac{3}{4}T^4_{\text{eff}}\left[\tau_R(m)\left(1 - \frac{\tau_R(m)}{2\tau_{\text{tot}}}\right) + \frac{1}{\sqrt{3}} + \frac{1}{3\epsilon\tau_{\text{tot}}}\frac{w(m)}{\overline{w}}\right], \tag{3}$$

where $\tau_R(m)$ is the Rosseland optical depth at level m, τ_{tot} is the total Rosseland optical depth at the central plane, $w(m)$ is the kinematic viscosity at level m, $\overline{w}$ is the depth-averaged viscosity, and ϵ is defined by $\kappa_B(m) = \epsilon\kappa_R(m)$, where $\kappa_B(m)$ is the Planck mean opacity per unit mass and $\kappa_R(m)$ is the Rosseland mean opacity per unit mass. Roughly, ϵ is an average of κ_ν/χ_ν, where κ_ν is the line opacity and χ_ν is the total opacity, line plus scattering. Thus, viscosity appears explicitly in the calculation of the vertical temperature profile, but the T_{eff} value depends on R according to (1).

For an optically thick annulus, $\tau_{\text{tot}} \gg 1$, $\epsilon \sim 1$ and (3) reduces to the expression for a stellar model atmosphere temperature with the Hopf function taken at the upper boundary, in the Eddington approximation [50]. In the case of an optically thin annulus the last two terms in (3) dominate. As discussed in H90, if the viscosity is independent of depth ($w(m) = \overline{w}$), then ϵ apparently approaches zero and the last term increases without limit, producing the "thermal catastrophe". To avoid this effect, the `TLUSTY` parameters ζ_0 and ζ_1 control the vertical viscosity profile in an annulus and are set to appropriate default values.

2.1 Irradiation

Hubeny [26] discusses external irradiation of accretion disk annuli and shows that the temperature structure is given by

$$T^4(m) = \frac{3}{4}T^4_{eff}\left[\tau_R(m)\left(1 - \frac{\tau_R(m)}{2\tau_{\text{tot}}}\right) + \frac{1}{\sqrt{3}} + \frac{1}{3\epsilon\tau_{\text{tot}}}\frac{w(m)}{\overline{w}}\right] + WT^4_*, \tag{4}$$

where T_* is the effective temperature of the irradiating star and W is the dilution factor. The dilution factor is

$$W = \frac{1}{4}[1 - \sqrt{1-y^2} + \arcsin y - y\sqrt{1-y^2}], \tag{5}$$

where $y = R_*/R$, and R, R_* are as in (1). The expression for W differs from that in [26] and represents a new derivation by I. Hubeny (private communication). Note that (4) differs from (3) only in the added irradiation term.

2.2 An Accretion Disk Model

Table 1 illustrates the preparation of an accretion disk model by an initial calculation of a series of annulus models. Each line in the table represents a separate annulus. The column headed by m_0 is the column mass above the central plane. The column headed by T_0 is the boundary temperature. Compare with $T_{\rm eff}$. The log g values are at an optical depth of 0.9. The z_0 column gives the height of the annulus in cm. The accretion disk radius at the tidal cutoff boundary is 2.75×10^{10}cm. The Ne column is the electron density at the upper annulus boundary. The $\tau_{\rm Ross}$ column is the Rosseland optical depth at the central plane.

It is important to calculate annulus models closely enough spaced in $R/R_{\rm wd}$ that interpolation, to be performed in BINSYN, is smooth. In the example described here, the individual annulus models used the default values of ζ_0 and ζ_1, and included irradiation by a 35,000K *WD*. The inner annuli with $T_{\rm eff}$ values above 12,000K included only H and He as explicit atoms in

Table 1. Properties of accretion disk with mass transfer rate $\dot{M} = 3.0E(-9)M_\odot\mathrm{yr}^{-1}$ and WD mass of $1.0M_\odot$.

$r/r_{\rm wd}$	$T_{\rm eff}$	m_0	T_0	log g	z_0	Ne	$\tau_{\rm Ross}$
1.36	66333	8.967E3	54213	7.16	4.63E7	1.11E15	1.48E4
2.00	59475	1.095E4	48448	6.90	8.01E7	6.75E14	1.76E4
3.00	48093	1.070E4	39159	6.60	1.35E8	4.14E14	2.08E4
4.00	40423	9.959E3	32916	6.38	1.94E8	2.90E14	2.39E4
5.00	35062	9.251E3	28556	6.21	2.55E8	2.24E14	2.71E4
6.00	31106	8.639E3	25342	6.07	3.18E8	1.82E14	3.06E4
7.00	28058	8.115E3	22867	5.95	3.81E8	1.52E14	3.39E4
8.00	25630	7.665E3	20894	5.84	4.47E8	1.31E14	3.77E4
10.00	21986	6.934E3	17395	5.67	5.84E8	1.02E14	4.58E4
12.00	19367	6.365E3	15812	5.53	7.24E8	8.21E13	5.42E4
14.00	17381	5.907E3	14201	5.40	8.68E8	6.74E13	6.25E4
16.00	15816	5.529E3	12933	5.30	1.01E9	5.71E13	7.00E4
18.00	14547	5.210E3	11905	5.20	1.16E9	4.96E13	7.60E4
20.00	13495	4.937E3	11054	5.12	1.31E9	4.36E13	7.98E4
22.00	12606	4.700E3	10333	5.04	1.44E9	3.78E13	8.13E4
24.00	11843	4.492E3	9717	4.97	1.59E9	3.25E13	8.10E4
30.00	10086	3.993E3	8287	4.78	2.02E9	1.49E13	7.52E4
34.00	9213	3.734E3	7572	4.68	2.32E9	6.68E12	7.17E4
40.00	8188	3.420E3	6744	4.61	3.13E9	1.87E12	6.97E4
45.00	7516	3.207E3	6202	4.39	2.68E9	5.27E11	6.69E4
50.00	6961	3.026E3	5759	4.18	2.36E9	1.59E11	7.48E4

the calculated models. The TLUSTY models for 12,000K and lower $T_{\rm eff}$ values used H, He, C, Mg, Al, Si, and Fe as explicit atoms and the remaining first 30 atomic species as implicit atoms. Inclusion as an implicit atom means that the atomic species contributes to the total number of particles and the total charge, but not to the opacity. SYNSPEC is used to calculate a synthetic spectrum for each annulus, and the array of annulus synthetic spectra, along with the corresponding synthetic spectrum arrays for the primary and secondary stars and the accretion disk rim, constitute the data base used to calculate a system synthetic spectrum. All of the models are on the hot branch of the disk instability model (DIM) [9, 53], are optically thick in the continuum, and show absorption line spectra. We measure the continuum optical thickness with the Rosseland optical depth. TLUSTY provides two options for the viscosity parameter–either the Shakura/Sunyaev α or a Reynolds number.

Figure 1 shows the synthetic spectrum for the annulus at $R/R_{\rm wd} = 6.0$, using the option to calculate only the continuum plus H lines. The corresponding line in Table 1 provides some of the detailed astrophysical information available from TLUSTY for this annulus. Note that the Balmer lines are very shallow for this annulus ($T_{\rm eff} = 31,106$K), and that there is essentially zero Balmer jump. The flux is in Eddington flux units. The calculated synthetic spectrum extends from 800Å to 10,000Å.

By contrast, Fig. 2 shows the synthetic line spectrum corresponding to the Table 1 annulus $R/R_{\rm wd} = 30.0$. The synthetic spectrum looks like a normal stellar model atmosphere spectrum, but the point deserves emphasis that this is a synthetic spectrum corresponding to an annulus model as described in

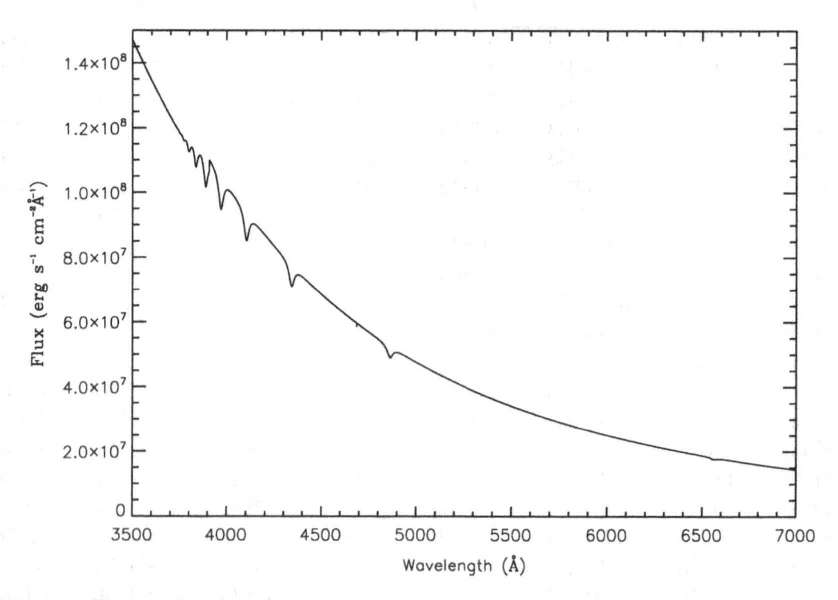

Fig. 1. Synthetic continuum spectrum for annulus $r/r_{\rm wd} = 6.0$.

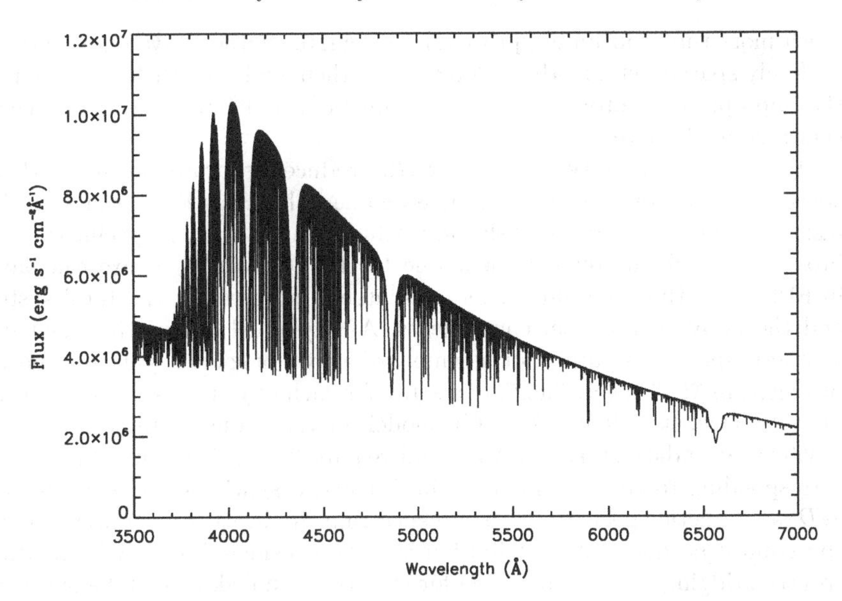

Fig. 2. Synthetic spectrum line for annulus $r/r_{\rm wd} = 30.0$.

H90. From Table 1, the Rosseland optical depth at the midplane is $7.52{\times}10^4$. At the temperatures involved, ϵ in (4) is close to 1.0 and the final term in the brackets is small compared with the first term, so the analogy to a model stellar atmosphere is justified. Irradiation by a 35,000K *WD*, included in the specific model calculated, amounts to a very small perturbation. As (5) indicates, the dilution factor, W, changes from annulus to annulus.

A further point about the synthetic spectrum: The `SYNSPEC` output optionally provides either specific intensities (erg cm^{-2} s^{-1} Hz^{-1} sterad^{-1}) for a specified number of zenith angles, or Eddington flux (erg cm^{-2} s^{-1} Å^{-1}). With the latter choice, an intensity normal to the local photosphere is calculated, followed by an intensity in the direction of the line of sight by application of an assignable limb darkening coefficient.

2.3 Calculation of Synthetic Spectra for *CV* Systems

The `BINSYN` software suite [40] is used to calculate system synthetic spectra for comparison with observational data. The program can also calculate synthetic light curves at specified effective wavelengths, based on a black body approximation for the radiation characteristics of the system objects. The accretion disk shadows part of the secondary star from irradiation by the primary star. The software evaluates this effect as well as irradiative heating of the secondary star by the faces and rim of the accretion disk. The program also calculates irradiation of the accretion disk rim by the secondary star. Synthetic photometry is also available using a more fundamental but much more time-consuming procedure: calculate synthetic system spectra at a closely-spaced set of orbital longitudes, then multiply each spectrum by the appropriate photometric response function, and integrate to produce a synthetic light curve.

For an accretion disk system, `BINSYN` produces separate output synthetic spectra for the total system, the mass gainer, the mass loser, the accretion disk face, and the accretion disk rim (including a hot spot, if specified). Line-broadening and line displacement due to the Doppler effect are calculated in `BINSYN` as the programs generate synthetic spectra for the total system and the individual system components. As explained above, the calculated synthetic spectra require prior assembly of arrays of source synthetic spectra produced in `TLUSTY+SYNSPEC`, one array for each object (mass gainer, accretion disk face, etc.). Thus, for a *CV* model corresponding to Table 1 and with a 3500K secondary star, `BINSYN` would require 21 synthetic annulus spectra corresponding to the annulus radii in Table 1, a synthetic spectrum for the *WD*, a synthetic spectrum (a single spectrum duplicating the outer annulus spectrum if no hot spot contribution is to be considered; otherwise multiple spectra bridging the hot spot $T_{\rm eff}$) for the accretion disk rim, and a synthetic spectrum for the secondary star. A single synthetic spectrum meets the formal requirement for the latter object, if its contribution is negligible. Otherwise, if the secondary star fills its Roche lobe, a two-dimensional array of synthetic spectra is required, bridging the photospheric variation in $T_{\rm eff}$ and $\log g$.

The output spectrum for a given object starts with interpolation within its array of fiducial synthetic spectra to produce, in effect, a local synthetic spectrum appropriate to the local $T_{\rm eff}$ value at each photospheric segment. (A grid of points defining the photosphere divides each object into a large number of segments.) Integration over the segments, with proper allowance for Doppler shifts and sources of line broadening, then produces the synthetic spectrum for the object (star, accretion disk, rim) at the particular orbital inclination and longitude under consideration. Each photospheric segment has an associated visibility key. If the particular segment is hidden from the observer, either by eclipse or by being below the observer's horizon, the value of the visibility key prevents inclusion of that segment in integration over the object in question. The system synthetic spectrum is the sum of contributions from the separate objects in the system. The synthetic light curve procedure uses a black body curve at the local $T_{\rm eff}$ value, replacing the local synthetic spectrum at each segment.

Stellar objects are represented by the Roche model, including allowance for rotational distortion up to critical rotation. The standard bolometric albedo formalism determines effects of mutual irradiation for both stellar components.

The accretion disk is represented by a specified number (typically 32) of concentric annuli with fixed radial width but of increasing thickness, up to a specified rim height. The rim height typically is set equal to the separately-calculated theoretical rim height [14], but may be set to a different value in

BINSYN. Each annulus photosphere is divided into azimuthal segments (typically 90) to permit evaluation of eclipse effects, if they exist. The number of annuli specified within BINSYN typically may exceed the number of calculated annulus models. This feature is necessary to provide adequate geometric resolution in calculating, e.g., eclipse effects. BINSYN calculates a T_{eff} value for each internally specified annulus (the 32 mentioned above); these are standard model [14] values by default, but an option permits assigning an individual T_{eff} to each internally specified annulus. This feature allows evaluation of an arbitrary temperature profile for the accretion disk, including an isothermal model. The synthetic spectrum for a given annulus specified by BINSYN is calculated by interpolation within the array of annulus spectra produced by SYNSPEC. BINSYN optionally permits calculation of irradiative heating of the accretion disk by the *WD*, again based on a bolometric albedo formalism. The spectra output from BINSYN are in the same mode as the annulus spectra from SYNSPEC–either line spectra or continuum spectra.

BINSYN requires input specification of the inner and outer radii of the accretion disk. This feature provides important flexibility to truncate the disk at an inner radius and/or to set the outer radius short of the tidal cutoff radius if the latter step is necessary to place all of the accretion disk on the hot branch of the *DIM*.

Figure 3 shows a system synthetic spectrum for the *CV* SDSS J0809 [42], with the accretion disk as listed in Table 1. The accretion disk is not truncated and extends to the tidal cutoff radius. The orbital inclination is 65°. The $0.3M_{\odot}$, 3500K (polar) secondary star makes an insignificant contibution

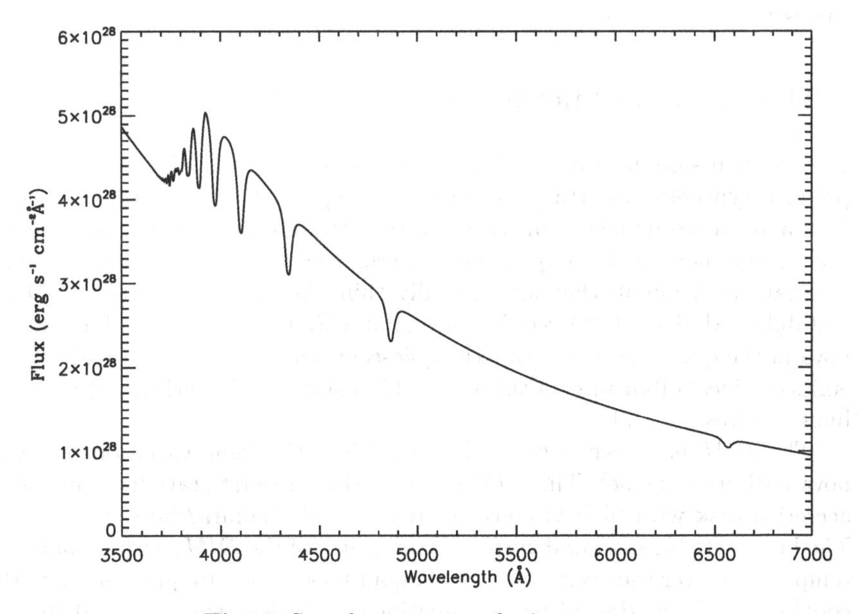

Fig. 3. Sample system synthetic spectrum.

to the system synthetic spectrum. The flux is the 4π value integrated over the system. In a comparison with an observed spectrum, an empirically determined divisor superposes the synthetic spectrum on the calibrated observed spectrum, permitting determination of the system distance, assuming no interstellar absorption.

It is straightforward to apply the procedures described herein and calculate a model for a *CV* system with a standard model accretion disk. Actual application to two systems [19, 41, 42] showed wide departures from the standard model. Both cases used the flexibility of `BINSYN` to depart from the standard model $T(R)$ relation to produce a successful model. Modeling of MV Lyr [19, 41] demonstrated that neither the intermediate state nor high state spectrum of that system could be fit by a standard model accretion disk. A fit was achieved with a much flatter temperature profile than the standard model. Similarly, standard model fits to the probable SW Sex system SDSS J0809 required such large truncation radii that the models were not credible. (The fits required a *WD* $T_{\rm eff}$ of, at most, 20,000K, a highly unlikely condition given the fairly large mass transfer rate, and a result that would be inconsistent with known properties of SW Sex stars.) Based on a likely similarity to the temperature profile of SW Sex itself [57], a reasonable fit could be achieved to *FUSE*, *HST*, and *SDSS* spectra with a temperature profile that is isothermal in an inner region and thereafter follows a standard model to the tidal cutoff radius. The reader may consult the separate references for details. Simulation of other *CV* systems are projected. However, the most important remaining work of the present simulation capability is to produce emission lines. We devote the remainder of the paper to a consideration of this issue.

3 The Emission Line Issue

*CV*s are emission line objects [55, 74, 75]. A fully successful *CV* model should produce synthetic spectra that accurately represent the observed emission lines and should be based on self-consistent physics. No current model meets that requirement. A basic question concerns whether observed emission lines associate with annuli that are optically thin. As discussed below, the Disk Instability Model (*DIM*) requires a Shakura/Sunyaev $\alpha \simeq 0.01$ for a dwarf nova in the quiescent state. But the quiescent state is associated with strong emission lines, often in contrast to the high state, and synthesis of emission lines requires $\alpha \simeq 1.0$.

The *DIM* has been successful in modeling the light variation of dwarf nova outbursts [17, 38]. The *DIM* explains the outburst state in terms of an accretion disk with high viscous dissipation and Shakura/Sunyaev $\alpha \simeq 0.1$. The high state is associated with the hot branch of the *DIM*, with a minimum temperature over the accretion disk of about 6300K [64]. In quiescence, on the cool branch of the *DIM*, viscous dissipation is much less, with $\alpha \simeq 0.01$ [16,38]. The *DIM* models showing the transition profile between the high and low

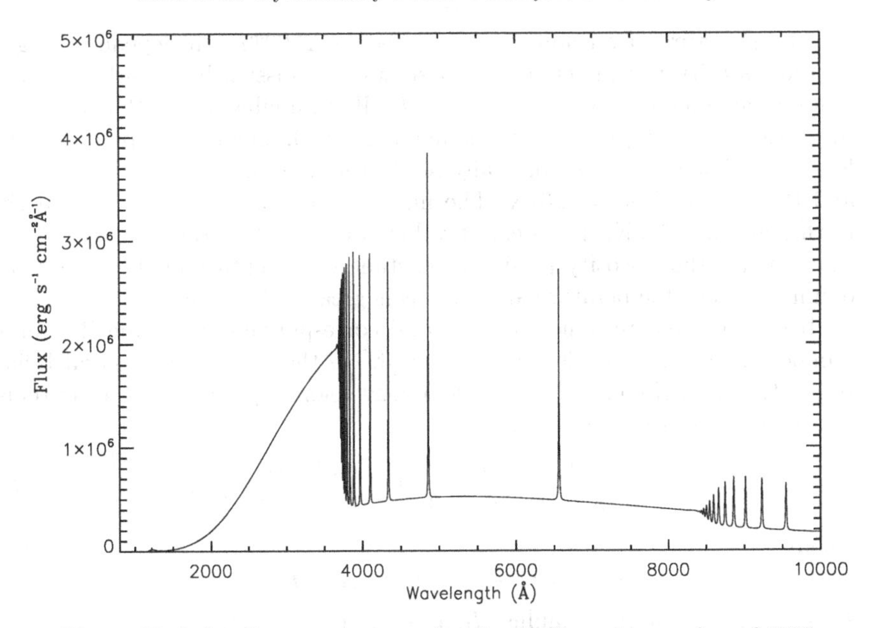

Fig. 4. Emission line spectrum produced with Reynolds number of 5000.

states (the "S" curve) have column mases (Σ) in the range 300−1000 gm cm^{-2} in the high state and $20 - 50$ gm cm^{-2} in the low state. Cannizzo & Wheeler [11] provide a detailed study of accretion disk structure as function of α and mass transfer rate.

It is often stated that an accretion disk in the low state is optically thin, and this condition explains the nearly universal presence of emission lines in the low state of *CV* systems. Mason et al. [44] provide spectroscopic evidence for an optically thin accretion disk in WZ Sge. Horne & Marsh [22] discuss emission line formation in accretion disks with optically thin continua and show that the shapes of double-peak profiles indicate whether the emission lines themselves are optically thin or thick. The authors also show that emission is anisotropic. Williams & Shipman [76] present non-*LTE* models for optically thin (continuum) annuli and apply their results to the Balmer decrement. Marsh & Horne [43] develop a mapping technique to use the line profile variation with phase to determine two-dimensional imaging of the disk, including detection of the impact region from the mass transfer stream. However, as the studies of Williams [75] and Tylenda [70] show, the production of emission lines in the presence of optically thin continua associates with large values of α, of order 1.0, rather than values of order 0.01 appropriate to the quiescent state. Thus, there is a conflict between the *DIM* and emission line synthesis calculations conflict that requires resolution. Cannizzo [9] discusses this point and states that annuli are optically thick in the quiescent state. The following discussion includes more recent publications bearing on this issue.

The structure of an annulus calculated by TLUSTY can depend strongly on the viscosity parameter. Figure 4 shows an emission line synthetic spectrum for an annulus centered on a $0.69M_{\odot}$ *WD*, annulus radius $R/R_{\rm wd} = 4.0$, $\dot{M} = 1.0 \times 10^{-11} M_{\odot} {\rm yr}^{-1}$, Reynolds number=5000. The central plane Rosseland optical depth is 0.103 (optically thin!), the column mass is 0.15 gm cm^{-2}, and the annulus $T_{\rm eff} = 6261$K. The internal annulus temperature is nearly isothermal, at 7500K. The default values of ζ_0 and ζ_1 set $w(m) = \overline{w}$, so that, in (3), the viscosity profile is constant with depth and the final term dominates. The temperature at all levels is greater than $T_{\rm eff}$.

It is useful to determine the value of α corresponding to a given Reynolds number. The conversion is described in [30] in the context of General Relativity. Křiž and Hubeny [37] provide a conversion applicable to *CV* systems in the nonrelativistic limit. It is

$$\alpha = \frac{1}{Re\dot{M}^{1/4}} \frac{m_H}{2k} \left(\frac{8\pi\sigma}{3}\right)^{1/4} \frac{(GM_{\rm wd})^{3/4}}{R_{\rm wd}{}^{1/4}} q(r), \quad (6)$$

where

$$q(r) = [r(1 - 1/\sqrt{r})]^{-1/4}, r = R/R_{\rm wd}. \quad (7)$$

R is the radius of the annulus, $R_{\rm wd}$ is the radius of the *WD*, $M_{\rm wd}$ is the *WD* mass, $\dot{M}$ is the mass transfer rate, G is the gravitation constant, σ is the Stefan-Boltzmann constant, k is the Boltzmann constant, m_H is the mass of the H atom, and Re is the (dimensionless) Reynolds number. The α corresponding to the Fig. 4 parameters is 6.36. TLUSTY tests, with Fig. 4 an example, confirm values of $\alpha \sim 1.0$ or more, and correspondingly small values of Σ, for optically thin annuli with emission lines, in disagreement with the *DIM*. Menou [47] emphasizes the same discrepancy. Idan et al. [33] used the Shaviv & Wehrse [61] code to model HT Cas, a SU UMa-type *CV*, and found it necessary to adopt an $\alpha \sim 1.0$ to produce emission lines, in agreement with Fig. 4. They found a mass transfer rate of $\dot{M}_0 = 2.5 \times 10^{-11} M_{\odot}$ yr^{-1}. The authors explain why an optically thin accretion disk, *i.e.*, an accretion disk with $\Sigma < 1.0$ gm cm^{-2}, implies $\alpha \geq 1.0$. The problem with use of optically thin continuum models is the requirement for high viscosity in a very low column mass. A physical basis for optically thin accretion disks in quiescent *CV* systems with low mass transfer rates remains obscure.

The $T_{\rm eff}$ value of 6261K for the Fig. 4 model would place this annulus on the hot branch of the *DIM*, implying an optically thick model. Changing the viscosity parameter input to TLUSTY to $\alpha = 0.1$, and keeping all other parameters the same, produces the spectrum shown in Fig. 5. This optically thick model has a column mass of $\Sigma = 319$ gm cm^{-2} and a midplane Rosseland optical depth of 7.4×10^4. The column mass is in agreement with the requirements of the *DIM*. These two examples demonstrate that the midplane Rosseland optical depth is a strong function of the viscosity parameter, as are the other physical properties of the annulus.

Tests with *WD* masses of $0.69M_{\odot}$ and greater, and mass transfer rates of $1.0 \times 10^{-11} M_{\odot}/{\rm yr}$ and greater, using $\alpha = 0.1$, show that all annuli extending

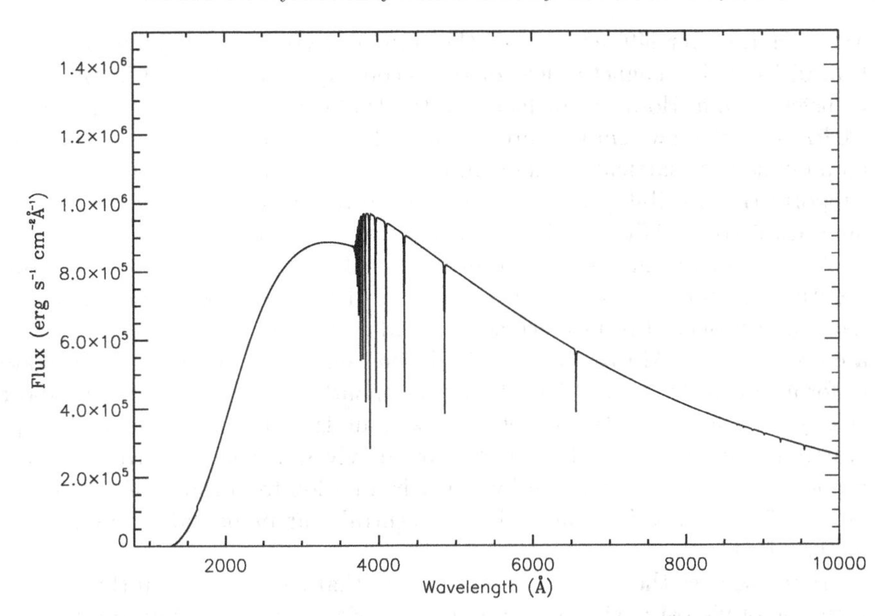

Fig. 5. Absorption line spectrum produced with $\alpha = 0.1$. Other parameters are the same as in Fig. 4.

either to a typical tidal cutoff radius or the end of the *DIM* hot branch at ~6300K are optically thick. Annulus column masses are of order $10^2 - 10^3$ gm cm^{-2}.

Nova-like *CVs* have mass transfer rates of order $1.0 \times 10^{-9} M_\odot$ yr^{-1}, and are optically thick to the outer edge of the accretion disk. Yet many of these *CVs* show strong emission lines. Robinson et al. [56] provide a detailed discussion. An example is SDSS J0809 [42], which has a mass transfer rate of $\sim 3.0 \times 10^{-9} M_\odot$ yr^{-1} but has an exclusively emission line optical spectrum and no Balmer jump. Its *FUSE* spectrum shows no P Cygni features, and the inclination is of order 65°, so it is unlikely that the emission lines are produced in a wind. Table 1 contains annulus models for this system, each calculated with $\alpha = 0.1$. All of the annulus spectra are absorption line spectra, and each annulus lies on the hot branch of the *DIM*. The observed emission line spectrum of this system cannot be explained by optically thin outer annuli.

There is general agreement that viscosity in the high state of dwarf novae, or for nova-like systems, is a result of the magnetorotational instability (*MRI*) [6,7]. Tout [68] discusses the connection of the *MRI* with the requirements of dwarf novae outbursts. Balbus [4] and references therein indicate that numerical *MRI* simulations lead to α values for fully ionized disks that are of order 0.1, in agreement with *DIM* results. At low temperatures ohmic resistance tends to suppress the magnetic field with consequent absence of viscosity. Menou [46] describes the viscosity problem associated with suppression of the magnetic field in the quiescent state [15]. Meyer &

Meyer-Hofmeister [49] argue that the source of viscosity in quiescent accretion disks is the magnetic field of the secondary star entrained in the mass transfer stream. However, inclusion of the Hall effect [5] appears to maintain (*MRI*) into the low temperature regime. The vertical viscosity profile within an annulus is of particular importance. (*MRI*) simulation by Stone et al. [67] supports the possibility of vertically concentrated viscous dissipation in an annulus. Also see Miller & Stone [51] and Hirose, Krolik, & Stone [18].

Production of emission lines from an optically thick annulus requires a temperature inversion in the outer atmosphere: a hot corona. Several mechanisms have been proposed to produce coronae. Meyer & Meyer-Hofmeister [48] use a coronal siphon; Murray & Lin [52] depend on sound waves that accelerate to form shocks; Liang & Price [39] propose nonthermal processes to transport energy vertically from the accretion disk. Cannizzo [10] uses a thermal evaporative instability originally identified by Shaviv & Wehrse [59]. Ko et al. [36] propose accretion disk heating by X-ray irradiation from the boundary layer. Smak [65] proposes irradiation by the central star or boundary layer, with further details in [56].

H90 discusses the details of a hot corona that can result from the vertical variation of viscosity within an annulus, a variation which, if improperly controlled, may lead to the "thermal catastrophe". Parallel results are discussed in Shaviv & Wehrse [59] and Adam et al. [1–3]. Thus, annulus models with prescribed viscosity profiles present an attractive possibility for production of emission lines from optically thick annuli, and recent theoretical (*MRI*) studies support a possible physical basis for such models. The emission line annulus models would substitute directly for the current annulus synthetic spectra, and their use with `BINSYN` would proceed exactly according to the current procedure.

References

[1] Adam, J., Störzer, H., Shaviv, G., & Wehrse, R. 1988, Astron. & Astroph., 193, L1

[2] Adam, J., Störzer, H., & Duschl, W.J. 1989a, Astron. & Astroph., 218, 205

[3] Adam, J., Innes, D.E., Shaviv, G., Störzer, H., & Wehrse, R. 1989b, in Theory of Accretion Disks ed. F. Meyer et al. (Dordrecht:Kluwer), p. 403

[4] Balbus, S.A. 2002, in The Physics of Cataclysmic Variables and Related Objects, ASP Conference Series vol. 261, ed. B.T. Gänsicke, K. Beuermann and K. Reinsch (San Francisco:Astron. Soc. Pacific), p. 356

[5] Balbus, S.A. 2005, in The Astrophysics of Cataclysmic Variables and Related Objects, ASP Conference Series vol. 330, ed. J.-M. Hameury and J.-P. Lasota (San Francisco: Astron. Soc. Pacific), p. 185

[6] Balbus, S.A. & Hawley, J.F. 1991, ApJ, 376, 214

[7] Balbus, S.A. & Hawley, J.F. 1998, Rev. Mod. Phys., 70, 1

[8] Buat-Ménard, V., Hameury, J.-M., & Lasota, J.-P. 2001, Astron. & Astroph., 366, 612

[9] Cannizzo, J.K. 1993, in Accretion Disks in Compact Stellar Systems ed. J.C. Wheeler (Singapore:World Scientific), p. 6
[10] Cannizzo, J.K. 2000, ApJ, 534, L35
[11] Cannizzo, J.K. & Wheeler, J.C. 1984, ApJS, 55, 367
[12] la Dous, C. 1989a, Astron. & Astroph., 211, 131
[13] la Dous, C. 1989b, MNRAS, 238, 935
[14] Frank, J., King, A., & Raine, D. 1992, Accretion Power in Astrophysics (Cambridge: Univ. Press)
[15] Gammie, C. F. & Menou, K. 1998, ApJ, 492, L75
[16] Hameury, J.-M., Menou, K., Dubus, G., Lasota, J.-P., & Huré, J.M. 1998, MNRAS, 298, 1048
[17] Hameury, J.-M. 2002, in The Physics of Cataclysmic Variables and Related Objects, ASP Conference Series vol. 261 ed. Gänsicke, Beuerman, and Reinsch (San Francisco:Astr.Soc.Pacific), p. 377
[18] Hirose, S., Krolik, J.H., & Stone, J.M. 2006, ApJ, 640, 901
[19] Hoard, D.W., Linnell, A.P., Szkody, P., Fried, R.E., Sion, E.M., Hubeny, I., & Wolfe, M.A. 2004, ApJ, 604, 346
[20] Hoard, D.W., Linnell, A.P., Szkody, P., & Sion, E.M. 2005, AJ, 130, 214
[21] Horne, K. 1993, in Accretion Disks in Compact Stellar Systems ed. J. Craig Wheeler (Singapore:World Scientific), p. 117
[22] Horne, K. & Marsh, T.R. 1986, MNRAS, 218, 761
[23] Huang, M., Sion, E.M., Hubeny, I., Cheng, F.H., & Szkody, P. 1996, ApJ, 458, 355
[24] Hubeny, I. 1988, Comp. Phys. Comm., 52, 103
[25] Hubeny, I. 1990, ApJ, 351, 632
[26] Hubeny, I. 1991, in Structure and Emission Properties of Accretion Disks, ed. C. Bertout, S. Collin, J-P. Lasota, J. Tran Thanh Van (Singapore:Fong & Sons), p. 227
[27] Hubeny, I., Stefl, S., & Harmanec, P. 1985, Bull. Astron. Inst. Czechosl. 36, 214
[28] Hubeny, I. & Lanz, T. 1995, ApJ, 439, 875
[29] Hubeny, I. & Hubeny, V. 1997, ApJ, 484, L37
[30] Hubeny, I. & Hubeny, V. 1998, ApJ, 505, 558
[31] Hubeny, I., Agol, E., Blaes, O., & Krolik, J.H. 2000, ApJ, 533, 719
[32] Hubeny, I., Blaes, O., Krolik, J.H., & Agol, E. 2001, ApJ, 559, 680
[33] Idan, I., Lasota, J.-P., Hameury, J.-M., & Shaviv, G. 1999, PhR, 311, 213
[34] Kiplinger, A.L. 1979, ApJ, 234, 997
[35] Kiplinger, A.L. 1980, ApJ, 236, 839
[36] Ko, Y.-K., Lee, Y.P., Schlegel, E.M., & Kallman, T.R. 1996, ApJ, 457, 363
[37] Kříž, S. & Hubeny, I. 1986, Bull. Astron. Inst. Czechosl., 37, 129
[38] Lasota, J.-P. 2001, New Astronomy Reviews, 45, 449
[39] Liang, E.P.T. & Price, R.H. 1977, ApJ, 218, 247
[40] Linnell, A.P. & Hubeny, I. 1996, ApJ, 471, 958
[41] Linnell, A.P., Szkody, P., Gänsicke, B., Long, K.S., Sion, E.M., Hoard, D.W., & Hubeny, I. 2005, ApJ, 624, 923
[42] Linnell, A.P., Hoard, D.W., Szkody, P., Long, K.S., Hubeny, I., Gänsicke, B. & Sion, E.M. 2006, ApJ, in press
[43] Marsh, T.R. & Horne, K. 1987, ApSpSci, 130, 85
[44] Mason, E., Skidmore, W., Howell, S.B., Ciardi, D.R., Littlefair, S., & Dhillon, V.S. 2000, MNRAS, 318, 440

[45] Mayo, S.K., Wickramasinge, D.T., & Whelan, J.A.J. 1980, MNRAS, 193, 793
[46] Menou, K. 2000, Science, 288, 2022
[47] Menou, K. 2002, in The Physics of Cataclysmic Variables and Related Objects, ASP Conference Series vol.261 ed. Gänsicke, Beuerman, and Reinsch (San Francisco:Astr.Soc.Pacific), p. 387
[48] Meyer, F. & Meyer-Hofmeister, E. 1994, Astron. & Astroph., 288, 175
[49] Meyer, F. & Meyer-Hofmeister, E. 1999, Astron. & Astroph., 341, L23
[50] Mihalas, D. 1978, Stellar Atmospheres, 2nd ed. (San Francisco:Freeman)
[51] Miller, K.A. & Stone, J.M. 2000, ApJ, 534, 398
[52] Murray, S.D. & Lin, D.N.C. 1992, ApJ, 384, 177
[53] Osaki, Y. 1996, PASP, 108, 39
[54] Pringle, J. 1981, Ann. Rev. Astron. Astrophys., 19, 137
[55] Robinson, E.L. 1976, Ann. Rev. Astron. Astrophys., 14, 119
[56] Robinson, E.L., Marsh, T.R., & Smak, J.I. 1993, in Accretion Disks in Compact Stellar Systems, ed. J.C. Wheeler (Singapore:World Scientific), p. 75
[57] Rutten, R.G.M., van Paradijs, J., & Tinbergen, J. 1992, Astron. & Astroph., 260, 213
[58] Schwarzenberg-Czerny, A., & Różyczka, M. 1977, Acta Astron., 27, 429
[59] Shaviv, G. & Wehrse, R. 1986, Astron. & Astroph., 159, L5
[60] Shaviv, G. & Wehrse, R. 1989, in Theory of Accretion Disks, eds. F. Meyer et al. (Dordrecht:Kluwer), p.419
[61] Shaviv, G. & Wehrse, R. 1991, Astron. & Astroph., 251, 117
[62] Skidmore, W., Mason, E., Howell, S.B., Ciardi, D.R., Littlefair, S., & Dhillon, V.S. 2000, MNRAS, 318, 429
[63] Shakura, N.I. & Sunyaev, R.A. 1973, Astron. & Astroph., 24, 337
[64] Smak, J. 1982, Acta Astron., 32, 199
[65] Smak, J. 1989, Acta Astron., 39, 201
[66] Smak, J. 2002, Acta Astron., 52, 263
[67] Stone, J.M., Hawley, J.F., Gammie, C.F., & Balbus, S.A. 1996, ApJ, 463, 656
[68] Tout, C.A. 2000, New Astronomy Reviews, 44, 37
[69] Tylenda, R. 1977, Acta Astron., 31, 267
[70] Tylenda, R. 1981, Acta Astron., 31, 127
[71] Wade, R.A. 1984, MNRAS, 208, 381
[72] Wade, R.A. 1988, ApJ, 335, 394
[73] Wade, R.A. & Hubeny, I. 1998, ApJ, 509, 350
[74] Warner, B. 1995, Cataclysmic Variable Stars (Cambridge:University Press)
[75] Williams, R. 1980, ApJ, 235, 939
[76] Williams, G.A. & Shipman, H.L. 1988, ApJ, 326, 738

Part IV

Modeling Short-Period Eclipsing Binaries

Distance Estimation for Eclipsing X-ray Pulsars

R.E. Wilson[1], H. Raichur[2], and B. Paul[2]

[1] University of Florida, Astronomy Department, Gainesville, FL 32611, USA
wilson@astro.ufl.edu
[2] Raman Research Institute, Bangalore, India
sharsha@rri.res.in, bpaul@rri.res.in

Distance of an X-ray binary can be computed from a rigorous flux scaling law that connects model stellar atmosphere output with observed standard magnitudes of the optical star via either of two standard magnitude calibrations that agree within 4 percent. Accordingly the corresponding distance disagreement (due to the calibrations only) is only 2 percent, which is negligible compared to several other error sources. The flux-distance scaling is not the usual one for spherical stars but preserves directional (i.e. aspect) information, and therefore is not limited to well detached binaries. Bolometric corrections are not needed, so errors in their estimation are avoided. The procedure also models dependence of system brightness and spectroscopically observable temperature on orbital phase and inclination due to tides, irradiance, and eccentric orbits, although those effects cause only minor distance uncertainties for most X-ray binaries. Not taken into account, due to their largely stochastic nature, are radial velocity variations caused by dynamical tides. Expressions are given for derivatives $\partial d/\partial p$, of distance with respect to various parameters. Some of the derivatives are entirely analytic while others are partly numerical. Upper and lower limits to the relative radius, $r = R/a$, of an X-ray binary's optical star can be measured, although actual r and inclination are otherwise uncertain. An application to the High Mass X-ray Binary Vela X-1/GP Vel, based on archival pulse arrival times and radial velocities, finds a distance of about 2.2 kiloparsecs and also finds distance uncertainties due to estimated magnitude, interstellar extinction, metallicity, orbit size, optical star size, surface temperature, and surface gravity.

1 Introduction

Published distances based on optical stars in X-ray binaries are affected by a variety of errors. Estimates scaled according to brightness vs. luminosity usually assume spherical symmetry, even for stars that very nearly or actually

E.F. Milone et al. (eds.), Short-Period Binary Stars: Observations, Analyses, and Results, 179–189.

fill limiting lobes, thereby neglecting aspect effects due to tides, rotation, and eccentric orbits. Also the connection between surface brightness and temperature may be handled by empirical calibrations of modest accuracy although model stellar atmospheres establish that connection to the order of 1 percent. Most X-ray binary distances published to date lack uncertainty estimates and essentially all lack formal standard errors. Here we (1) address these problems following ideas on rigorous flux scaling for light curves that effectively are in absolute units (Wilson 2004, 2005, 2007a, 2007b), (2) adapt the general plan to X-ray binaries, and (3) list expressions for partial derivatives of distance with respect to observationally estimated quantities such as standard magnitude, interstellar extinction, metallicity, effective temperature and surface gravity. The derivatives allow computation of formal standard errors for distance or, more specifically, components of standard errors due to the various parameter uncertainties. Both the flux scaling and the derivatives are very simple, but still need to be written out and evaluated if distance uncertainties are to be quantified. The central ideas are to:

1\. ...operate with stellar atmosphere radiative output (not blackbody, not bolometric) integrated over standard bands (*U, B, V, R, I*, etc.), with input of T_{eff}, $\log g$, and $[Fe/H]$. Note that, at 25,000 K, a blackbody is about 33% brighter in V band than a real star of equal size, corresponding to about 15% distance error.

2\. ...avoid scaling via simple spheres, thus removing any limitation to detached binaries (DB). Rather than the traditional scaling of flux from luminosity, the idea is to preserve directional information by scaling observable flux from emergent intensity at a (star) surface reference point. For the general *EB* case (not necessarily an X-ray binary), resulting distances are at least as accurate for semi-detached (*SD*) and over-contact (OC) binaries as for *DB*'s. Also, *SD*'s and especially OC's are relatively efficient in use of light curve data, as their (distance information bearing) eclipses occupy larger fractions of the orbital cycle than *DB* eclipses.

2 X-ray Binaries – a Special Case with Special Difficulties

Because the optical stars of most X-ray binaries are tidally and rotationally distorted, with local temperature varying over the surface, observed temperature and observed magnitude depend on orbital phase and inclination, while other surface variations (*e.g.* reflection effect) are significant in some X-ray binaries. Accordingly three effective temperatures enter the distance problem – temperature seen by an observer (averaged) over the instantaneously visible surface (T_{obs}), temperature at a surface reference point (T_{ref}, usually T_{pole}), and globally averaged temperature (T_{ave}). The logical solution parameter is T_{ave}, which is not a distribution but a single number for a given star. However T_{ave} is not directly observable but must be inferred from a fitted model.

Table 1. Comparison of Johnson and Bessell Calibrations in $erg \cdot \sec^{-1} \cdot cm^{-3}$

Band	Johnson (1965)	Bessell (1979)	Johnson/Bessell
U	$4.35 \cdot 10^{-1}$	$4.19 \cdot 10^{-1}$	1.038
B	$6.88 \cdot 10^{-1}$	$6.60 \cdot 10^{-1}$	1.042
V	$3.78 \cdot 10^{-1}$	$3.61 \cdot 10^{-1}$	1.047
R	$1.85 \cdot 10^{-1}$	...	...
R_C	...	$2.25 \cdot 10^{-1}$	...
I	$8.99 \cdot 10^{-2}$	...	...
I_C	...	$1.22 \cdot 10^{-1}$	...
J	$3.40 \cdot 10^{-2}$	...	...
K	$3.90 \cdot 10^{-3}$	$4.01 \cdot 10^{-3}$	0.973
L	$8.04 \cdot 10^{-4}$	...	...
M	$2.16 \cdot 10^{-4}$	...	...
N	$1.24 \cdot 10^{-5}$	...	...

It can be computed from T_{obs} and two dimensionless ratios derivable from the system model,

$$T_{ave} = \left(\frac{T_{ave}}{T_{pole}}\right)\left(\frac{T_{pole}}{T_{obs}}\right)T_{obs}, \tag{1}$$

where the first ratio is easy and comes from eqn. 8 of Wilson (1979) while the second is much more intricate computationally (Wilson 2007a, 2007b). The current model applies Kurucz (1993) atmospheres via the Van Hamme & Wilson (2003) Legendre polynomials that represent *cgs* fluxes in standard photometric bands. Utilization of *cgs* flux allows Direct Distance Estimation (*DDE*, see §3), which will be from the optical star alone for typical X-ray binaries. The calibrations of standard magnitudes to physical units by Johnson (1965) and Bessell (1979) disagree by only 4 percent in *U*, *B*, and *V*, corresponding to only a 2 percent difference in distance. They have been converted to $erg \cdot \sec^{-1}\ cm^{-3}$ in Table 1, where bands R_C and I_C are on the Cousins (1971, 1973, 1980, 1981) system.

3 Light/Velocity Curve Solutions in Physical Units

Simultaneous light curve and radial velocity (*RV*) curve solutions for *EB*'s now can be carried out with light in standard physical units of $erg \cdot \sec^{-1} \cdot cm^{-3}$ (Wilson 2007a, 2007b). Distance can be derived along with its standard error via *DDE* by *W-D*'s Differential Corrections (*DC*) program, in contrast to conventional global scaling with assumption of spherical stars. Flux to distance scaling that is tied to a surface reference point can utilize all local effects of the *W-D* model. If one already has a *W-D* solution, say from the literature, a quicker way to find distance is to extract a few numbers from a *W-D* light curve run (*viz.* scaling relation below) and write a small program that only does the intensity to flux scaling. Both ways, full solution and scaling of an existing solution, are founded on the relation (Wilson 2004):

$$F_d^{abs} = 10^{-.4A} \left[F_{a,1}^{prog} \left(\frac{I_1^{abs}}{I_1^{prog}} \right) + F_{a,2}^{prog} \left(\frac{I_2^{abs}}{I_2^{prog}} \right) \right] \left[\frac{a}{d} \right]^2 . \tag{2}$$

Subscript a means that F_a is flux for a hypothetical observer at distance a (length of the semi-major axis), of course *without* compensation for finite distance effects (*i.e.* the flux is for a *large* distance with a simple inverse square law conversion). F_d^{abs} will be in standard physical units because $I_{1,2}^{abs}$ is in standard physical units and the equation simply converts from intensity to flux by applying *EB* geometry and radiative physics. Only one of the two terms will be needed for an X-ray binary, assuming negligible light from the X-ray star, so subscripts 1, 2 are deleted from this point on. Basically, local effective temperature, $\log g$, and chemical composition jointly fix I^{abs} at the reference point, then I^{abs} fixes F_d^{abs} via eqn. 2 at d_{pc} (distance in parsecs) through the *EB* model (which supplies I^{prog} and F^{prog}), and then d_{pc} becomes an ordinary parameter in the Differential Corrections (*DC*) solution algorithm. Alternatively, the solution algorithm could be other than *DC*, although only *DC* is in the present *DDE* program. So the idea is to compare theoretical fluxes with observed fluxes by the Least Squares criterion, all in *cgs* units. *RV*'s are needed for absolute scale and mass information, so either the solution must be simultaneous in light and *RV* or – if in light only – it must be be given the ordinary *RV* parameters as fixed input.

4 Pulsed X-ray Binaries

Pulse arrival times establish the absolute scale of an X-ray star's orbit (*i.e.* $a_x \sin i$) via the light-time effect, and also e and ω (eccentricity and argument of periastron), which may be poorly known from optical *RV*'s. However optical[1] star *RV*'s are needed to complete the orbital scale by furnishing $a_o \sin i$, although with large systematic and random errors for most High Mass X-ray Binaries (*HMXB*'s). Optical star *RV*'s, weakly measured as they may be, are the only link to systemic *RV*, as pulse frequency in the neutron star rest frame is not directly known.

4.1 Lower and Upper Limits on R_o/a

DDE operates with local geometry and radiative physics that are integrated into synthesized observables, but from an intuitive viewpoint *DDE* effectively requires knowledge of mean optical star size, R_o/a, for which lower and upper

[1] Subscript o is for the optical star.

limits can sometimes be assessed[2]. For the lower limit, note that a neutron star's point-like nature eliminates the possibility of finding inclination (i) from eclipses – all we know is that the neutron star disappears for a fraction of the orbit cycle set by R_o/a, i, e, ω, and a rotation parameter (or possibly more than one)[3]. Given an X-ray eclipse, R_o/a must be large enough to account for the eclipse duration in the extreme case of an edge-on orbit ($i = 90°$). Since sufficient light, *RV*, spectral, and pulse information may exist to estimate e, ω, and rotation, a lower limit to R_o/a can come from X-ray eclipse duration under the edge-on assumption. A fast-converging iterative scheme based on formal relations among these parameters (Wilson, 1979) can generate R_o/a for given i. Eclipse width (in phase) can be introduced as a constraint that eliminates otherwise possible solutions (Wilson & Wilson, 1976; Wilson, 1979). For an upper limit, relative lobe size for the optical star (R_{lobe}/a) depends on mass ratio, rotation, and e, so boxing of R_o/a (and R_o) within a narrow range set by these lower and upper limits may be possible. In overview, we have far less than the usual eclipse information for an X-ray pulsar, but can assess T_{obs} from spectra, generate R_o/a and R_o limits from eclipse width and mass ratio (above), and set the absolute scale from the pulses and *RV*'s.

4.2 Solutions for Distance

Equation 2 gives F_d^{abs} for an observer at distance d, according to values of F_d^{prog} and I^{prog}, along with orbit size, a, polar normal intensity in physical units, I^{abs}, and bandpass interstellar extinction, A. The Cardelli, Clayton, and Mathis (1989) tabulation converts V extinction to extinction in other bands. Radiative behavior is set by a stellar atmosphere model that supplies bandpass-integrated intensity as a function of T_{pole}, $\log g$, chemical composition, and direction (*viz.* Van Hamme & Wilson, 2003 for a fast radiative scheme). *DC* output will include a distance with a standard error. *HMXB*-related difficulties usually include large and uncertain interstellar extinction, roughly known (perhaps sub-synchronous) rotation, and essentially stochastic effects of dynamical tides.

5 Vela X1 (GP Vel) Results from RXTE Data

The binary composed of the 283^s X-ray pulsar Vela X-1 and the *B0.5Ib* supergiant GP Velorum can serve as an example of *DDE*. The supergiant's optical light curves show essentially ellipsoidal variation that surely must be

[2] Of course the actual star-size parameter is a surface potential that can be converted to local or mean R_o/a.

[3] The situation can be more difficult if the optical star's edge is not sharp, as with atmospheric or wind eclipses.

due to dynamical tides driven by the eccentrically orbiting neutron star. However a large chaotic component is superposed on the regular variation and, as chaotic tides are beyond present *EB* model capabilities, we analyze only *RV* and pulse data (the chaotic behavior also affects *RV*'s, but not so seriously as it does the light curves). Wilson & Terrell (1994, 1998) developed a procedure for simultaneous analysis of optical star light curves, *RV*s, and X-ray pulse arrival times (t_a), following ideas in Wilson (1979) while deriving analytic derivatives $\partial t_a/\partial p$ for *DC* solutions (p means parameter). Although the X-ray pulse capability was never integrated into the general (*i.e.* public) *W-D* model because of the scarceness of suitable data, we can find a distance for Vela X-1 in two steps – first applying the program of Wilson & Terrell (1998) simultaneously to pulses and *RV*s for evaluation of binary parameters, and then applying the scaling logic of §3. We also corrected the observed optical star temperature, T_{obs}, for aspect so as to estimate T_{ave} according to eqn. 1, although the correction is much smaller than the uncertainty in T_{obs} for GP Vel. The pulse data are from the Proportional Counter Array (*PCA*) of the Rossi X-ray Timing Explorer satellite (*RXTE*) and the optical *RV*s are from Barziv, et al. (2001).

Pulsations from Vela X-1 were discovered by McClintock, et al. (1976). The Compton Gamma Ray Observatory's Burst and Transient Source Experiment (*BATSE*) found alternations between spin-up and spin-down, with no long-term trends in spin frequency (Bildsten, et al., 1997). The pulse profiles vary with energy and time, being complex below 6 *kev* and simpler at higher energies, although double-peaked. We used *RXTE/PCA* archival data from January of 2005 to estimate arrival times. A template profile was made by averaging all pulses. We then made averaged pulse profiles by combining 10 consecutive pulses and cross-correlated each such profile with the template. An advance/delay from the cross-correlation was then converted to arrival time. *RXTE*'s All Sky Monitor (*ASM*) X-ray light curve yielded an eclipse semi-duration of $0^P.0940$ that *DC* used to constrain possible solutions, according to the logic in Wilson (1979). That is, only solutions that produce eclipses of the specified width are allowed. The supergiant's angular rotation was assumed to be half the mean orbital rate (*cf.* estimate from line widths by Zuiderwijk, 1995). Further work on the difficult observational problem of the supergiant's rotation would be helpful. Simultaneous pulse-*RV* fitting then gave binary system parameter estimates for input to the distance step. Figures 1 and 2 respectively show the pulse arrival and radial velocity fits. Numerical results are in the next section.

These and related data, plus a T_{ave} that is representative of published spectroscopic estimates and an assumption that $[Fe/H] = 0.0$ for the optical star, are our inputs to the distance estimation step, with the *W-D* light curve program computing F^{prog} and I^{prog} for insertion into eqn 2. Output is d_{pc}.

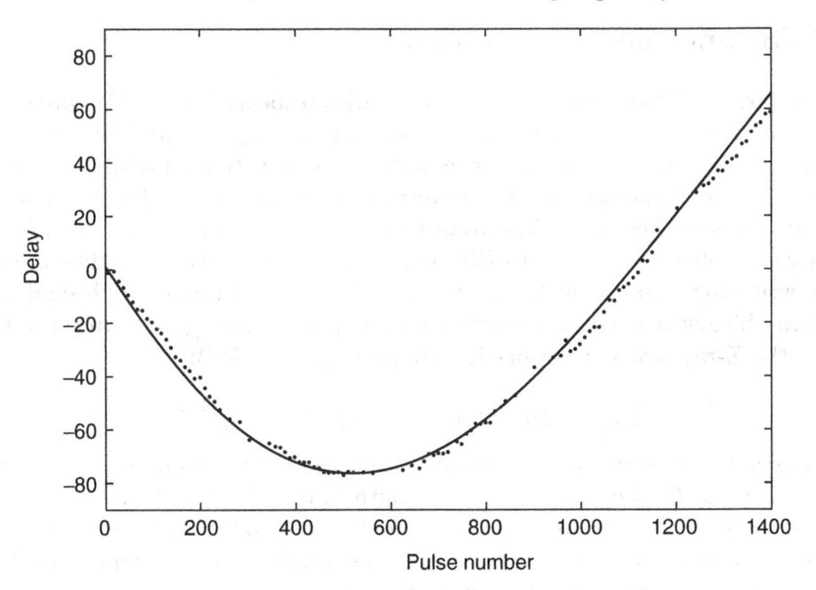

Fig. 1. Vela X1 advance-delay (advances being negative delays) in pulse arrival over about 4.6 days, with dots for RXTE data and a continuous curve for our Least Squares solution. The pulse frequency was changing by large amounts on time scales shorter than the orbit period, as shown by short-term biases in advance-delay that made the solution difficult.

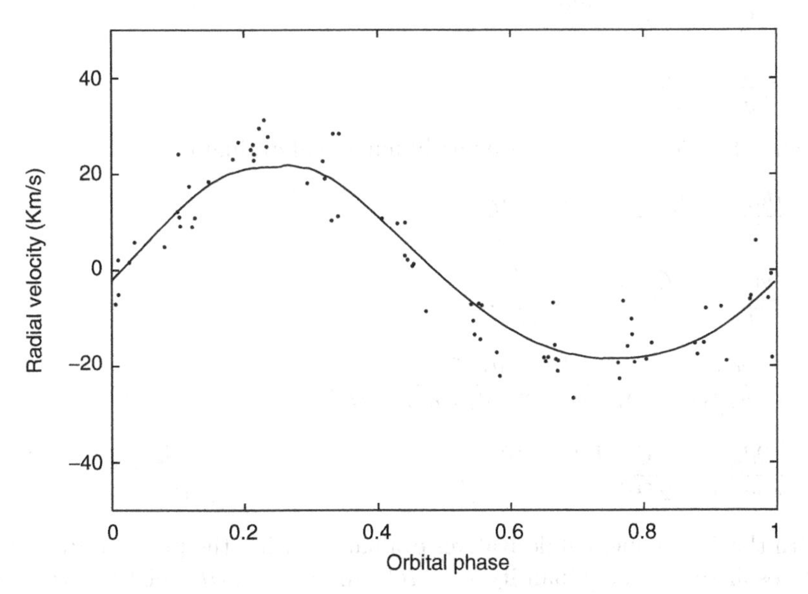

Fig. 2. Radial velocity *vs.* phase for the supergiant companion (GP Vel) to Vela X1, showing a clear signature of orbital eccentricity. Dots are Barziv, *et al.* (2001) data and the curve is our Least Squares solution (simultaneous in RV and pulse arrival time).

6 Distance and Uncertainties

The largest distance errors for X-ray binaries (especially *HMXB*'s) are caused by uncertainties in standard magnitude (*e.g.* V), T_{obs}, a, and interstellar extinction, A, where needed improvements are observational problems. Distance errors depend not only on those uncertainties but also on factors that convert errors of these quantities to distance error, such as $\partial d/\partial T_{obs}$ and $\partial d/\partial a$. Given an effectively absolute *EB* model such as *W-D*, the derivatives are easy to write down and to evaluate, although that has not previously been done in X-ray binary papers. So re-write eqn. 2 explicitly for d_{pc} and drop the term for the X-ray star (subscripts 1, 2 are no longer needed):

$$d_{pc} = 10^{-.2A} aC \left[F_a^{prog}\left(I^{abs}/I^{prog}\right)\right]^{1/2}. \tag{3}$$

Constant C converts the length unit from the orbital semi-major axis, a, to 1 parsec, so $C = a_{pc}/a$. The *W-D* length unit is 1 solar radius so, for *W-D* applications, C is $R_\odot$ in parsecs, or about $2.256 \cdot 10^{-8}$. With conversion from flux to magnitude (here the V system is assumed so as to exemplify a definite magnitude system), analytic differentiation gives

$$\frac{\partial d_{pc}}{\partial V} = 0.2 \ln 10 \cdot d_{pc} \approx 0.46052\, d_{pc},$$

$$\frac{\partial d_{pc}}{\partial A_V} = -\frac{\partial d_{pc}}{\partial V}, \text{ and}$$

$$\frac{\partial d_{pc}}{\partial a} = \frac{d_{pc}}{a}.$$

Four other derivatives require partly numerical evaluation:

$$\frac{\partial d_{pc}}{\partial r} = \frac{d_{pc}}{2}\frac{dF^{prog}/dr}{F^{prog}} \approx \frac{d_{pc}}{r},$$

$$\frac{\partial d_{pc}}{\partial T} = \frac{C}{2I^{1/2}}\frac{F^{prog}}{I^{prog}}\frac{dI^{abs}}{dT},$$

$$\frac{\partial d_{pc}}{\partial [Fe/H]} = \frac{C}{2I^{1/2}}\frac{F^{prog}}{I^{prog}}\frac{dI^{abs}}{dT}\frac{dI^{abs}}{d[Fe/H]}, \text{ and}$$

$$\frac{\partial d_{pc}}{\partial \log g} = \frac{C}{2I^{1/2}}\frac{F^{prog}}{I^{prog}}\frac{dI^{abs}}{d\log g},$$

with the I^{abs} numerical derivatives evaluated within the program from stellar atmosphere output. Quantity $r = R/a$ in the $\partial d_{pc}/\partial r$ relations is a mean dimensionless radius for the optical star. The *DC* solution gave $r = 0.56$, $a = 51.92\ R_\odot$, $e = 0.124$, $\omega = 3.003$ radians, and $\log g = 2.80$. For input to the distance step we adopt $T_{pole} = 25{,}929\ K$ and $V = 6^m.93$. Literature values of V band interstellar extinction, A_V, are typically around $1^m.8$ to $2^m.2$. Here we estimated the color excess, E_{B-V}, to be $0^m.60$ by differencing $B-V = +0^m.38$

Table 2. Evaluated Derivatives and Distance Standard Errors (S.E.)

Derivative	Value	Parameter S.E. (estimated)	Distance S.E.
$\partial d_{pc}/\partial r$	$+3738\ pc$	0.03	$112\ pc$
$\partial d_{pc}/\partial V$	$+993\ pc/mag.$	$0^m.1$	$99\ pc$
$\partial d_{pc}/\partial A_V$	$-993\ pc/mag.$	$0^m.3$	$298\ pc$
$\partial d_{pc}/\partial a$	$+41.5\ pc/R_\odot$	$3R_\odot$	$125\ pc$
$\partial d_{pc}/\partial T$	$+0.074\ pc/K$	$1000\ K$	$74\ pc$
$\partial d_{pc}/\partial [Fe/H]$	$+95.6\ pc/\log_{10}$	0.5	$48\ pc$
$\partial d_{pc}/\partial \log g$	$-21.3\ pc/\log_{10}$	0.1	$2\ pc$

from the SIMBAD data base and $(B-V)_0 \approx -0^m.22$ from the Flower (1996) calibration, assuming $T_{ave} \approx 26{,}000\ K$. Multiplication by 3.1 (assumed ratio of V band extinction to $E(B-V)$) then gave $A_V = 1^m.86$. Although i must be in the vicinity of 90° to have eclipses, no reasonably accurate value can be found for typical X-ray binaries for reasons given by Wilson & Terrell (1998). However the supergiant cannot be kept within its limiting lobe if $\sin i$ is much below 1.00, so we assumed $i = 90°$. The V magnitude is a consensus of several published values and refers to phase $0^P.20$. We assume solar metallicity for lack of contrary information, so $[Fe/H] = 0.00$. The Ω_2 potential parameter (output from the DC solution) that sets the supergiant's size was 18.818. Values for F_a^{prog}, I^{abs}, and I^{prog} are taken from a W-D run, whereupon eqn. 3 yields $d_{pc} = 2155\ pc$. Then, as an example,

$$\frac{\partial d_{pc}}{\partial r} = 3738\ pc$$

(from the approximate version of $\frac{\partial d_{pc}}{\partial r}$, which is good enough for demonstration purposes), so the distance error for 0.03 uncertainty in r is $112\ pc$. Seven such contributions to distance standard errors are similarly evaluated in Table 2, ranging from a negligible 2 pc for $\log g$ to several hundred pc for A_V. While the distance standard errors of column 4 depend on roughly estimated parameter standard errors (column 3), they can be refined by improved estimates for column 3.

If, for definiteness, we assume the input parameter uncertainties to be realistic and the errors to be independent, the root mean square uncertainty in d_{pc} from these seven error sources is 367 pc (dominated by the A_V error). One should remember that all these formulas pertain to a definite aspect (*i.e.* phase and inclination).

7 Summary

In an eccentric, near lobe-filling binary such as GP Vel = Vela X1, large errors in absolute orbit size (a) and consequently also star size ($R/R_\odot$) result from systematic RV errors caused by chaotic tidal dynamics, for which no simple remedy seems apparent. Although $r = R_{\odot}/a$ cannot be computed very accurately from typical X-ray binary light curves, r can be boxed between an upper limit imposed by lobe filling and a lower limit imposed by X-ray eclipse duration. Other uncertainties come from roughly measured temperature, standard magnitude, interstellar extinction, and metallicity. However if reasonable estimates of such errors can be made, their propagation into distance estimates can be computed via the derivatives of §6. The derivatives are elementary in form, but do need to be evaluated so as to quantify the relative influence of parameter errors on d_{pc} uncertainty, as well as the overall standard error of d_{pc}. One also can take care of systematic errors due to inaccurate or neglected conversion from T_{obs} to T_{ave}, neglect of eccentric orbit and asynchronous rotation effects, or even (as in most X-ray binary distance estimates) complete neglect of aspect effects by assumption of spherical symmetry. One can thereby quantify error contributions and know which errors cause the most and least distance uncertainty. In the case of GP Vel, the uncertainty due to interstellar extinction appears most important, orbit size and relative star size turn out to be moderately important, and $\log g$ is unimportant. We find the distance of this *HMXB* to be $2155 \pm 367(S.E.)$ *pc*.

8 Acknowledgments

We thank W. Van Hamme for valuable suggestions on the manuscript. We made extensive use of the *SIMBAD* and *NASA ADS* web sites. The work was supported by U.S. National Science Foundation grant 0307561.

References

1. M.S. Bessell 1979, PASP, 102, 1181
2. O. Barziv, L. Kaper, M.H. Van Kerkwijk, J.H. Telting, J. Van Paradijs 2001, A&A, 377, 925
3. L. Bildsten, D. Chakrabarty, J. Chiu, M.H. Finger, D.T. Koh, R.W. Nelson, T.A. Prince, B.C. Rubin, D.M. Scott, M. Stollberg, B.A. Vaughan, C.A. Wilson, R.B. Wilson 1997, ApJS, 113, 367
4. P.E. Boynton, J.E. Deeter, F.K. Lamb, G. Zylstra 1986, ApJ, 307, 545
5. J.A. Cardelli, G.C. Clayton, J.S. Mathis 1989, ApJ, 345, 245
6. A.W. Cousins 1971, Roy. Obs. Ann., No. 7
7. A.W. Cousins 1973, Mem. Roy. Astron. Soc, 77, 223
8. A.W. Cousins 1980, S. Africa Astron. Obs. Circ., 1, 166
9. A.W. Cousins 1981, S. Africa Astron. Obs. Circ., 6, 4
10. J.E. Deeter, P.E. Boynton, N. Shibazi, S. Hayakawa, F. Nagase, N. Sato 1987, AJ, 93, 877
11. P.J. Flower 1996, ApJ, 469, 355
12. H.L. Johnson 1965, Contrib. Lunar & Planetary Lab., 3, 73
13. R. Kurucz 1993, in Light Curve Modeling of Eclipsing Binary Stars, ed. E.F. Milone, Springer Publ. (New York), p. 93

14. J.E. McClintock, S. Rappaport, P.C. Joss, H. Bradt, J. Buff, G.W. Clark, D. Hearn, W.H.G. Lewin, T. Matilsky, W. Mayer, F. Primini 1976, ApJ, 206, 99
15. W. Van Hamme, R.E. Wilson 2003, ASP Conf. Ser., 298, 323
16. R.E. Wilson 1979, ApJ, 234, 1054
17. R.E. Wilson 2004, New Astr. Rev., 48, 695
18. R.E. Wilson 2005, ApSS, 296, 197
19. R.E. Wilson 2007a, ASP Conf. Ser., 362, 3
20. R.E. Wilson 2007b, ApJ, in press
21. R.E. Wilson, D. Terrell 1994, Am. Inst. Phys. Conf. Proc. 308, ed. S.S. Holt & C.S. Day, p. 483
22. R.E. Wilson, D. Terrell 1998, MNRAS, 296, 33
23. R.E. Wilson, A.T. Wilson 1976, ApJ, 204, 551
24. E.J. Zuiderwijk 1995, A&A, 299, 79

The Tools of the Trade and the Products they Produce: Modeling of Eclipsing Binary Observables

Eugene F. Milone[1] and Josef Kallrath[2]

[1] University of Calgary, Dept. of Physics & Astronomy, 2500 University Dr., NW, Calgary, AB T2N 1N4 Canada
milone@ucalgary.ca
[2] University of Florida, Dept. of Astronomy, Gainesville, FL 32611
kallrath@astro.ufl.edu

1 Introduction

In this volume we have been discussing the significance, and observational and theoretical state of knowledge of, short-period binaries, binary stars that encompass the entire close-binary realm from unevolved binaries to black hole systems. The purpose of this portion of the volume is to report on results from systems less evolved than black–holes, neutron stars, and white dwarfs, and in doing this to emphasize both the current modeling techniques developed to analyze them, and those on the horizon. Some of our suggestions for future development may allow the tools to be used for systems with more evolved components.

In this paper, we discuss some parameter–fitting techniques for eclipsing binary light curves[1] making primary use of the Wilson–Devinney model in and with various programs. These include the simplex method, damped least squares, Powell's direction methods, and metaheuristic search techniques such as simulated annealing and genetic algorithms. We describe the results of applications of some of these techniques to eclipsing binaries across a range of contact configurations and evolutionary states, and, by a not quite trivial extension, to planetary eclipses in extra–solar systems.

2 Parameter Estimation of Eclipsing Binary Data

Eclipsing binary parameters are computed by fitting an eclipsing binary model to the observed data. The set of parameters includes the mass ratio, the Roche

[1] The term *light curves* is used here in the general sense of eclipsing binary observables (see KM99, pp. 49–51).

E.F. Milone et al. (eds.), Short-Period Binary Stars: Observations, Analyses, and Results, 191–214.

potentials representing the shape of the components, the inclination, the temperature of at least one component, epoch and orbital period, the semi-major axis and the systemic radial velocity. A major improvement in a future version of the Wilson–Devinney program (Wilson 2007) is the calculation of both distance and third body adjustable parameters. For general light curve analysis, what is used is a maximum likelihood estimator, for which we can thank Boscovic (*Theorie Philosophia Naturalis*, Vienna 1758). If the measurement errors of the observed data are independent and normally distributed we usually want to solve a least squares problem; this is possible thanks to Legendre (1805) and Gauß (1809). Mathematicians call this norm ℓ_2. Thus far, nearly all light curve analyses have been based on this norm. If the errors follow a Laplace distribution, ℓ_1 is the appropriate choice; in this norm the absolute values of the difference between observed and calculated observables are compared.

There are various methods to solve light curve least–squares problems. For detailed discussions we refer the reader to Kallrath & Milone (1999), hereafter *KM99* and also Wilson (2007, §4). Some of the methods (differential corrections, Levenberg–Marquardt damped least-squares) use derivatives, others (Simplex, direction set methods) do not. As an extensive history of the development of light curve modeling code can be found in KM99 and in the on-line proceedings of an AAS topical session in 1997, http://www.ucalgary.ca/~milone/AAS97, we summarize developments only from the beginning of exploration of derivative-free techniques to those most recently exploited.

The first application of simplex to eclipsing binary analysis was carried out by Kallrath (1987), in conjunction with the Wilson–Devinney program (Wilson and Devinney 1971) and the Light Synthesis Program of Linnell (1984) as reported in Kallrath & Linnell (1987). This implementation is close to that of Spendley, *et al.* (1962). The simplex algorithm provides the benefit of exploring parameter space and good starting values for derivative based methods. The first application of damped least squares (Levenberg-Marquardt) engines to light curve solution was by Kallrath, *et al.* (1995) and this same work also illustrated the operation of self-iterative procedures in the package `WD95`; that package was followed successively by `WD98`, `WD2002`, `WD2005`, `WD2006`, and, most recently, `WD2007`. The Levenberg-Marquardt scheme has now been implemented in recent versions of Wilson's base code, and in `PHOEBE`, developed by Andrej Prša (2005) at the University of Lubljana. `PHOEBE` also contains a variant of the simplex algorithm, similar to that found in Press, *et al.* (1992). Recently, Prša [Prša & Zwitter (2005)] reported positive results from applications of a direction set method (*cf.*, Press, *et al.* 1992, p.406) that is also implemented in their light curve analysis tool `PHOEBE`. They use Powell's direction set method with appropriate conjugate directions preserving the derivative free nature of the method. Both `PHOEBE` and `WD2007` contain a version of simulated annealing discussed in Section 4.1. The attempts to use metaheuristics are partly motivated by our long-term goal to find a robust method to uncover the global minimum for the parameter fit problem. However, both applications and results thus far have been somewhat limited, as we note below.

Although the available methods are sufficient to support analyses of individual eclipsing binaries, the focus is now shifting more to the analysis of large sets of eclipsing binaries obtained by surveys. Neural networks (Devinney, *et al.*, 2006) as well as the archive approaches suggested by Wyithe & Wilson (2001), and Devor & Charbonneau (2006) are being discussed. In the approach of Wyithe & Wilson, initial parameter guesses are taken from a large precalculated library or archive of eclipsing binary curves. This approach was, in fact, taken in preparation of a search for planet transits in the globular cluster 47 Tuc (Milone, *et al.*, 2004a). For their study of the eclipsing binaries found in that 47 Tuc survey, Milone, *et al.* (2004b), are making use of the specific, non–solar isochrones produced for that cluster by Bergbusch & VandenBerg (1992). Devor & Charbonneau similarly employ isochrones [the Yonsei-Yale isochrones with solar metallicity (Kim, *et al.*, 2002)], to attempt to discern the properties of both components of eclipsing systems.

Several programs or software packages are available to the eclipsing binary community or researchers in other areas of astrophysics as well as to amateur astronomers. Some of those with which we are familiar are described in §3.

3 Getting to the Bottom of it: Software for Eclipsing Binaries Analysis

KM99 provides an overview of existing software. In this section, we comment on four software programs which have seen active developments over the years since then: the original Wilson–Devinney program (`WD`), `WD2007` by Josef Kallrath, Bradstreet's `Binary Maker`, and `PHOEBE`.

3.1 The BASE Software: The Wilson–Devinney Program (WD)

This program is continually improved and new versions are produced aperiodically. Major releases appeared in 1971 (Wilson & Devinney 1971), 1979 (Wilson 1979), 1998, 2003, and 2007 (Wilson 2007). Documentation and updates are available from `ftp://ftp.astro.ufl.edu/pub/wilson/`. The Wilson–Devinney (*WD*) program is used to study eclipsing binary light curve observables, through:

- Computation of synthetic curves from a model and suite of parameters with the Light Curve (*LC*) routine;
- Analysis, to yield a set of least-squares determined parameetrs with the Differential Corrections (*DC*) routine.

The next Wilson–Devinney program will include, among other improvements, two major new features:

- Adjustable *third body parameters* (ephemerides, heliocentric reference time or epoch $T_{0,3b}$ and period $P_{0,3b}$, eccentricity, e_{3b}, and argument of the pericenter, ω_{3b}, and orbital semi-major axis, a_{3b}) (Wilson 2007, §3.1), and
- *Direct distance estimation* (Wilson 2007, §5).

Both features are important additions to complete the picture of eclipsing binary star analysis. The most problematic part of the analysis is the determination of the orbit period of the third body, mainly due to the inadequate time coverage of most light curve data sets.

Direct distance estimation is based on absolute calibration of light curves. In favorable cases, the temperatures of both components can be obtained. The goal here is the comparison of the computed, distance–dependent flux with the observed flux. Although this feature produces stellar distances that are limited by the appropriateness of the models, calibration accuracy, and by the assumed interstellar extinction, the production of self–consistent distances in a single software program marks an important advance for binary analysis.

3.2 The `WD2007` Package

Here we describe the latest version of a package of light curve analysis programs, the first version of which was `WD95`. All of these versions were developed and maintained by one of us (JK), with testing and analyses performed episodically by the other (EFM). `WD2007` runs under the `LINUX` operating system as well as in Microsoft `Windows`, under `Win95`, `win98`, `WinNT`, `Win2000`, and `XP` operating systems, and is expected to run also under such other `UNIX` operating systems as `cygWin`. `WD2007` enables the user to make use of the `WD` program to fit as well as to compute eclipsing binary light curves.

In fact, `WD2007` contains the `WD` program and several additional features and programs. Unlike what is done in a "wrapping" approach, it couples directly to the Wilson–Devinney program so whenever a new version of the base Wilson–Devinney program, is issued, the package needs to be revised accordingly. The documentation has been improved significantly from that available in 2005, and is freely available from the authors. A word of caution is appropriate here, however. It is not generally appreciated that modeling eclipsing binaries and solving inverse problems is a major research effort and requires expertise to use the software effectively. Great effort has been invested to make the software as stable as possible, but using it successfully is still an art; reading the manual is therefore a necessary first step to mastering the texhnique.

`WD2007` has been developed within the framework of several preceding software programs and packages, namely `LCCTRL`, `wd98k93`, `WD95`, `WD98`, and `WD2002`. `WD2002` was the first release that combined all previous developments and included the stellar atmospheres improvements added by Milone, Stagg, and Kurucz (1992), as well as other improvements described by Milone, *et al.* (1988, 1993, 1997) and Stagg and Milone (1993) and it was also the first release which ran under `LINUX` and other `UNIX` operation systems.

WD2007 incorporates simulated annealing as a minimization algorithm; it produces tables of absolute dimensions, and, in particular, LaTeX output tables. It also supports and embeds a Fortran subroutine for the on–the–fly computation of limb–darkening coefficients coded by Walter Van Hamme and based on the atmosphere models of Robert L. Kurucz (1993). During the iterations of least squares runs, the limb-darkening coefficients are automatically computed as a function of temperature, log g, and effective wavelength of the light– and RV–curve passbands.

The procedure for creating the flux files is described in the appendix along with changes to the atmospheres option in WD programs. The use of the flux ratio files that make use of Kurucz atmospheres models are coded in the subroutine ATM2000 (which replaces subroutine ATM developed by C. R. Stagg for a stand–alone, non–iterating program, wd98k93), to run on UNIX platforms.

In short, WD2007 is an eclipsing binary simulation and optimization tool that enables the modeler to

- carry out simulations using the orginal Wilson–Devinney programs LC and DC;
- use Kurucz atmospheres to model better the stellar spectral energy distribution;
- fit eclipsing binary observables using the simplex method, differential corrections, a Levenberg-Marquardt scheme, or simulated annealing;
- do automatic iterations;
- produce gnuPlot graphics files; and to
- obtain best-fit solutions for grids or tables over some fixed parameters which otherwise may be poorly determined.

The program WD2007 is a stand–alone Fortran77 program called from a UNIX script file or a DOS batch file, and connects to the main subroutines of our version of the Wilson–Devinney program, namely, LC and DC. These are fed the input data through the following WD2007 input files:

1. An information file that contains files names and procedural switches that tell WD2007 how to proceed, **.inf*;
2. A file containing upper and lower bounds on the adjustable parameter to be fitted, **.con*;
3. A DC input file that has the light and radial velocity curves and input parameters for the Wilson–Devinney subroutine DC. This file is nearly identical to most WD input files; it is called **.dci*;
4. An LC input file controlling the Wilson–Devinney subroutine LC to evaluate light and radial velocity curves and other observables for a grid of time or phase values, **.lci*;
5. A set of flux files providing the ratio between the Kurucz atmospheres and a black body radiator; examples: *Rflux*, *Bflux*, *Vflux*, *Vm05*, *Im10* [The latter two files involve atmosphere models which have non-solar metallicities]; and

6. An ascii file containing `gnuPlot` commands to produce graphics, *fit.gnu.*

The main parts of the program `WD2007`, itself, are:

1. Subroutines to read the **.inf* and **.con* input files;
2. A subroutine to interface with and call `LC` and `DC`;
3. Subroutines that control the solution to the inverse problem through the simplex method, the Levenberg-Marquardt scheme, or by simulated annealing;
4. A subroutine to implement a variant of the simplex algorithm (Nelder & Mead, 1965; Kallrath & Linnell, 1987), also described in KM99;
5. A subroutine to implement a variant of the Levenberg-Marquardt scheme, described in Kallrath, *et al.* (1998), and KM99;
6. A subroutine to implement a simulated annealing algorithm [see Aarts & Korst, 1989; Aarts, *et al.*, 1997; Eglese, 1990, Kirkpatrick, *et al.*, 1983; (Julia) Kallrath, 2004]; and
7. Several subroutines to handle character strings in Fortran.

3.3 Binary Maker

`Binary Maker` is a commercially available software package developed by David Bradstreet (Eastern College, Pennsylvania) to visualize light and radial velocity curves, and the appearance of the system itself with varying phase. It permits the plotting of the Roche potentials and the outer and inner Lagrangian surfaces in either Kopal or modified Kopal potentials, and the calculation of several quantities. Depending on what is given, these computed quantities are: the radii of the components in back, side, pole and point facings, surface areas and volumes, mean densities, locations of the Lagrangian points L_1^{p} and L_2^{p}, and the fill-out factor. Most importantly, it computes and plots synthetic light and radial velocity curves. It can also be used to derive the mass ratio from observed radial velocity curves, and other parameters, through trial approximations of light curve fittings to data.

`Binary Maker` can plot observed data for comparison with computed light and radial velocity curves. It also computes and plots spectral line profiles at selected phases. Generated light curve and radial velocity curve data points can be simultaneously displayed with the projected three-dimensional view of the system model replete with star spots and in the correct orientation toward the observer. The surface elements may be shown in both spherical and cylindrical coordinates. One of its most interesting features is the demonstration of the effects of star spot regions on the light and radial velocity curves and on the profiles.

`BM3`, the version (3.0) current at this writing, does not make use of the latest Wilson–Devinney or `WD93K93` program features such as asynchronous rotation, nonlinear limb-darkening, multiple reflection, or Kurucz atmosphere calculations. However, most of the features available in the pre-1979 version of the Wilson–Devinney program are included. Unlike BM2, however, eccentric orbits are treated and BM3 runs on a variety of platforms, including Windows on PCs, Unix, and Macintosh computers.

The program has many uses for both research and teaching. It is perhaps most valuable for demonstrating the effects on the synthetic light and radial velocity curves and spectral profiles of changes of particular parameters. The manual has been updated and defines many of the classical Wilson–Devinney program features.

3.4 PHOEBE

PHOEBE (PHysics Of Eclipsing BinariEs) by A. Prša (Prša & Zwitter (2005) is a modeling package built on top of the WD program. This paper describes most of the uses for and extensions to PHOEBE, especially to the Wilson–Devinney program which is at its center. These include:

The incorporation of observational spectra of eclipsing binaries into the solution-seeking process; the extraction of individual temperatures from observed color indices of main-sequence stars and treatment of interstellar reddening. Among its innovations are: suggested improvements to WD Differential Corrections, inclusion of the new Nelder & Mead's downhill Simplex method) and various technical structures (*e.g.*, back–end scripter, graphical user interface[2]). Although PHOEBE retains 100% WD compatibility, its add–ons enhance WD. The usefulness of these extensions is demonstrated for a synthetic main–sequence test binary. Applications to real data are to be published in follow-up papers. PHOEBE is released under the GNU General Public License, which guaranties it to be free.

PHOEBE is best understood as a wrapper on top of the WD model. However, it provides several physical enhancements such as:

- color indices as indicators of individual temperatures (color constraints);
- spectral energy distribution as an independent data source;
- main sequence constraints; and
- interstellar and atmospheric extinction.

This program is currently used on LINUX systems. At present, the convenient graphical user interface has not been ported to Microsoft systems.

[2] In the late 1990's Dirk Terrell (now at the Southwest Research Institute, Boulder, CO) distributed a copy of the Wilson–Devinney program with a user–friendly I/O interface running under Microsoft Windows. Part of this interface has been integrated into PHOEBE.

4 Newer Investigatory Approaches

4.1 Simulated Annealing

Our WD2007 implementation of simulated annealing (*SA*) is based on the Corana, *et al.* (1987) simulated annealing algorithm for multimodal and robust optimization as implemented and modified by Goffe, *et al.* (1994). The implementation also contains a multimodal example from Judge, *et al.* (1985). Simulated annealing is a function-evaluation based improvement method often used to solve optimization problems. It is capable of escaping local minima and can, in principle, find the global optimium although one can never be sure about this, unless one knows the function to be minimized thoroughly. Corana, *et al.* (1987) test *SA* with 2– and 4–dimensional Rosenbrock (1960) classical unimodal test functions:

$$f_2(x_1,x_2) = 100(x_2 - x_1^2)^2 - (1 - x_1)^2,$$

and

$$f_4(x_1,x_2,x_3,x_4) = \sum_{k=1,3}^{3} [100(x_{k+1} - x_k^2)^2 - (1 - x_k)^2]$$

Corana, *et al.* (1987) compared the results of the *SA* tests with those of an application of the Nelder and Mead simplex (*NMS*) and an Adaptive Random Search (*ARS*) algorithm. They ran the experiment over a wide range of independent variable values and with a variety of starting points. The experiment showed that *SA* never failed to find the global minimum, whereas *ARS* failed to find the minimum in three cases for the 2-dimensional function test, and stopped at a saddle point in the 4-dimensional function test. *NMS* stopped twice in the saddle point of the functions. In these tests, therefore, *SA* proved the most robust of the three functions, whereas simplex was by far the most efficient, requiring less than 0.2% of the number of iterations required for *SA* to achieve convergence (less than 1000 compared to ∼500,000). These authors also tested *SA* with multimodal functions, one of which was a parabola with square–edged flat–bottomed cut–out regions, that required 10 parameters to describe; the central minimum was the deepest, whereas the cutout regions closest to the central minimum were nearly as deep. This multi-minimum problem proved to be difficult even for *SA*, which found the global minimum 8 out of 10 times; in the other two cases, it found the deep minimum closest to the central, absolute minimum solution. Apparently, *SA*, too, is not fool proof; however, compared to the other methods, it certainly proved far more robust. To emphasize this point, in the Corana *et al.* (1987) experiments the *ARS* method failed to find the global minimum in a two–dimensional test after 5 million evaluations, whereas simplex found the global minimum after 240 restarts and 150,000 evaluations. In 4–dimensions, neither *NMS* nor *ARS* found the global solution, whereas in 10–dimensions, with a

more retricted range, *NMS* was able to find a good, but not global, solution with 6500 restarts and ~10 million evaluations.
The basic requirements are an initial starting point, a "neighborhood concept" and a cooling scheme involving a few tuning parameters. For continuous optimization problems, the neighborhood concept leads to parameter variations of the form:

$$x_i \rightarrow x'_i = x_i + r \cdot v_i$$

where x_i is a parameter, r is a random number in the range [-1, +1], and v_i is the step length for this parameter. The function to be minimized, $f(x)$, is then computed for the new values of the parameter, to give the value $f'(x')$. In our program, this function is the sum of squares of the weighted residuals. If $f' < f$, the function is said to have moved downhill, and the values x' are taken as the new optimum values. All downhill steps are accepted and the process repeats from this new point. An uphill step may be accepted, allowing the parameter values a chance to escape from local minima. The acceptance of uphill movements (for example, a value f' such that $f' > f$), is controlled by application of the Metropolis criterion:
An "acceptance probability", p, is defined such that

$$p = e^{(f-f')/T}$$

where f is the previously found minimum value, f' is the most recently computed value of the function, and T is the current value of the "temperature," the thermodynamic analogue, not a stellar light curve parameter. If p is larger than a randomly selected number between 0 and 1, the new point is accepted. Both v_i and T are changed over time by the program. The step length is changed such that a fixed percentage (this is one of the tuning parameters mentioned above) of the new evaluations are accepted. When more points are accepted, v_i is increased, and more points may be "out of bounds," that is, beyond the parameter limits that are established in a constraints file. The temperature is decreased after a specified number of times through the iteration procedure. For this set of trials, this number was 20. After this number of iterations has occurred, the temperature is changed to

$$T' = r_T \cdot T$$

where $0 \leq r_T \leq 1$, where r_T is another tuning parameter; in our experiments, r_T = 0.85. As the optimization process proceeds, the step length decreases and the algorithm terminates ideally close to the global optimum.

Recently we performed a number of experiments, in which we solved for suites of three and four parameters of a single passband synthetic light curve based roughly on the V781 Tau data set (Kallrath, *et al.*, 2006). Only a single, phase-ordered light curve was placed in the *DC* input file, and only three or four parameters were adjusted for any one run. The `WD` package that we used (a slightly older version of `WD2007`) permits several programs to be run in succession. For most of the trials, we ran the simulated annealing (*SA*) program first, followed by our damped least squares version of the Wilson–Devinney program, and/or *DC* and *LC*, the solver and light curve simulator programs, respectively, of the Wilson–Devinney Program. All of these *WD* versions have been modified as per our `WD98` package, described elsewhere (Kallrath, *et al.*, 1998). An initial temperature (tuning parameter number 3) must be specified. This is typically a very large number to permit adequate ranges of excursion across parameter space. Usually, we used 9×10^4 K for this value but changed this in a series of trials to see how much it affected the search.

The parameter suites were:

1. $\Omega_{1,2}$, L_1;
2. i, Ω_1, L_1;
3. i, T_1, Ω_1;
4. $pshift$, T_1, L_1;
5. i, T_2, L_1;
6. i, Ω_2, L_1;
7. i, Ω_1, Ω_2;
8. i, T_1, Ω_1; L_1;
9. i, T_2, Ω_1, L_1

The three-parameter suites usually converged and produced tight solutions, and parameters to higher precision, typically, than the other methods. The runs on the 9th suite, performed after most of the other experiments had been run, and with more highly refined parameter values thereby established, did converge, and to high precision. The latter fitting, particularly, suggests that simulated annealing algorithms may be of some practical benefit in light curve modeling on very fast platforms.

On the down side, our experiments indicate that solving for only three or four variables often requires very long running times, depending on the initial temperature, the initial parameters, and on the stopping criteria; the desired precision of the fitting must be specified in the *.inf* file, in the sense that if this is set to a value that exceeds the precision of the data themselves (such as, say 0.1 mmag when the mean standard error of the light curve is no better than 0.01 mag), this criterion cannot be satisfied and the program will stop only when another criterion (such as a limiting temperature or a maximum number of function evaluations), is met. There is no disguising that there is a heavy price to be paid for an attempt to find the absolute extremum in parameter space with this algorithm, compared with the simplex algorithm that we have used for the past two decades. In the case of Suite 9, the time taken to run 6 iterations with the simplex program took only 139 seconds to achieve convergence on the same data set and initial values and run on the same platform (a 2GHz laptop PC with `XP`) as used to test the *SA* program. Given the criteria and initial values for the run, *SA* required more than 17 hours to fulfill the criteria (which did result in convergence).

Therefore, refinement of the criteria to improve the efficiency of the *SA* run is essential. The silver lining in this is that the results need not, after all, be particularly precise if the program does the one thing that is required of it most of all: to locate at least approximately the region of deepest minimum of the function (in our case, the sum of squares of the weighted residuals). If this is done approximately and the multiparameter terrain is not highly variable in this part of parameter space, the more efficient simplex program can then determine the solution to better precision, and *WD* can be found to follow the gradients to extract the optimum parameters and their true uncertainties. Examples are provided below.

The details of the fittings with *SA* are shown in Table 1. The second line of each set of parameters or other data gives the subsequent *WD* determination of that parameter, always determined after the *SA* part of the run, that is, with the *SA* results as input. The parameter uncertainties, which are not computed in the *SA* program, are enclosed in parentheses and are in units of the last digit of the parameter values. The exceptions to this structure begin with the sum of squares of the residuals, *SSR*, on the 15th line. Other quantities in Table 1 are as follows:

- SSR is the sum of squares of residuals in the simulated annealing program;
- $\sigma_{fit} = \sqrt{\Sigma wr^2/(n-1)}$, the standard error of a single observation of average weight, computed in the damped least squares program, *DDC*);
- Σwr^2, *SSR*, but as computed in *DC*;
- The quantity T_{final} is the final temperature;
- The quantity t_{SA} is the full running time of the package. All but a few minutes, at most, is due to the simulated annealing part of the run.

The initial values of the adjusted parameters for the Suite 1 fitting were:

- $\Omega_1 = 4.0100$
- $\Omega_2 = 5.0000$
- $L_1 = 7.0000$

The results for this trial were not as precise as for the others, because the criteria were met before the step interval decreased sufficiently to permit achievement of high precision for all the parameters. The refinement produced by the damped least–squares *WD* program (*DDC*) that followed the *SA* trial and the resulting *DC* solution were not quite converged as a consequence, but sufficently close so that with more iterations they clearly would have been. The fitting can be seen in Figure 1. When suite 1 was run with initial T_1 and $\Omega_{1,2}$ values closer to the solution, namely,

- $\Omega_1 = 4.0704$
- $\Omega_2 = 4.4925$
- $L_1 = 6.93053$

and was allowed to run long enough, it converged. This trial took more than 4^h ($15,699^s$), but also achieved greater precision: [SSR $(SA) = 0.00217$,

Table 1. Fittings of Simulated Annealing Tests

Parameter	Suite 1	2	3	4	5	6	7	8	9
p_{shift} (ph)	0.0000[a]	0.0000[a]	0.0000[a]	$8.34 \cdot 10^{-6}$	0.0000[a]	0.0000[a]	0.0000[a]	0.0000[a]	0.0000[a]
p_{shift} (ph)	...	...	...	$0.0(2) \cdot 10^{-4}$	...	...	...	...	...
i (deg.)	79.983[a]	79.977	80.019	79.983[a]	79.978	79.984	79.983	79.964	79.967
i (deg.)	...	79.98(1)	80.02(1)	...	79.98(1)	79.98(1)	79.98(1)	79.98(2)	79.98(1)
T_1 (K)	4500[a]	4500[a]	4501	4503	4502[a]	4500[a]	4500[a]	4503	4500
T_1 (K)	...	...	4500(1)	4502(1)	...	...	...	4502(1)	...
T_2 (K)	4500[a]	4500[a]	4500[a]	4500[a]	4499.1	4500[a]	4500[a]	4500[a]	4499[a]
T_2 (K)	...	...	...	...	4499.8(5)	...	...	...	4499.8(5)
Ω_1	4.436	4.077	4.087	4.077[a]	4.077[a]	4.077[a]	4.077	4.076	4.076
Ω_1	4.027(3)	4.077(2)	4.079(1)	...	...	...	4.077(1)	4.076(1)	4.076(1)
Ω_2	5.222	4.493[a]	4.500[a]	4.492[a]	4.492[a]	4.493	4.492	4.492[a]	4.492[a]
Ω_2	4.524(7)	...	...	...	...	4.492(3)	4.492(1)	...	...
L_1 (4π)	7.1412	6.9305	6.9397[a]	6.9423	6.9428	6.9316	6.9305[a]	6.9473	6.9458
L_1 (4π)	7.069(20)	6.932(3)	...	6.940(2)	6.940(2)	6.931(5)	...	6.941(3)	6.941(3)
SSR	0.03349	0.00043	0.00051	0.00032	0.00032	0.00043	0.00043	0.00032	0.00031
σ_{fit}	0.01471	0.00038	0.00048	0.00033	0.00030	0.00037	0.00036	0.00030	0.00030
Σwr^2	$3.08 \cdot 10^{-4}$	$1.55 \cdot 10^{-6}$	$2.40 \cdot 10^{-6}$	$9.28 \cdot 10^{-7}$	$9.00 \cdot 10^{-7}$	$1.53 \cdot 10^{-6}$	$1.53 \cdot 10^{-6}$	$9.24 \cdot 10^{-7}$	$8.98 \cdot 10^{-7}$
T_{final} (K)	99250	$4.07 \cdot 10^{-5}$	$6.63 \cdot 10^{-5}$	$2.94 \cdot 10^{-5}$	$7.79 \cdot 10^{-5}$	99250	44038	$4.79 \cdot 10^{-5}$	$4.07 \cdot 10^{-5}$
t_{SA} (secs)	386	45,426	45,619	35,953	35,257	395	2,643	54,199	62,176

[a] indicates parameters that were fixed and unadjusted in particular runs.

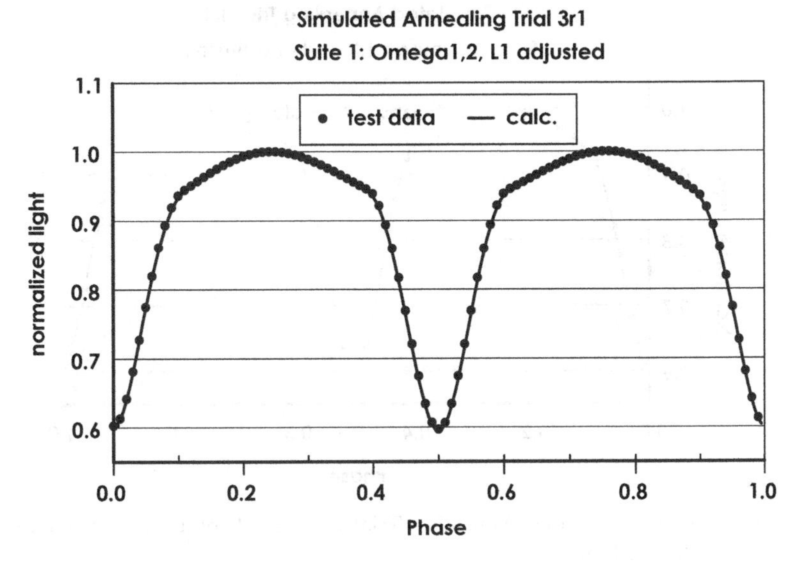

Fig. 1. The Simulated Annealing Trial for Suite 1, which did not converge before the termination criteria were met. Note that the fitting is good but not excellent; compare this fitting to that shown in Figure 2.

σ_{fit} $(DDC) = 0.00219$, and $SD_{<w>}$ $(DC) = 0.000340]$, than the suite 1 trial discussed above. The final *DC* results for the parameters in the converged solution were as follows:

- $\Omega_1 = 4.0739 \pm 0.0029$, $\Delta = +0.0017$
- $\Omega_2 = 4.4944 \pm 0.0043$, $\Delta =$ -0.0012
- $L_1 = 6.941 \pm 0.013$, $\Delta =$ -0.0061

The fitting for this trial are shown in Figure 2. Note the improvement in the fit compared to that in Figure 1.

To gauge the relative efficiencies as well as the robustness of the *SA* and simplex programs, we ran Suite 8 with the simplex program substituted for the simulated annealing program in the list of tasks within the *.inf* file. The simplex run was set to terminate after a fixed number of iterations, as well as at specific error limits of the parameters (fixed at 0.001) and the fitting. With a limit of only 5 iterations, the solution had not yet converged, and the fitting achieved a σ_{fit} of only 0.010050. After the maximum number of iterations was increased to 50, the fit error was reduced to 0.003859, achieved in 23 iterations, and convergence was nearly achieved. When the error criterion was reduced to 0.003 (from 0.007 for the other trials), after 31 iterations $\sigma_{fit} = 0.001796$. Finally, when this criterion was set to 0.001, $\sigma_{fit} = 0.000776$, in 41 iterations. It is noteworthy that only the final simplex run achieved (marginally) convergence[3] in all parameters, approximating the absolute minimum. These

[3] For one parameter, $\Omega_1 = 4.07491 \pm 0.001136$, $\Delta = +0.001146$.

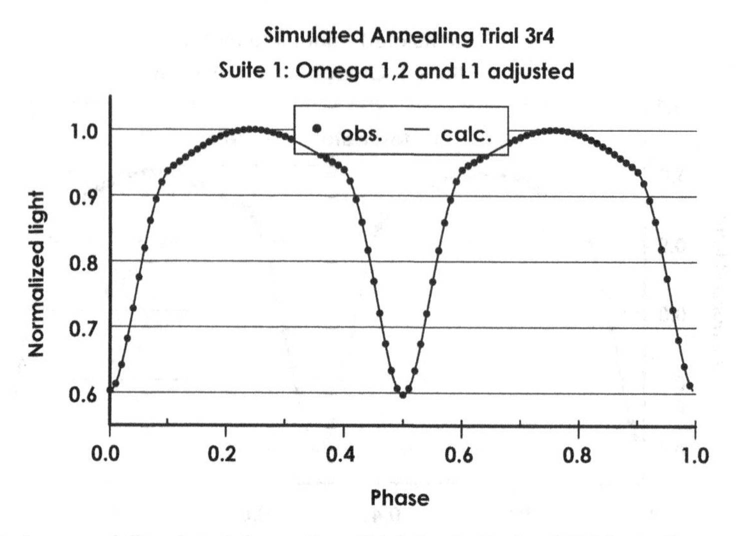

Fig. 2. Improved Simulated Annealing Trial for Suite 1, of Table 1. Compare this fitting with that in Figure 1.

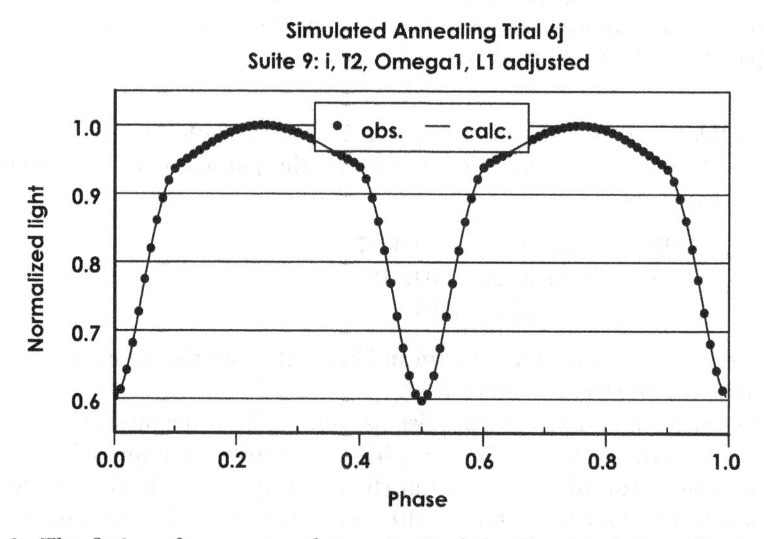

Fig. 3. The fitting of parameter data suite 9 of the Simulated Annealing experiments. See Table 1 for the parameters and fitting errors and the text for details.

results tend to confirm the relative robustness of *SA* as well as the efficiency of simplex.

Figure 3 shows the converged fitting of the Suite 9 parameters.

We conclude from these experiments that given enough CPU time, and optimized choices for the initial temperature, step length, and tuning parameters, simulated annealing, as coded in the `WD2007` package, is a robust tool to explore parameter space to insure that the solution found through more efficient algorithms cannot be improved. We found that with even a limited exploration, one can have reasonable confidence that the solution is sufficiently close to the correct region of parameter space that the solution can be honed to high precision by our simplex and damped least squares. *SA* runs could be used to test any "final" solution that is found with other methods.

4.2 Genetic Algorithms (*GA*)

Genetic Algorithms were introduced by J. H. Holland (1975), who coded data in strings and applied transformations to alter the strings so as to produce improvements. These are the two most important principles. Useful expositions of *GA* in an astronomical context is given by Charbonneau (1995; 2002). The basic idea behind it is the adjustment of a parameter through the alteration of numerical strings that represent adjustable parameters. This is accomplished by swapping a portion of one string with another, in analogy to the exchanging of genetic material in a chromosome. The question is then: is the new generation an improvement over the previous generation or not? The longevity of the traits of this new generation is decided by whether a better set emerges in the future. This is a type of "natural selection," if the word "natural" can be defended in this context; after all, the modeler, not mother nature, allows the fittest parameters to survive, although chance (as employed in the algorithms) plays a role.

The characteristics of *GA* are basically as follows:

- Technique: Stochastic and iterative;
- Purpose: global optimization;
- Application: low dimensionality problems;
- Process: points in multi-dimensional parameter space coded as finite–length strings, transformations are applied to improve fitting Operators;
- Operators: selection, reproduction (crossover), mutation; new generation.

There have been very few light curve solutions found with the aid of *GA*s, despite a discussion of the value of *GA* for eclipse mapping purposes by Bobinger (2000), who studied the CV IP Peg. Metcalfe (1999) used *GA* to analyze BH Cas, an over–contact system. However, there have been many other astronomical applications of *GA*, and some of these have involved optimization of orbital parameters in binary star systems: Torres (2001) determined the rate of change of the inclination with time of the orbit of the variably-eclipsing binary, SS Lac; Noyes, *et al.* (1997). Many more *GA* applications have been to solar and stellar astrophysics; one of the latter is a spectroscopic treatment of polar accretion disk emission, carried out by Hakala (1995). Charbonneau's (1995) exposition includes an example of uncovering the periods of the δ Scuti variable θ^2 Tau. Many others are listed in the website mentioned in the next paragraph.

A program implementing *GA* is *Pikaia*, named for the first chordate, and sole back–boned creature found in the ancient Burgess Shale deposit in British Columbia, *Pikaia gracilens*. Whether this annelid survived through natural selection or, as per the theory of Stephen J. Gould, by pure chance, we leave to others to argue. The program, developed by Paul Charbonneau and Barry Knapp at the High Altitude Observatory (and mentioned first in Charbonneau, 1995) is currently available in several computing languages from: http://whitedwarf.org/index.html?research/archive/&0 from which there is a link to the "Parallel PIKAIA homepage,"PIKAIA of Travis Metcalfe: where two parallel versions are available. These make use of the message–passing routines of *Parallel Virtual Machines* (*PVM*) and *Message Passing Interface* (*MPI*) software. As for *SA*, *GA*, too requires extensive computing time, so massively parallel computing is a very good idea if metaheuristic methods are to be used a great deal.

4.3 Neural Networks

Another potential biological analogue for modeling eclipsing binary observables is that of neural networks (*NN*). The analogy is to the pattern of nerves and neural activity in the human body. The use of networks of nodes of binary decision–making, or "neurons," was first suggested by McCulloch & Pitts (1943). A recent update of *NN* development is given by Müller, Reinhardt, & Strickland (1995).

Basically, *NN* is a decision-making procedure with progressive branching. It requires a critical suite of features and an accurate and comprehensive "training set." Devinney, *et al.* (2006) describe the use of neural nets to identify types of light curves and to obtain starting parameters for follow-up analyses.

This technique is used to sort objects for further analysis, and has been widely used in other contexts, such as the identification of objects in CCD images. Its increased use for preliminary analysis of the thousands of light curves expected from deep, ground–based and space surveys can be expected over the next few years.

4.4 Hybrid Methods

An optimization method that incorporates *SA*, *GA*, and other stochastic methods, such as Monte Carlo and Tabu searches, was suggested by Fox (1993, 1995), and further developed by Besnard, *et al.* (1999) in a hybrid optimization method called *Advanced Simulated Annealing* (*ASA*). These authors advocate easing of some *GA* procedures in order to improve efficiency. They suggest the use of "intelligent moves," that restrict searches in a domain around previously visited locations in parameter space. Their method makes use of the *SA* finite probability of accepting a less optimum state to escape from local minima; it also uses the *GA* coded string concept, but differs from it in using a Markov chain that retains a kind of "DNA" history of previous "generation," and, from *Tabu* searches, it takes the idea of a penalty function. It uses an algorithm to simulate the time required to jump to a new state, reducing the number of rejected moves in the later stages, when the "temperature" has fallen to very low values. Besnard, *et al.* (1999) apply *ASA* to problems in computational fluid dynamics. It would be interesting to test such a hybrid method on light curve analysis.

5 Future Directions

Here we list briefly a few areas for inclusion in the next generation of modeling code: Inclination variation adjustment; Tomography (now carried out in separate programs); Zeeman Doppler image modeling of magnetic field structures in interacting binaries; and the incorporation of a wider range of atmospheric models (in T and $\log g$ grids) than is currently available in most light curve modeling codes, for more adequate modeling of:

(a) Orbits of EBs in 3-star systems (see Torres 2001);
(b) Binary systems with greatly disparate components, involving components such as white, red, or brown dwarfs or planets;
(c) The narrow neck region of over–contact systems; and
(d) Low temperature starspots on stars with active regions.

5.1 Astronomical Tomography

Astronomical tomography can permit the detailed modeling of structures that extend beyond the photospheric limb, and that are normally seen in spatially and spectrally discrete emission features.

The software package, `SHELLSPEC` was developed starting about 2004 by Budaj & Richards (2004). Its capabilities are very useful to complement the diagnostic capabilities of *WD*. A set of output images from `SHELLSPEC` is seen in Figure 4. The input values for the model that is illustrated were manipulated so that a contribution from each feature is clearly visible. These features are:

- Both component stars are assumed to be spherical;
- The primary star has limb–darkening;
- A secondary star filling its Roche lobe, with gravity darkening
- A Keplerian disk around the primary star;
- An inclined jet;
- A slowly expanding shell surrounding the system;
- The centers of the jets and shell have no net space velocity and that of the disk center corresponds to that of the primary star;
- the jet precesses with orbital period.

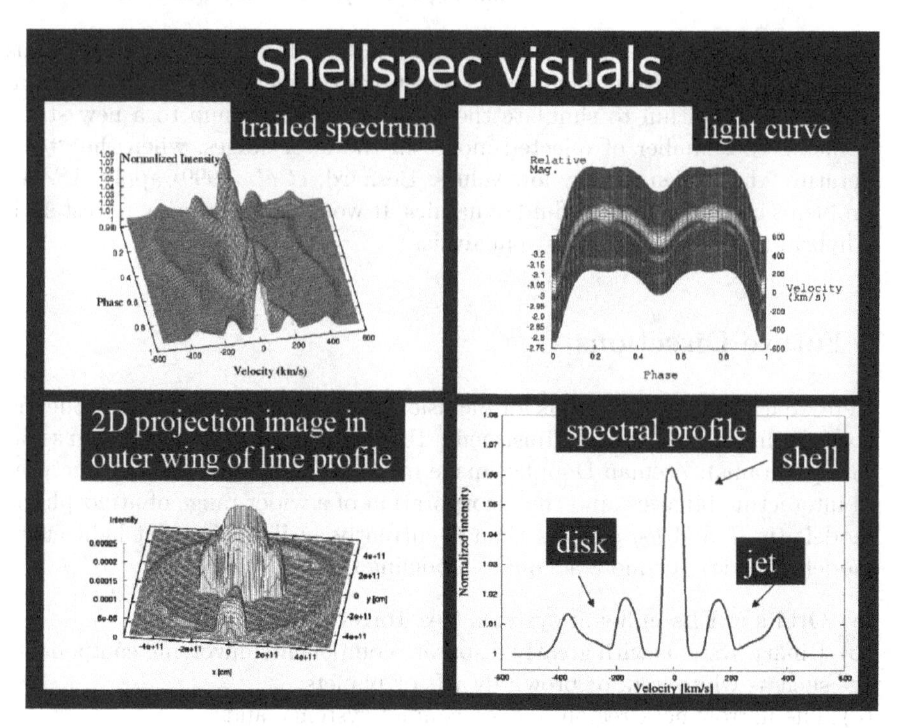

Fig. 4. Features of an Algol system model, that contains a disk around the primary, a Roche-lobe filling secondary star, and a precessing jet. The features are visualized with the software package SHELLSPEC developed by Budaj & Richards (2004). See the text for further details of the model and software.

Formerly, spherical symmetry was assumed in the modeling of the limb. The advantage of use of the Wilson–Devinney program is that *WD* avoids such approximations, and thereby yields improved physical models of the stars, themselves. In fact, however, recent versions of SHELLSPEC (*e.g.*, version 07) make use of Roche geometry. SHELLSPEC thus allows valid model comparisons with Doppler signatures in observed spectra of the stellar, disk, and optically thin circumstellar material, making use of commercial products (*e.g.*, *gnuplot* and *gimp*) for visualization (Budaj, private communication, 2007). Recent applications can be found in Budaj, Richards, and Miller (2005) and Miller, *et al.* (2007).

5.2 Improved Synthetic Spectral Energy Distributions

The limitations of atmospheric models are of ongoing concern for light curve observables modeling, but over the past two decades much progress has been achieved in producing models for individual stars (*e.g.*, by Hauschildt's group and associates, with PHOENIX; see Hauschildt, *et al.*, (2003), for a relatively recent summary of its capabilities), and we expect to be able to improve the temperature range of available atmospheric models to be able to handle the narrow neck region of over–contact systems, as well as the generally cooler stellar and substellar objects that are being observed in increasing numbers. Both `ATLAS` (Nesvacil, *et al.*, 2003) and `PHOENIX` have been used to model a broad range of spectral types. As Hauschildt states on his website, "The atmospheres and spectra of novae, M dwarfs & giants, brown dwarfs, L/T dwarfs, T Tauri's, SNe, basically everything that radiates." The `PHOENIX` website also states that "spectra can be calculated for any desired resolution (standard are 20000 to 500000 wavelength points spread from the UV to the radio)." We note also the development of `BINSYN` and `SYNSPEC` used extensively for modeling CV spectra and described elsewhere in these proceedings.

5.3 Potential for High–Precision Infrared Light Curves

The availability of more refined atmospheric models should permit the adequate analyses of light curves obtained, for example, in the improved infrared passbands developed by the Infrared Working Group (*IRWG*) of IAU Commission 25. These passbands (the near–IR components of which are illustrated along with the Earth's atmospheric transmission windows in Figure 5) promise to deliver truly millimagnitude precision *and* accuracy to light curve acquisition by careful placement and shaping of passband profiles to fit within the spectral transmission windows of the Earth's atmosphere. See Milone and Young (2005; 2007) for the evidence and discussion of the implications for ground–based infrared photometry from observatories at all elevations.

5.4 Synoptic Observations and Surface Modeling

The infrared is especially useful for photometry of late–type stars and for cool regions of hotter stars. Currently little light curve work has been done in infrared passbands, mainly because the observatories with the best IR facilities tend to be those at high–elevation, optimally dry sites where observing time is at a premium and rarely granted to obtain IR light curves. The *IRWG* passband set may permit many more observatories to attempt at least differential IR photometry. Consequently, long series of synoptic multi–λ and spectroscopic data have been obtained almost exclusively in optical passbands. Nevertheless, with such data, Lanza, *et al.* (1998; 2002) have shown how effectively Tikhonov and maximum entropy techniques can be applied, in their case, to studies of the RS CVn–type systems AR Lac and RT Lac.

5.5 Spectropolarimetry

Yet another promising area of light curve observables modeling is spectropolarimetry. The use of polarization curve modeling has already been discussed by Wilson (1993), and but the more recent availability of high resolution

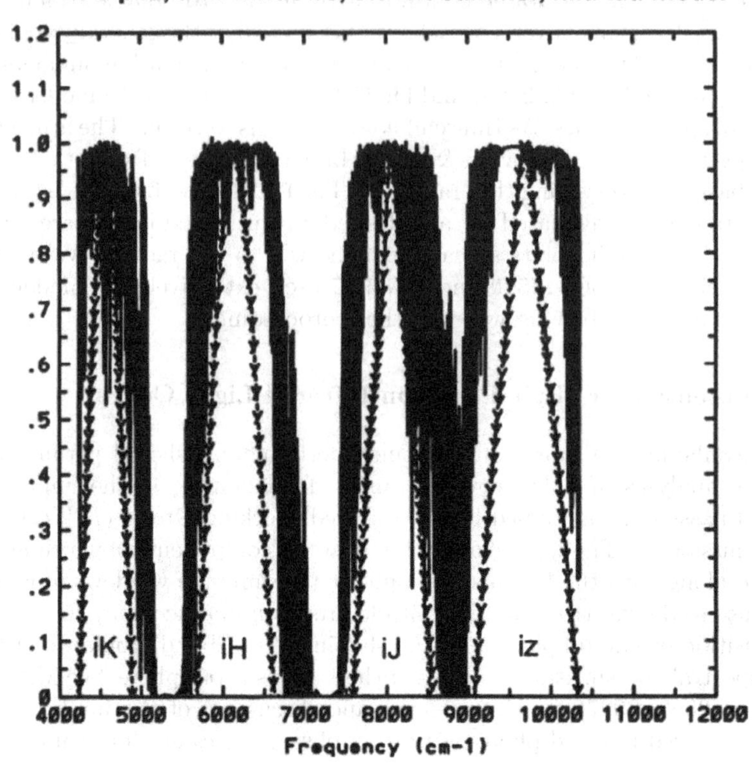

Fig. 5. The near–infrared suite of IRWG passbands, the shape and placement of which in the Earth's atmospheric windows permits less dependence on water vapor transmission variations and thus improved precision and accuracy in IR photometry The atmospheric transmission is computed for a site at 0.3–km elevation, and for a moist summer, mid-latitude model atmosphere. See Milone & Young (2005; 2007) for further details.

spectropolarimetric data from instruments such as ESPaDOnS (for **E**chelle **S**pectro**P**olarimetry **D**evice for the **O**bservatio**n** of **S**tars), makes this work all the more relevant. See Landstreet & Mathys (2000), Petit, *et al.* (2004), and Pointer *et al.* (2002) for illustrations of how Zeeman Doppler imaging can provide magnetic field maps of magnetic Ap stars, an FK Com star (HD 199178), and a K0 dwarf (AB Dor), respectively.

5.6 Planetary Eclipse Modeling

Finally, we mention the modeling of planetary eclipses. The modeling of planetary transits can be and have been already carried out with versions of the Wilson-Devinney code (*e.g.*, by Williams 2001), notwithstanding a recent unsubstantiated comment about the unsuitability of *WD* programs for modeling planetary transits (Giménez 2006). Planetary transits can be modeled readily with the Wilson–Devinney program, *if* the number of grid elements across the stellar disks is set high enough within the program to provide sufficient phase resolution. This is exactly what was done by Williams (2001), in our stand–alone *WD* program. `wd98k93h`. Figure 6 indicates how successful his fitting of the transit of HD 209458b across the face of its parent star has been. Of course, planetary occultations, necessarily observed in infrared passbands, could be modeled with *WD* with appropriate atmospheric flux models.

Because the discovery of many transit cases were anticipated, a grid of models was developed by Williams (Milone, *et al.*, 2004a), for preliminary use with the planetary transit survey of the globular cluster 47 Tuc, where, however, *none* were found; the grid may eventually be useful, but thus far the sparseness of planetary host stars that are cluster members argues against clusters of any kind as propitious settings for planetary systems. The eclipsing binary studies of that cluster have continued, however, and the grid of stellar eclipse models developed for that study, continues to be used.

The experimental work described here was supported in part by Canadian *NSERC* grants to EFM.

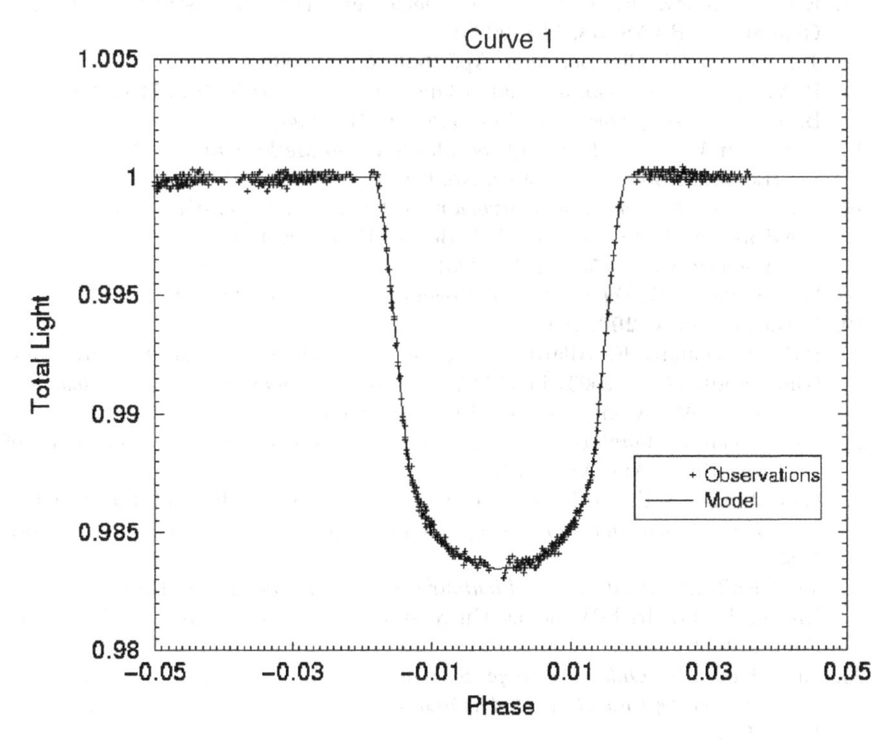

Fig. 6. The fitting of HD 209458b HST transit data obtained from David Charbonneau (personal communication to M.D. Williams) with a *wd98k93h* fitted to a model produced by M.D. Williams (2001).

References

1. E.H.L. Aarts and J.H.M. Korst, eds.: *Simulated Annealing and Boltzmann Machines, Stochastic Approach to Combinatorial Optimization and Neural Computing* (Wiley, Chichester, UK 1989)
2. E.H.L. Aarts, J.H.M. Korst and P.J.M. van Laarhoven: In *Simulated Annealing and Boltzmann Machines*, eds. E.H.L. Aarts, and J.H.M. Korst, pp. 91-120 (1997)
3. P.A. Bergbusch and D.A. VandenBerg: ApJS, **81**, 163 (1992)
4. E. Besnard, N. Cordier–Lallouet, and O. Kural, H.–P. Chen: *Bulletin, Amer. Inst. Aeronautics & Astronautics*, Paper 99–0186 (1999)
5. A. Bobinger: A&A, **357**, 1170 (2000)
6. J. Budaj and M.T. Richards: "A Description of the SHELLSPEC Code," Contrib. Astron. Obs. Skalnate Pleso, **34**, 167-196 (2004)
7. R. Budaj, M.T. Richards, and B. Miller: ApJ, **623**, 411 (2005)
8. P. Charbonneau: ApJS, **101**, 309 (1995)
9. P. Charbonneau: *An Introduction to Genetic Algorithms for Numerical Optimization*, (NCAR/TN-450 + 1A, 74pp 2002)
10. A. Corana, M. Marches, C. Martini, and S. Ridella: Transactions in Mathematical Software **13**, 262 (1987)
11. E.J. Devinney, E. Guinan, M. DeGeorge, D.H. Bradstreet, and J.M. Giammarco: BAAS, **38**, 119 (2006)
12. J. Devor and D. Charbonneau: ApJ, **653**, 647 (2006)
13. R.W. Eglese: European Journal of Operational Research, **46**, 271 (1990)
14. B. Fox: *Annals of Operation Research*, **41**, 47 (1993)
15. B. Fox: in *Monte Carlo and Quasi Monte Carlo Methods in Scientific Computing* (Lecture Notes in Statistics, No. 6 1995)
16. C.F. Gauß: *Theoria Motus Corporum Coelestium in Sectionibus Conicus Solem Ambientium*, (F. Perthes and J.H. Besser, Hamburg 1809)
17. A. Giménez: A&A, **450**, 1231 (2006)
18. W.L. Goffe, G.D. Ferrier, and J. Rogers: J. of Econometrics, **60**, 650 (1994)
19. P. Hakala: A&A, **296**, 164 (1995)
20. P.H. Hauschildt, F. Allard, E. Baron, J. Aufdenberg, and A. Schweitzer: (Hauschildt, *et al.*, 2003) in *GAIA Spectroscopy, Science and Technology*, ed. U. Munari, ASP Conf. Ser. Vol. **298**, 179 (2003)
21. J.H. Holland: *Adaptation in Natural and Artificial Systems* (University of Michigan Press, Ann Arbor 1975)
22. G.G. Judge, W.E. Griffiths, R. Carter Hill, H. Lütkepohl, and Tsoung-Chao Lee: *The Theory and Practice of Econometrics* 2nd edn. (Wiley, New York 1985)
23. Josef Kallrath: *Analyse von Lichtkurven enger Doppelsterne*, Diploma thesis, Rheinische Friedrich-Wilhelms Universität Bonn, Sternwarte der Universität Bonn (1987)
24. Julia Kallrath: *Online Storage Systems and Transportation Problems with Applications: Optimization Models and Mathematical Solutions* (Springer, New York 2004)
25. Josef Kallrath, E.F. Milone: *Eclipsing Binary Stars: Modeling and Analysis* (Springer, New York Berlin Heidelberg 1999)
26. Josef Kallrath and A.P. Linnell: ApJ, **313**, 346 (1987)

27. Josef Kallrath, E.F. Milone, D. Terrell, and A.T. Young: ApJS, **508**, 308 (1998)
28. Josef Kallrath, E.F. Milone, R.A. Breinhorst, R.E. Wilson, A. Schnell, and A. Purgathofer: A&A, **452**, 959 (2006)
29. Y. Kim, P. Demarque, S.K. Yi, and D.R. Alexander: ApJS, **143**, 499 (2002)
30. S. Kirkpatrick, C.D. Gelatt, and M.P. Vecchi: Science, **220**, 671–680 (1983)
31. R.L. Kurucz: in *Light Curve Modeling of Eclipsing Binary Stars*, ed. E.F. Milone, (Springer–Verlag, New York), pp. 93–101 (1993)
32. J.D. Landstreet and G. Mathys: A&A, **359**, 213 (2000)
33. A.F. Lanza, M. Rodonò, and R. Rosner: A&A, **332**, 541 (1998)
34. A.F. Lanza, S. Catalano, M. Rodonò C. Ibanoglu, S. Evren, G. Tas, Ö. Çakirli, and A. Devlen: (Lanza, *et al.*, 2002) A&A, **386**, 583 (2002)
35. A.M. Legendre: *Nouvelles Methodes pour la Determination des Orbites des Comètes*, (Courcier, Paris 1805)
36. A.P. Linnell: ApJS, **54**, 17 (1984)
37. W.S. McCulloch and W. Pitts: Bulletin of Mathematical Biophysics, **5**, 115 (1943)
38. T.S. Metcalfe: AJ, **117**, 2503 (1999)
39. B. Miller, J. Budaj, M. Richards, P. Koubsk, and G.J. Peters: (Miller, *et al.*, 2007), ApJ, **656**, 1075 (2007)
40. E.F. Milone: in *Critical Observations versus Physical Models for Close Binary Systems*, ed. K.–C. Leung, (Gordon and Breach Science Publishers, Inc., Montreux, Switzerland 1988), pp. 139–149
41. E.F. Milone: in *New Frontiers in Binary Star Research*, eds., K.–C. Leung and I.–S. Nha, ASP Conference Series, **38**, pp. 172–180 (1993)
42. E.F. Milone: in *The Third Pacific Rim Conference on Recent Development on Binary Star Research*, ed., K.–C. Leung, ASP Conference Series, **130**, pp. 87–93 (1997)
43. E.F. Milone and A.T. Young: PASP, **117**, 485 (2005)
44. E.F. Milone and A.T. Young: in *The Future of Photometric, Spectrophotometric and Polarimetric Standardization*, ed. C. Sterken. ASP Conference Series, **364**, pp. 387–407 (2007)
45. E.F. Milone, M.D. Williams, C.R. Stagg, M.L. McClure, B. Desnoyers Winmill, T. Brown, D. Charbonneau, R.L. Gilliland, J. Kallrath, G.W. Marcy, D., Terrell, and W. VanHamme: (Milone, *et al.*, 2004a) in *Planetary Systems in the Universe: Observation, Formation and Evolution*, eds, A. Penny, P. Artymowicz, A.-M. Lagrange, and S. Russell. IAU Symposium 202 Proceedings, pp. 90–92 (2004)
46. E.F. Milone, J. Kallrath, C.R. Stagg, and M.D. Williams: (Milone, *et al.*, 2004b) in *The Environment and Evolution of Double and Multiple Stars*, C. Scarfe and C. Allen, eds., Revista Mexicana AA (SC), **21**, 109–115 (2004)
47. B. Müller, J. Reinhart, and M.T. Strickland: Neural Networks (Springer–Verlag, Berlin 1995)
48. J.A. Nelder and R. Mead: Computer J., **7**, 308 (1965)
49. R.W. Noyes, S. Jha, S.G. Korzennik, M. Krockenberger, P. Nisenson, T.M. Brown, E.J. Kennelly, S.D. Horner: (Noyes, *et al.*, 1997) ApJ, **483**, L111 (1997)
50. P. Petit, J.–F.Donati, J.M. Oliviera, M. Aurière, S. Bagnulo, J.D. Landstreet, F. Lignières, T. Lüftinger, S. Marsden, D. Mouillet, F. Paletou, S. Strasser, N. Toquè, and G.A. Wade: (Petit, *et al.*, 2004) MNRAS, **351**, 826 (2004)
51. G.R. Pointer, M. Jardine, A. Collier Cameron, and J.–F. Donati: (Pointer, *et al.*, 2002) MNRAS, **330**, 160 (2002)

52. W.H. Press, B.P. Flannery, S.A. Teukolsky, and W.T. Vetterling: *Numerical Recipes — The Art of Scientific Computing*, (Cambridge University Press, Cambridge 1992)
53. N. Nesvacil, Ch. Stütz, and W.W. Weiss: (Nesvacil, *et al.*, 2003) in *GAIA Spectroscopy, Science and Technology*, ed. U. Munari, ASP Conf. Ser. Vol. **298**, 173 (2003)
54. A. Prša and T. Zwitter: ApJ, **628**, 426 (2005)
55. H.H. Rosenbrock: Computer J., **3**, 175 (1960)
56. W. Spendley, G.R. Hext, and F.R. Himsworth: Technometrics, **4**, 441 (1962)
57. C.R. Stagg and E.F. Milone: in *Light Curve Modeling of Eclipsing Binary Stars*, ed. E.F. Milone, (Springer–Verlag, New York), pp. 75–92 (1993)
58. G. Torres: ApJ, **121**, 2227 (2001)
59. M.D. Williams: *Shadows of Unseen Companions: Modeling the Transits of Extra–Solar Planets* (University of Calgary, MSc Thesis 2001)
60. R.E. Wilson and E.J. Devinney: ApJ, **166**, 605 (1971)
61. R.E. Wilson: ApJ, **234**, 1054 (1979)
62. R.E. Wilson: in *Light Curve Modeling of Eclipsing Binary Stars*, ed. E.F. Milone, (Springer–Verlag, New York), pp. 7–25 (1993)
63. R.E. Wilson: Close Binary Star Observables: Modeling Innovations 2003-06, in W. I. Hartkopf, E. F. Guinan, & P. Harmanec (eds.), *Binary Stars as Critical Tools & Tests in Contemporary Astrophysics*, No. 240 in Proceedings IAU Symposium, Kluwer Academic Publishers, Dordrecht, Holland (2007, in press)
64. J.S.B. Wyithe and R.E. Wilson: ApJ, **559**, 260 (2001)

The Closest of the Close: Observational and Modeling Progress

W. Van Hamme[1] and R.E. Cohen[1,2]

[1] Department of Physics, Florida International University,
Miami, Florida 33199, USA
vanhamme@fiu.edu
[2] Astronomy Department, University of Florida, Gainesville, Florida 32611, USA
cohenr@astro.ufl.edu

Two of today's most critical unresolved problems in binary star research are the formation and structure of W Ursae Majoris binaries. Demands imposed by nearly equal effective temperatures for the two stars lead to a grossly different mass–luminosity relation from that of single main sequence stars. That only non-equilibrium models can account for many W UMa properties is widely recognized. One such model is the thermal relaxation oscillator model, in which systems oscillate between states of good thermal contact and broken contact. Notwithstanding decades of theoretical research, W UMa's have not been fitted into a comprehensive structural theory, and the formation mechanism also needs to be resolved. The recent idea that all W UMa's are part of triple (or multiple) systems and formed through dynamical interactions with third bodies may well answer that question. On the observational front, we have seen in the last decade a vast increase in the number of double-lined systems with complete radial velocity curves that leads to parameters of improved reliability, including absolute dimensions. After a brief review of observational efforts, we look at recent modeling progress, describe a project aimed at determining broken-contact binary parameters, and comment on specific challenges with W Crv as an example. We then discuss finding third light from simultaneous light and velocity solutions for W UMa's suspected of having tertiary components. Showing the specific example of V2388 Oph, we explore the prospects of this method for finding W UMa companions.

1 Introduction

In this contribution we focus on the group of short-period binaries (periods less than 1 day) with non-degenerate components and spectral types in the range F to K. The majority of these are W UMa-type overcontact (OC) systems that are very common among main sequence stars. For example, already in

E.F. Milone et al. (eds.), Short-Period Binary Stars: Observations, Analyses, and Results, 215–230.

the 1940s, Shapley (1948) indicated that W UMa binaries are 20 times as numerous as all other types of eclipsing binaries combined. More recently, for stars in the galactic disk toward Baade's window, Rucinski (1998) derived a W UMa frequency relative to other main sequence stars of 1/130. In a 7.5 magnitude-limited sample of stars, Rucinski (2002) found a somewhat lower but still impressive relative frequency of 1/500, with a corresponding space density of $1.02 \pm 0.24 \times 10^{-5}$ W UMa's per cubic pc. Very recently, this result was confirmed by Gettel, Geske & McKay (2006) who derived a space density of $1.7 \pm 0.6 \times 10^{-5}$ pc^{-3} W UMa binaries extracted from *ROTSE-I* sky patrol data. The high incidence of W UMa binaries is thus well established.

A comprehensive theory for the structure of W UMa binaries has yet to emerge. Observational evidence indicates that W UMa systems consist of main sequence stars of *unequal* mass but *nearly equal* effective temperature. Furthermore, the Roche geometry imposes a ratio of radii proportional to the mass ratio raised to a power of about 0.46, and consequently, a luminosity ratio $L_1/L_2 = (M_1/M_2)^\alpha$, with α about 0.9. For main sequence stars, α is roughly around 4. This leads to the inevitable conclusion that considerable energy generated in the more massive component must flow to the less massive star, as pointed out by Kuiper (1948) more than half a century ago. Over the years, various energy transfer mechanisms have been proposed, and for specific references we refer the reader to a number of excellent review papers such as Smith (1984), Rucinski (1985), Eggleton (1996), and Webbink (2003). That a sharp drop in W UMa's occurs around spectral types where convective outer layers disappear indicates that convection likely plays a role in the energy transfer.

It is important to note that there are two distinct categories of W UMa binaries, A-types and W-types, originally defined on observational grounds by Binnendijk (1965, 1970). These two types appear to be structurally different as well. Wilson (1978, 1994) argues that A–types, which have larger masses, longer periods, smaller mass ratios and larger degrees of overcontact, are more evolved within the main sequence and are in equilibrium. W–types have smaller (often marginal) degrees of overcontact, are essentially on the zero age main sequence, and show signs of not being in equilibrium (Lucy, 1976; Flannery, 1976; Robertson & Eggleton, 1977). According to *Thermal Relaxation Oscillator (TRO)* theory, W-types undergo cyclic oscillations between overcontact and broken-contact states, with the latter stage much shorter in duration than the former (Robertson & Eggleton, 1977). In the broken–contact stage, thermal contact is lost, the two stars acquire substantially different surface temperatures and the two eclipses have much different depths. The primary, more massive component continues to fill its limiting lobe, but the less massive star becomes detached. Although the number of systems showing broken-contact light curves is expected to be small compared to that of genuine *OC* W UMa's, a number of papers have focused on finding genuine broken-contact systems (Kałużny, 1986; Hilditch, King & McFarlane, 1988; Rucinski & Lu, 2000; Van Hamme *et al.*, 2001).

W UMa's are convenient targets for photometric observing with modest-sized telescopes at smaller observatories, as complete light curves can be obtained in just a few nights. Light curve solutions, however, can be problematic because many systems have strongly asymmetric or distorted light curves. Also, there is no firm handle on the mass ratio for partially eclipsing cases that lack radial velocities (viz. Terrell & Wilson, 2005). An important observational development of the last decade has been the effort to obtain radial velocities for many W UMa binaries, and this has improved the mass ratio situation dramatically. For example, the David Dunlap Observatory radial velocity program by Rucinski and collaborators[1] has produced radial velocities for about 100 W UMa and near-contact (*NC*) binaries. The list of Van Hamme, *et al.* (2001) of double-lined systems with good radial velocities that would benefit from new solutions with updated models has grown considerably (for example, see Table 2 in Yakut & Eggleton, 2005). Many research groups are currently producing light and velocity solutions for these systems, including Kreiner, *et al.*, (2003); Baran, *et al.*, (2004); Zola, *et al.*, (2004); Gazeas, *et al.*, (2005); Zola et al., (2005); and Gazeas, *et al.* (2006).

In Sect. 2 we discuss W Corvi, one of the very first broken-contact candidates to be discussed in the literature. Lucy & Wilson (1979) modeled light curves obtained in 1967–1968 by Dycus (1968) and found geometrical overcontact but poor thermal contact, with a temperature difference between the two stars of $\approx 650\,\mathrm{K}$. Dycus observed with a small telescope with the star low in the sky, making the light curves fairly noisy and the solution inconclusive. Light curves by Odell (1996) are better and clearly show variable asymmetries and distortions. Primary and secondary minima have unequal widths, and the maximum after secondary fluctuates in height (see Figs. 1–3, see also Odell & Cushing, 2004). The section around primary minimum is likely to be the clean part and relatively unaffected by the source of distortions (Odell, 1996; Odell & Cushing, 2004; Rucinski & Lu, 2000). Small (0.01–0.02 mag) epoch-to-epoch variations in primary minimum may be due to long-term intrinsic variability of the comparison star (Odell & Cushing, 2004), but probably do not affect general conclusions. Based on eclipse timing analysis, Rucinski & Lu, (2000) argue against W Crv being in broken contact. The timing diagram shows a period increase and, if caused by conservative mass transfer, mass flow from the less massive to the more massive star. The opposite is expected for systems in the *TRO* broken contact stage, as the more massive star should be the lobe filler and mass should flow to the less massive (detached) component. Here we show that a unique set of binary parameters can be found that describe all available light and velocity curves. Light curve distortions are modeled by surface spots that vary from one epoch to another, but leave the underlying binary parameters unaffected. The only satisfactory fit to the light curves is for a genuine *OC* system, but with a large temperature difference between the components. Consequently, one of the main questions regarding

[1] See http://www.astro.utoronto.ca/DDO/research/binaries_prog.html

W Crv remains unanswered – how can a system that appears to be genuinely *OC* be in such poor thermal contact?

In Sect. 3 we present a light and velocity solution for V2388 Oph, a system whose formation is discussed by D'Angelo, van Kerkwijk, & Rucinski (2006). In an earlier paper of the same series, Pribulla & Rucinski (2006) find at least 42% of 151 W UMa stars brighter than 10th magnitude to be part of multiple systems. For a subsample of better observed systems, this fraction increases to 59%. This hints at the intriguing idea that *all* W UMa binaries may be part of triple (or multiple) systems, which clears the way for a formation scenario that involves bringing two stars, originally formed farther apart, closer together through gravitational interactions with another star (or stars). The authors present an impressive array of observational techniques that can be used to detect the presence of third bodies. One such technique is finding third light in light curve solutions. Pribulla and Rucinski consider this technique to be of limited use because possible correlations with other parameters (inclination and mass ratio in particular) can render third light unreliable. In order to explore this idea further, we started a program to determine third light from light and velocity solutions of systems which are found by D'Angelo, van Kerkwijk & Rucinski (2006) to have tertiaries. We present V2388 Oph as a first example. Because this binary is known to be part of a visual system, with the third star only 0.″088 away from the eclipsing pair, we expect to be able to easily extract third light from the system's light curves. We solve light and velocity curves simultaneously, with and without third light as a fitted parameter. The solution without third light is sufficiently inferior to the other that it would be rejected, even if no other evidence for the third star existed. Serendipitously, we find that the radial velocity fit significantly improves if we allow for the presence of another star that is *not* the visual companion. This star orbits the eclipsing system with a 311 d period at a distance of 1.25 AU, assuming a circular orbit coplanar with that of the eclipsing pair.

2 W Crv

Light and radial velocities were fitted with the physical model of Wilson (1979, 1990), with results given in Sect. 2.1. Star spots (hot or cool) were invoked to handle light curve asymmetries. We exploited the benefits of the bandpass-based stellar atmosphere approach outlined in Van Hamme & Wilson (2003). Time instead of phase was used as the independent variable and ephemeris parameters (reference epoch T_0, period P and the period time rate of change $\mathrm{d}P/\mathrm{d}t$) were determined as part of the solution. In Sect. 2.2 we compare this $\mathrm{d}P/\mathrm{d}t$ determined from whole light curves with that obtained from a traditional eclipse timing analysis.

2.1 Datasets and Solutions

Table 1 lists the light and velocity curves analyzed in this paper and the standard deviations used in assigning curve-dependent weights. The R band observations of 2002–03 were obtained with the Southeastern Association for Research in Astronomy (*SARA*) 0.9-m telescope and are described in Cohen & Van Hamme (2003). Individual weights inversely proportional to the square root of light level were applied (for a general discussion of weights see Wilson, 1979). We adopted a logarithmic limb darkening law with coefficients x, y from Van Hamme (1993). The detailed reflection treatment of Wilson (1990) was used with double reflection. Initial experiments showed that there was sufficient consistency among solutions of epoch-specific sets of light curves that indicated the existence of a single set of physical parameters (excluding spot parameters) that could reasonably fit all curves. We then applied this fixed parameter set to the light curves at each epoch and adjusted spot parameters only, to account for the specific light curve shape at that epoch. Spot effects were then subtracted from each of the light curves and non-spot parameters refitted to all curves combined. The procedure was repeated until no further significant improvement was obtained. The result was a *unique* set of physical parameters (excluding spot parameters) that fitted *all* data sets, including the radial velocities. No acceptable fits could be obtained for detached or semidetached configurations, i.e., the light curves do not support a broken-contact or Algol-type status, as favored by Rucinski & Lu (2000). W Crv appears to be a genuine *OC* binary. Parameters obtained from the global multi-epoch solution are listed in Table 2, with absolute dimensions in Table 3. Solutions are shown in Figs. 1–4.

Table 1. W Crv Data Sets

Year	Band	σ^a	Reference
1967	U	0.0455	Dycus (1968)
1967	B	0.0275	Dycus (1968)
1967	V	0.0282	Dycus (1968)
1981	V	0.0104	Odell (1996)
1982	B	0.0233	Odell (1996)
1982	V	0.0118	Odell (1996)
1988	B	0.0133	Odell (1996)
1988	V	0.0180	Odell (1996)
1993	B	0.0188	Odell (1996)
1993	V	0.0185	Odell (1996)
1995	RV1	9.77 km/s	Rucinski & Lu (2000)
1995	RV2	17.3 km/s	Rucinski & Lu (2000)
2002	R	0.0154	Cohen & Van Hamme (2003)
2003	R	0.0052	Van Hamme & Branly (2003)

[a]For light curves, in units of light at phase $0\overset{P}{.}25$

Table 2. W Crv Multi-epoch Light and Velocity Solution

Parameter	1981–2003 Global
T_0 (HJD)	$2449108.79127 \pm 0.00010$ s.e.
P_0 (d)	$0.3880816023 \pm 0.0000000037$
$\mathrm{d}P/\mathrm{d}t$	$2.746 \pm 0.057\, 10^{-10}$
$a\,(R_\odot)$	2.820 ± 0.039
V_γ (km/s)	-17.7 ± 2.0
i	$83\overset{\circ}{.}428 \pm 0\overset{\circ}{.}059$
g_1, g_2	0.32, 0.32
T_1 (K)[a]	5700
T_2 (K)	4813 ± 4
A_1, A_2	0.5, 0.5
Ω_1	3.1244 ± 0.0039
Ω_2	3.12442
$f = \frac{\Omega_{\mathrm{L}_1} - \Omega}{\Omega_{\mathrm{L}_1} - \Omega_{\mathrm{L}_2}}$	0.183
$q = M_2/M_1$	0.6731 ± 0.0016
$L_1/(L_1 + L_2)_V$ (1981)	0.7838 ± 0.0011
$L_1/(L_1 + L_2)_V$ (1982)	0.7838 ± 0.0018
$L_1/(L_1 + L_2)_V$ (1988)	0.7838 ± 0.0016
$L_1/(L_1 + L_2)_V$ (1993)	0.7838 ± 0.0024
$L_1/(L_1 + L_2)_B$ (1982)	0.8254 ± 0.0033
$L_1/(L_1 + L_2)_B$ (1988)	0.8254 ± 0.0014
$L_1/(L_1 + L_2)_B$ (1993)	0.8254 ± 0.0024
$L_1/(L_1 + L_2)_R$ (2002)	0.7549 ± 0.0010
$L_1/(L_1 + L_2)_R$ (2003)	0.7549 ± 0.0008
x_1, y_1 (bolo)	0.646, 0.204
x_2, y_2 (bolo)	0.639, 0.156
x_1, y_1 (V)	0.764, 0.226
x_2, y_2 (V)	0.801, 0.072
x_1, y_1 (B)	0.839, 0.132
x_2, y_2 (B)	0.850, −0.065
x_1, y_1 (R)	0.694, 0.248
x_2, y_2 (R)	0.743, 0.141
Auxiliary Parameters	
r_1 (pole)	0.4001 ± 0.0007
r_1 (side)	0.4248 ± 0.0008
r_1 (back)	0.4592 ± 0.0012
$< r_1 >$	0.4302 ± 0.0005
r_2 (pole)	0.3344 ± 0.0007
r_2 (side)	0.3516 ± 0.0009
r_2 (back)	0.3904 ± 0.0015
$< r_2 >$	0.3612 ± 0.0004

[a] Adopted from spectral type

Table 3. W Crv Absolute Parameters

Parameter	Star 1	Star 2
$M\,(M_\odot)$	1.192 ± 0.049	0.803 ± 0.033
$R\,(R_\odot)$	1.213 ± 0.017	1.018 ± 0.014
$M_{\rm bol}$	4.38 ± 0.08	5.50 ± 0.10
$\log(L/L_\odot)$	0.144 ± 0.033	-0.301 ± 0.039
$\log g$ (cm·s^{-2})	4.347 ± 0.030	4.327 ± 0.030

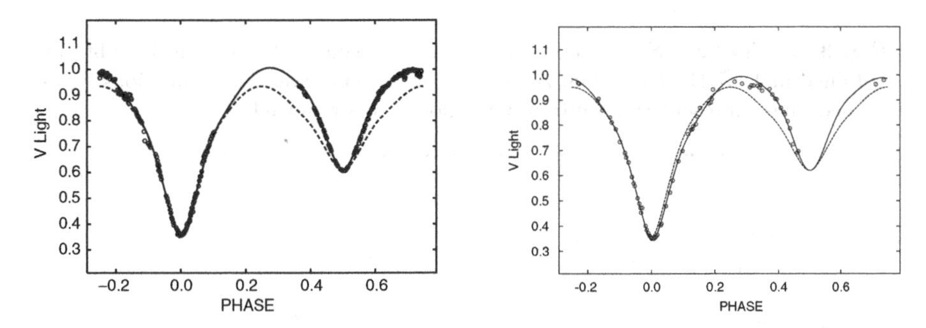

Fig. 1. W Crv 1981 (*left*) and 1982 (*right*) V observations of Odell (1996) and computed curves. Solution curves with spot effects removed are shown as dashed lines. Note the difference in eclipse widths in the observations.

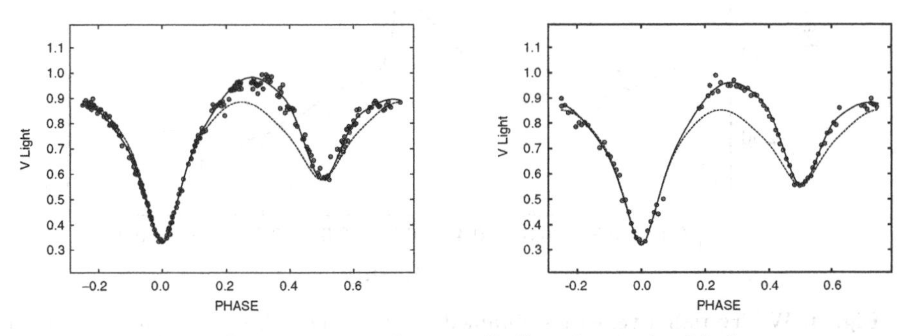

Fig. 2. Same as Fig. 1 for 1988 (*left*) and 1993 (*right*) data.

2.2 W Crv Period Behavior

The list of W Crv eclipse timings in Odell (1996), Rucinski & Lu (2000) and Odell & Cushing (2004) was updated with new minima (Table 4) determined from our 2002–03 observations (Sect. 2.1). We used the method of Kwee & Van Woerden (1956) to determine these minima. Relative weights of individual minima are needed before fitting ephemerides and analyzing $O-C$ residuals. These weights should scale with inverse standard errors squared. Unfortunately, authors do not always list uncertainties of eclipse timings. In order to estimate errors of previously published minima, and to compare typical errors

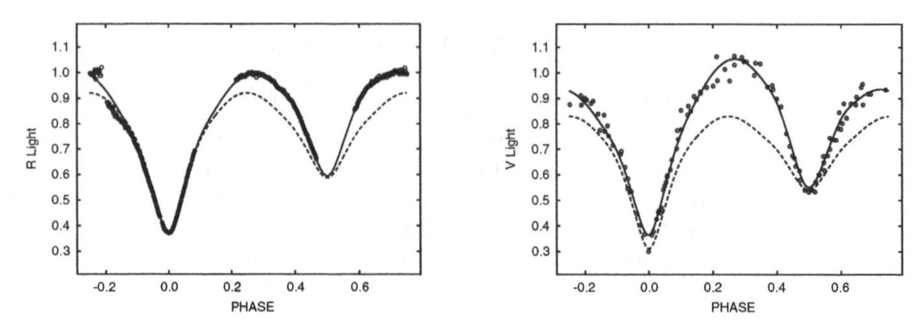

Fig. 3. W Crv 2003 *SARA* light curve in the R band (*left*) and the V light curve obtained in 1967 (Dycus, 1968, *right*), with solution curves. As in previous figures, dashed lines represent the solution with spot effects removed.

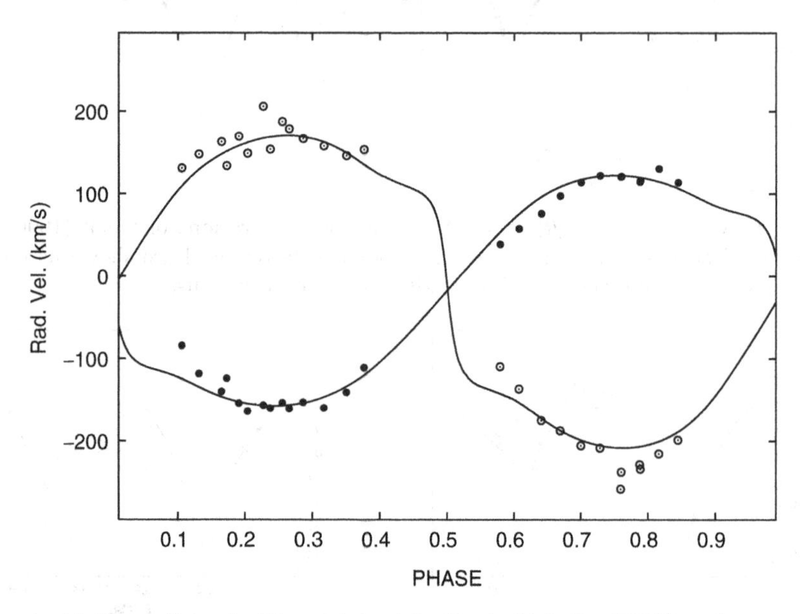

Fig. 4. W Crv radial velocities obtained by Rucinski & Lu (2000) and computed curves based on Table 2 parameters.

of photoelectric and visual minima, we redetermined those eclipse timings for which the original eclipse data are available in the literature. For each such minimum, we generated sets of 1000 light curves, sampled at the same Julian dates as the actual observations, and applying Gaussian noise with standard deviation typical for that set of observations. For visual observations, Gaussian noise of standard deviation 0.1 (in units of light at phase $0^{\mathrm{P}}_{\cdot}25$) was applied. A normal distribution was fitted to each of these sets of 1000 synthetic minima, and the standard deviation of this normal distribution was then considered to be the standard error of that particular minimum. On average, visual and photoelectric minima had errors of $\approx$0.005 d and 0.0005 d, respectively, which

Table 4. New Times of Minima for W Crv

$T_{\min}$ (HJD−2400000.0)	Standard Error
52438.72923	0.00016
52439.69876	0.00005
52701.85002	0.00065
52701.84950	0.00039
52701.85045	0.00044
52704.76055	0.00020
52704.76031	0.00018
52704.75971	0.00016
52723.77613	0.00020
52723.77631	0.00015
52723.77616	0.00015
52770.73369	0.00004

justified a weight ratio of 1:100 for both types of minima. In the least squares fits that follow, minima were assigned weights inversely proportional to their variance.

A general linear least squares fit to all minima (the earliest ones dating back to 1935) revealed a complex period structure. Therefore, we fitted ephemerides to minima in specific subintervals. Between 1935 and 1947, minima are fitted well by the linear ephemeris

$$T_{\min} = 2431180.23205(\pm 0.00084) + 0.38808128(\pm 0.00000018) \times E \qquad (1)$$

while minima between 1967 and 1976 are represented by the linear ephemeris

$$T_{\min} = 2442510.6362(\pm 0.0060) + 0.38808093(\pm 0.00000081) \times E\,. \qquad (2)$$

For minima obtained since 1981, there is a clear upward sloping parabolic trend (see also Rucinski & Lu, 2000) which continues to this date. A weighted quadratic least squares fit to all post-1981 minima yields the equation

$$\begin{aligned} T_{\min} = {} & 2449108.79129(\pm 0.00041) + 0.388081547(\pm 0.000000013) \times E \\ & + 5.53(\pm 0.43)\,10^{-11} \times E^2\,, \end{aligned} \qquad (3)$$

while the global solution of post-1981 light and velocities (Sect. 2.1) gives

$$\begin{aligned} T_{\min} = {} & 2449108.79127(\pm 0.00010) + 0.3880816023(\pm 0.0000000037) \times E \\ & + 5.33(\pm 0.11)\,10^{-11} \times E^2\,. \end{aligned} \qquad (4)$$

Some time around 1981–1982, W Crv's period decreased by a fractional amount $\Delta P/P$ of $2.75 \pm 0.46 \times 10^{-6}$. Since then, the period has increased steadily. Equations (3) and (4) correspond to $\mathrm{d}P/\mathrm{d}t$'s of $2.85 \pm 0.22 \times 10^{-10}$ and $2.746 \pm 0.057 \times 10^{-10}$ (see Table 2), respectively. The agreement between these two estimates, one from eclipse timings and the other from whole light and velocity curves, is excellent. Figure 5 shows post-1981 eclipse timing residuals with respect to a linear fit and the parabola corresponding to (3).

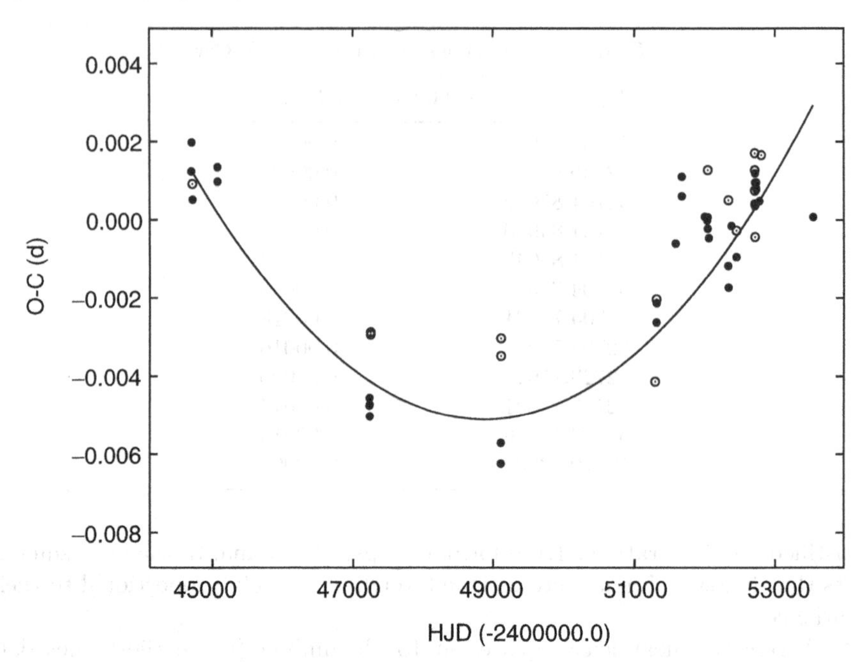

Fig. 5. W Crv post-1981 times of minima residuals for a linear fit and the parabola corresponding to eq.(3). Filled and open circles indicate primary and secondary minima, respectively.

2.3 W Crv Conclusions

All available W Crv light and velocity data can be described by a unique set of binary parameters, with individual epoch-to-epoch light curve changes successfully modeled with at most two spots. While individual spot parameters vary slightly from epoch to epoch, the general spot configuration and spot locations do not change dramatically. All light curves are characterized by a large hot spot on the smaller (cooler) star and positioned close to the connecting "neck" between the two stars (average co-latitude 50°, longitude 355°, radius 77° and temperature factor 1.15). Except for the 1967 and 1981 curves, a second, cool spot (presumably magnetic) is located on the trailing hemisphere of the larger (hotter) star (average co-latitude 80°, longitude 27°, radius 12° and temperature factor 0.71). Profiles of Hα lines in spectra obtained in 1995 by Odell (private communication) are consistent with the presence of a dark area on the primary star in the same general location of this dark spot. The global light and velocity solution has the system in genuine *OC*, with an 18% fillout. The *OC* state is quite robust as we failed to produce acceptable fits for detached or semidetached configurations. Because of the large temperature difference between the two components ($\approx$900 K), the *OC* configuration is puzzling, although W Crv is not unique in this respect. BE Cep (Kałużny, 1986) is another example of an *OC* system with a large

(900 K) temperature difference between components. We can only speculate on the cause of the hot spot on the secondary star. It could be the result of gas flow from the hotter to the cooler star, with the transferred matter causing a heated impact area on the accreting star. The recent period *increase*, however, indicates that this mass transfer cannot be conservative and that some of the transferred mass is leaving the system. W Crv deserves continued observational attention.

3 V2388 Oph

V2388 Oph is a bright (6th magnitude) eclipsing binary classified as a W UMa system by Rodríguez, *et al.* (1998) who obtained Strömgren u, v, b, y photometry. The system is a member of the visual system Fin 381 (HD 163151) whose orbital elements were determined by speckle interferometry by Hartkopf, Mason, & McAlister (1996). Improved elements (period 8.9 yr, separation $0\farcs088$, eccentricity 0.32) were published by Söderhjelm (1999). Rucinski, *et al.* (2002) obtained radial velocities and, based on broadening functions, found a magnitude difference of 1.75 ± 0.02 between the eclipsing system and the tertiary star, in agreement with Söderhjelm's estimate of 1.80 mag. Because V2388 Oph has total eclipses and the evidence in favor of a third star is very robust, we decided to use this system to make light and velocity solutions that include third light, and to test the validity of the claim by Pribulla & Rucinski (2006) that third light determined from light curves is of limited or no use in finding third components.

3.1 Light and Velocity Solutions

We made simultaneous solutions of the four light curves of Rodríguez, *et al.* (1998) and the velocities of Rucinski, *et al.* (2002), both without and with third light as an adjustable parameter. Not surprisingly, given the brightness of the companion, there is a marked improvement in the overall light curve fit for the solution with third light. The sum of squared light residuals drops between 10% and 20% depending on band. Comparing both panels in Figs. 6 and 7, note the much better fit of secondary minimum for the solution with third light. Parameters for the solution with third light are listed in Table 5, and corresponding absolute dimensions in Table 6. We find 24–25% third light in each of the bands. The source of this third light is likely the visual companion, which would be $\approx$1.2 mag fainter than the eclipsing pair. The amount of third light is slightly higher than the 20% estimated from spectra (Rucinski, *et al.*, 2002), but agrees well with the solution of Yakut, Kalomeni, & İbanoğlu (2004).

Graphs of radial velocity residuals versus time (Figs. 8–9) reveal the presence of an 8 km/s semi-amplitude oscillation, larger than expected given the precision of the individual velocities. This oscillation could be caused by the

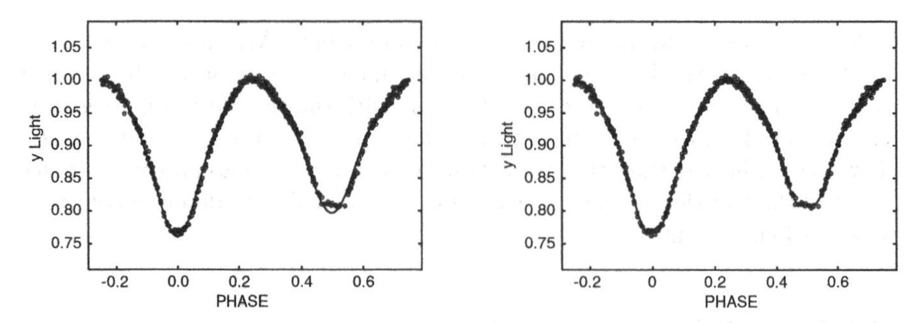

Fig. 6. V2388 Oph y light curve of Rodríguez, *et al.* (1998) and fitted curve for the solution without (*left*) and with (*right*) third light.

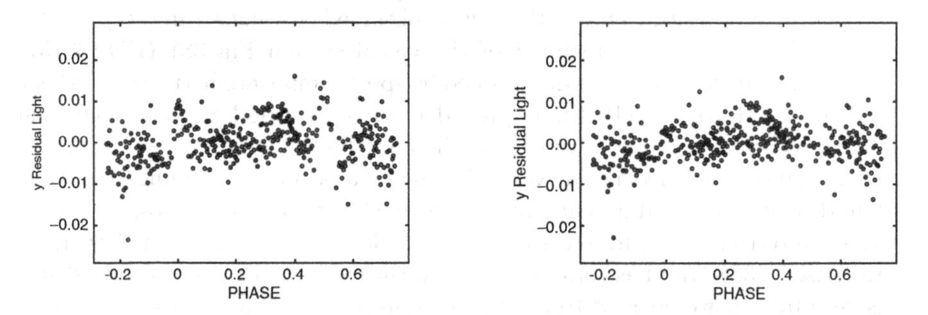

Fig. 7. V2388 Oph y light residuals for the solution without (*left*) and with (*right*) third light.

motion of the eclipsing pair around a third star. Clearly, this star cannot be the visible companion, as its period has to be considerably smaller than 8.9 yr. Using an experimental version of the Wilson program that includes third body orbital effects on light and velocities (Van Hamme & Wilson, 2005, 2007), we were able to obtain a substantial decrease in the root mean square radial velocity residuals. When third body effects were included, the rms residual velocity for the primary component dropped from 3.08 to 2.27 km/s, while it decreased from 5.90 to 3.43 km/s for the secondary. The orbital parameters of this third star derived as part of the light and velocity solution, and assuming a circular and nearly coplanar orbit, are included in Table 5 and carry the subscript "3b."[2] Because current velocities only span ≈1.5 orbital cycles and phase coverage is sparse, additional radial velocities are needed to confirm V2388 Oph's quadruple nature. If confirmed it would then be a system in which third light and orbital dynamical effects do not originate in the same component. On a final note, our experience tells us that including third light as a solution parameter in light curve solutions can be a valuable tool for determining multiplicity of W UMa systems.

[2] Note that the source of third light (l_3) is not the "3b" star but the visual companion instead.

Table 5. V2388 Oph Light and Velocity Solution

Parameter	Simultaneous RV, LC, plus 3b
T_0 (HJD)	$2450820.36719 \pm 0.00042$ s.e.
P_0 (d)	$0.80229664 \pm 0.00000034$
$\mathrm{d}P/\mathrm{d}t$	0.0
$a\,(R_\odot)$	4.707 ± 0.014
V_γ (km/s)	-18.4 ± 1.8
i	$77^\circ\!.09 \pm 0^\circ\!.23$
g_1, g_2	0.32, 0.32
T_1 (K)[a]	6450
T_2 (K)	6088 ± 7
A_1, A_2	0.5, 0.5
Ω_1	2.1708 ± 0.0024
Ω_2	2.17083
$f = (\Omega_{\mathrm{L}_1} - \Omega)/(\Omega_{\mathrm{L}_1} - \Omega_{\mathrm{L}_2})$	0.321
$q = M_2/M_1$	0.1909 ± 0.0011
$a_{3\mathrm{b}}\,(R_\odot)$	269.4 ± 6.8
$P_{3\mathrm{b}}$ (d)	311.0 ± 6.2
$i_{3\mathrm{b}}$	77°
$T_{3\mathrm{bconj}}$ (HJD)	2451586 ± 12
$L_1/(L_1+L_2)_u$	0.8638 ± 0.0082
$L_1/(L_1+L_2)_v$	0.8624 ± 0.0079
$L_1/(L_1+L_2)_b$	0.8534 ± 0.0075
$L_1/(L_1+L_2)_y$	0.8474 ± 0.0073
l_3 (u)	0.2342 ± 0.0061
l_3 (v)	0.2511 ± 0.0058
l_3 (b)	0.2536 ± 0.0055
l_3 (y)	0.2536 ± 0.0053
x_1, y_1 (u)	0.840, 0.271
x_2, y_2 (u)	0.868, 0.194
x_1, y_1 (v)	0.836, 0.204
x_2, y_2 (v)	0.851, 0.149
x_1, y_1 (b)	0.787, 0.243
x_2, y_2 (b)	0.807, 0.219
x_1, y_1 (y)	0.713, 0.274
x_2, y_2 (y)	0.736, 0.259
Auxiliary Parameters	
r_1 (pole)	0.5000 ± 0.0006
r_1 (side)	0.5481 ± 0.0010
r_1 (back)	0.5733 ± 0.0013
$< r_1 >$	0.5410 ± 0.0005
r_2 (pole)	0.2405 ± 0.0018
r_2 (side)	0.2516 ± 0.0022
r_2 (back)	0.2950 ± 0.0049
$< r_2 >$	0.2634 ± 0.0010

[a] Based on color indices (Rodríguez et al., 1998)

Table 6. V2388 Oph Absolute Parameters

Parameter	Star 1	Star 2	Star "3b"
$M\,(M_\odot)$	1.824 ± 0.016	0.348 ± 0.004	0.54 ± 0.11
$R\,(R_\odot)$	2.547 ± 0.008	1.240 ± 0.006	...
$M_{\rm bol}$	2.24 ± 0.03	4.05 ± 0.04	...
$\log(L/L_\odot)$	1.003 ± 0.014	0.278 ± 0.017	...
$\log g$ (cm·s^{-2})	3.888 ± 0.007	3.794 ± 0.009	...

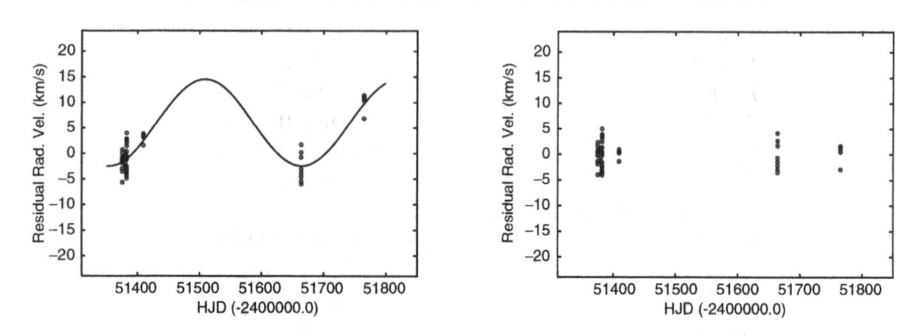

Fig. 8. V2388 Oph primary radial velocity residuals for the solution without (*left*) and with (*right*) the effects of a third star in a 1.25 AU, 311 d orbit around the eclipsing pair. The solid line in the left panel represents the oscillation due to this third star and is based on the "3b" parameters in Table 5.

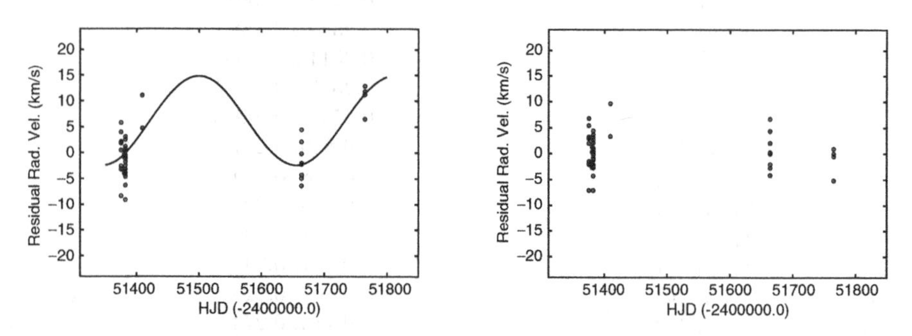

Fig. 9. Same as Fig. 8 for the secondary star.

4 Acknowledgments

REC acknowledges support from a Research Experiences for Undergraduates (*REU*) grant by the National Science Foundation (grant AST-0097616) to the Southeastern Association for Research in Astronomy. We made use of the *SIMBAD* database, operated at CDS, Strasbourg, France.

References

1. Baran, A., Zola, S., Rucinski, S. M., Kreiner, J. M., Siwak, M., & Drozdz, M. 2004, Acta Astronomica, 54, 195
2. Binnendijk, L. 1965, Kleine Veroff. Bamberg, 4, No. 40, 36
3. Binnendijk, L. 1970, Vistas Astr., 12, 217
4. Cohen, R. & Van Hamme, W. 2003, I.A.P.P.P. Comm. 91, 28
5. D'Angelo, C., van Kerkwijk, M. H., & Rucinski, S. M. 2006, AJ, 132, 650
6. Dycus, R. D. 1968, PASP, 80, 207
7. Eggleton, P. P. 1996, ASP Conf. Ser. 90, The Origins, Evolution, and Destinies of Binary Stars in Clusters, ed. E. F. Milone & J.-C. Mermilliod (San Francisco: ASP), 257
8. Flannery, B. P. 1976, ApJ, 205, 217
9. Gazeas, K. D., *et al.* 2005, Acta Astronomica, 55, 123
10. Gazeas, K. D., Niarchos, P. G., Zola, S., Kreiner, J. M., & Rucinski, S. M. 2006, Acta Astronomica, 56, 127
11. Gettel, S. J., Geske, M. T., & McKay, T. A. 2006, AJ, 131, 621
12. Hartkopf, W. I., Mason, B. D., & McAlister, H. A. 1996, AJ, 111, 370
13. Hilditch, R. W., King, D. J., & McFarlane, T. M. 1988, MNRAS, 231, 341
14. Kałużny, J. 1986, PASP, 98, 662
15. Kreiner, J. M., *et al.* 2003, A&A, 412, 465
16. Kuiper, G. P. 1948, ApJ, 108, 541
17. Kurucz, R. L. 1993, Light Curve Modeling of Eclipsing Binary Stars, ed. E. F. Milone (New York: Springer-Verlag), 93
18. Kwee, K. K. & Van Woerden, H. 1956, Bull. Astr. Inst. Netherlands, 12, 327
19. Lucy, L. B. 1976, ApJ, 205, 208
20. Lucy, L. B. & Wilson, R. E. 1979, ApJ, 231, 502
21. Odell, A. P. 1996, MNRAS, 282, 373
22. Odell, A. P. & Cushing, G. E. 2004, IBVS, 5514
23. Pribulla, T. & Rucinski, S. M. 2006, AJ, 131, 2986
24. Robertson, J. A. & Eggleton, P. P. 1977, MNRAS, 179, 359
25. Rodríguez, E., *et al.* 1998, A&A, 336, 920
26. Rucinski, S. M. 1985, Interacting Binary Stars, ed. J. E. Pringle & R. A. Wade (Cambridge: CUP), 113
27. Rucinski, S. M. 1998, AJ, 116, 2998
28. Rucinski, S. M. 2002, PASP, 114, 1124
29. Rucinski, S. M. & Lu, W. 2000, MNRAS, 315, 587
30. Rucinski, S. M., Lu, W., Capabianco, C. C., Mochnacki, S. W., Blake, R. M., Thomson, J. R., Ogłoza, W., & Stachowski, G. 2002, AJ, 124, 1738
31. Shapley, H. 1948, Harvard Obs. Monograph, No. 7, 249
32. Smith, R. C. 1984, QJRAS, 25, 405
33. Söderhjelm, S. 1999, A&A, 341, 121
34. Terrell, D. & Wilson, R. E. 2005, ApSS, 296, 221
35. Yakut, K. & Eggleton, P. P. 2005, ApJ, 629, 1055
36. Yakut, K., Kalomeni, B., & İbanoğlu, C. 2004, A&A, 417, 725
37. Van Hamme, W. 1993, AJ, 106, 2096
38. Van Hamme, W. & Branly, R. M. 2003, BAAS, 35, 1224
39. Van Hamme, W. & Wilson, R. E. 2003, ASP Conf. Ser. 298, GAIA Spectroscopy, Science and Technology, Ed. U. Munari (San Francisco: ASP), 323

40. Van Hamme, W. & Wilson, R. E. 2005, ApSS, 296, 121
41. Van Hamme, W. & Wilson, R. E. 2007, ApJ, 661, 1129
42. Van Hamme, W., Samec, R. G., Gothard, N. W., Wilson, R. E., Faulkner, D. R. & Branly, R. M. 2001, AJ, 122, 3436
43. Wilson, R. E. 1978, ApJ, 224, 885
44. Wilson, R. E. 1979, ApJ, 234, 1054
45. Wilson, R. E. 1990, ApJ, 356, 613
46. Wilson, R. E. 1994, PASP, 106, 921
47. Webbink, R. F. 2003, ASP Conf. Ser. 293, 3D Stellar Evolution, ed. S. Turcotte, S. C. Keller, & R. M. Cavallo (San Francisco: ASP), 76
48. Zola, S., *et al.* 2004, Acta Astronomica, 54, 299
49. Zola, S., *et al.* 2005, Acta Astronomica, 55, 389

Part V

Aspects of Short-Period Binary Evolution

Common Envelope Evolution Redux

Ronald F. Webbink

Department of Astronomy, University of Illinois
1002 W. Green St., Urbana, IL 61801, USA

Common envelopes form in dynamical time scale mass exchange, when the envelope of a donor star engulfs a much denser companion, and the core of the donor plus the dense companion star spiral inward through this dissipative envelope. As conceived by Paczynski and Ostriker, this process must be responsible for the creation of short-period binaries with degenerate components, and, indeed, it has proven capable of accounting for short-period binaries containing one white dwarf component. However, attempts to reconstruct the evolutionary histories of close double white dwarfs have proven more problematic, and point to the need for enhanced systemic mass loss, either during the close of the first, slow episode of mass transfer that produced the first white dwarf, or during the detached phase preceding the final, common envelope episode. The survival of long-period interacting binaries with massive white dwarfs, such as the recurrent novae T CrB and RS Oph, also presents interpretative difficulties for simple energetic treatments of common envelope evolution. Their existence implies that major terms are missing from usual formulations of the energy budget for common envelope evolution. The most plausible missing energy term is the energy released by recombination in the common envelope, and, indeed, a simple reformulation the energy budget explicitly including recombination resolves this issue.

1 Introduction

From the realization [26, 27, et seq.] that all cataclysmic variables (CVs) are interacting binary stars, their existence posed a dilemma for theories of binary evolution. The notion that close binary stars might evolve in ways fundamentally different from isolated stars was rooted in the famous 'Algol paradox' (that the cooler, lobe-filling subgiant or giant components among these well-known eclipsing binaries are less massive, but more highly evolved, than their hotter main-sequence companions). The resolution of that paradox invoked large-scale mass transfer reversing the initial mass ratios of these binaries [34].

E.F. Milone et al. (eds.), Short-Period Binary Stars: Observations, Analyses, and Results, 233–257.

Indeed, model calculations assuming conservation of total mass and orbital angular momenum are qualitatively consistent with the main features of Algol-type binaries. Even if quantitative consistency between models and observational data generally requires some losses of mass and angular momentum among Algol binaries (e.g., [8, 10, 21]), the degree of those losses is typically modest, and the remnant binary is expected to adhere closely to an equilibrium core mass-radius relation for low-mass giant stars (see, e.g., the pioneering study of AS Eri by Refsdal, Roth & Weigert [46]). Those remnant binaries are typically of long orbital period (days to weeks) in comparison with *CV*s, and furthermore typically contain helium white dwarfs of low mass, especially in the short-period limit. In contrast, CVs evidently contain relatively massive white dwarfs, in binary systems of much shorter orbital periods (hours), that is, with much smaller total energies and orbital angular momenta.

In an influential analysis of the Hyades eclipsing red dwarf/white dwarf binary BD $+16^{\circ}$ 516 (= V471 Tau), Vauclair [57] derived a total system mass less than the turnoff mass of the Hyades, and noted that the cooling age of the white dwarf component was much smaller than the age of the cluster. He speculated that V471 Tau in its present state was the recent product of the ejection of a planetary nebula by the white dwarf. Paczynski [41] realized that, immediately prior to that event, the white dwarf progenitor must have been an asymptotic giant branch star of radius $\sim$600 $R_{\odot}$, far exceeding its current binary separation $\sim 3\, R_{\odot}$. He proposed that the dissipation of orbital energy provided the means both for planetary nebula ejection and for the severe orbital contraction between initial and final states, a process he labeled 'common envelope evolution' (not to be confused with the common envelopes of contact binary stars). Discovery soon followed of the first 'smoking gun', the short-period eclipsing nucleus of the planetary nebula Abell 63 [3].

Over the succeeding three decades, there have been a number of attempts to build detailed physical models of common envelope evolution (see [55] for a review). These efforts have grown significantly in sophistication, but this phenomenon presents a daunting numerical challenge, as common envelope evolution is inherently three-dimensional, and the range of spatial and temporal scales needed to represent a common envelope binary late in its inspiral can both easily exceed factors of 10^3. Determining the efficiency with which orbital energy is utilized in envelope ejection requires such a code to conserve energy over a similarly large number of dynamical time scales.

Theoretical models of common envelope evolution are not yet capable of predicting the observable properties of objects in the process of inspiral. If envelope ejection is to be efficient, then the bulk of dissipated orbital energy must be deposited in the common envelope on a time scale short compared with the thermal time scale of the envelope, else that energy be lost to radiation. The duration of the common envelope phase thus probably does not exceed $\sim 10^3$ years. However general considerations of the high initial orbital angular momenta of systems such as the progenitor of V471 Tau, and the fact that most of the orbital energy is released the envelope only very late in the inspiral have led to a consensus view [32, 33, 50, 60, 65] that the planetary nebulae they eject should be bipolar in structure, with dense equatorial rings absorbing most of the initial angular momentum of the binary, and higher-velocity polar jets powered by the late release of orbital energy. Indeed, this appears to be a signature morphology of planetary nebulae with binary nuclei (e.g., [2]), although it may not be unique to binary nuclei.

2 The Energetics of Common Envelope Evolution

Notwithstanding the difficulties in modeling common envelope evolution in detail, it is possible to calculate with some confidence the initial total energy and angular momentum of a binary at the onset of mass transfer, and the corresponding orbital energy and angular momentum of any putative remnant of common envelope evolution.

Consider an initial binary of component masses M_1 and M_2, with orbital semimajor axis $A_{\rm i}$. Its initial total orbital energy is

$$E_{\rm orb,i} = -\frac{GM_1M_2}{2A_{\rm i}} . \tag{1}$$

Let star 1 be the star that initiates interaction upon filling its Roche lobe. If $M_{\rm 1c}$ is its core mass, and $M_{\rm 1e} = M_1 - M_{\rm 1c}$ its envelope mass, then we can write the initial total energy of that envelope as

$$E_{\rm e} = -\frac{GM_1M_{\rm 1e}}{\lambda R_{\rm 1,L}} , \tag{2}$$

where $R_{\rm 1,L}$ is the Roche lobe radius of star 1 at the onset of mass transfer (the orbit presumed circularized prior to this phase), and λ is a dimensionless parameter dependent on the detailed structure of the envelope, but presumably of order unity. For very simplified models of red giants – condensed polytropes [14, 16, 40] – λ is a function only of $m_{\rm e} \equiv M_{\rm e}/M = 1 - M_{\rm c}/M$, the ratio of envelope mass to total mass for the donor, and is well-approximated by

$$\lambda^{-1} \approx 3.000 - 3.816 m_{\rm e} + 1.041 m_{\rm e}^2 + 0.067 m_{\rm e}^3 + 0.136 m_{\rm e}^4 , \tag{3}$$

to within a relative error $< 10^{-3}$.

For the final orbital energy of the binary we have

$$E_{\rm orb,f} = -\frac{GM_{\rm 1c}M_2}{2A_{\rm f}} , \tag{4}$$

where $A_{\rm f}$ is of course the final orbital separation. If a fraction $\alpha_{\rm CE}$ of the difference in orbital energy is consumed in unbinding the common envelope,

$$\alpha_{\rm CE} \equiv \frac{E_{\rm e}}{(E_{\rm orb}^{\rm (f)} - E_{\rm orb}^{\rm (i)})} , \tag{5}$$

then

$$\frac{A_{\rm f}}{A_{\rm i}} = \frac{M_{\rm 1c}}{M_1}\left[1 + \left(\frac{2}{\alpha_{\rm CE}\lambda r_{\rm 1,L}}\right)\left(\frac{M_1 - M_{\rm 1c}}{M_2}\right)\right]^{-1} , \tag{6}$$

where $r_{\rm 1,L} \equiv R_{\rm 1,L}/A_{\rm i}$ is the dimensionless Roche lobe radius of the donor at the start of mass transfer. In the classical Roche approximation, $r_{\rm 1,L}$ is a function only of the mass ratio, $q \equiv M_1/M_2$ [7]:

$$r_{\rm 1,L} \approx \frac{0.49q^{2/3}}{0.6q^{2/3} + \ln(1+q^{1/3})} . \tag{7}$$

Typically, the second term in brackets in (6) dominates the first term.

As formulated above, our treatment of the outcome of common envelope evolution neglects any sources or sinks of energy beyond gravitational terms and the thermal energy content of the initial envelope (incorporated in the parameter λ). The justification for this assumption is again that common envelope evolution must be rapid compared to the thermal time scale of the envelope. This implies that radiative losses (or nuclear energy gains – see below) are small. They, as well as terminal kinetic energy of the ejecta, are presumably reflected in ejection efficiencies $\alpha_{\rm CE} < 1$. We neglect also the rotational energy of the common envelope (invariably small in magnitude compared to its gravitational binding energy), and treat the core of the donor star and the companion star as inert masses, which neither gain nor lose mass or energy during the course of common envelope evolution. One might imagine it possible that net accretion of mass by the companion during inspiral might compromise this picture. However, the common envelope is typically vastly less dense than the companion star and may be heated to roughly virial temperature on infall. A huge entropy barrier arises at the interface between the initial photosphere of the companion and the common envelope in which it is now embedded, with a difference in entropy per particle of order $(\mu m_{\rm H}/k)\Delta s \approx$ 4–6. The rapid rise in temperature and decrease in density through the interface effectively insulates the accreting companion thermally, and strongly limits the fraction of the very rarified common envelope it can retain upon exit from that phase [15, 62].

Common envelope evolution entails systemic angular momentum losses as well as systemic mass and energy losses. Writing the orbital angular momentum of the binary,

$$J = \left[\frac{GM_1^2M_2^2A(1-e^2)}{M_1+M_2}\right]^{1/2} , \tag{8}$$

in terms of the total orbital energy, $E = -GM_1M_2/2A$, we find immediately that the ratio of final to initial orbital angular momentum is

$$\frac{J_{\rm f}}{J_{\rm i}} = \left(\frac{M_{\rm 1c}}{M_1}\right)^{3/2}\left(\frac{M_{\rm 1c}+M_2}{M_1+M_2}\right)^{-1/2}\left(\frac{E_{\rm i}}{E_{\rm f}}\right)^{1/2}\left(\frac{1-e_{\rm f}^2}{1-e_{\rm i}^2}\right)^{1/2} . \tag{9}$$

Because $M_{1c} < M_1$ and we expect the initial orbital eccentricity to be small ($e_i \approx 0$), it follows that any final energy state lower than the initial state ($|E_f| > |E_i|$) requires the loss of angular momentum. The reverse is not necessarily true, so it is the energy budget that most strongly constrains possible outcomes of common envelope evolution.

3 Does Common Envelope Evolution Work?

As an example of common envelope energetics, let us revisit the pre-*CV* V471 Tau, applying the simple treatment outlined above. It is a member of the Hyades, an intermediate-age metal-rich open cluster (t = 650 Myr, [Fe/H] = +0.14) with turnoff mass $M_{TO} = 2.60 \pm 0.06\,M_\odot$ [28]. The cooling age of the white dwarf is much smaller than the age of the cluster ($t_{cool,WD} = 10^7$ yr [39] – but see the discussion there of the paradoxical fact that this most massive of Hyades white dwarfs is also the youngest). Allowing for the possibility of significant mass loss in a stellar wind prior to the common envelope phase, we may take M_{TO} for an upper limit to the initial mass M_1 of the white dwarf component. The current masses for the white dwarf and its dK2 companion, as determined by O'Brien et al. [39] are $M_{WD} = 0.84 \pm 0.05\,M_\odot$, $M_K = 0.93 \pm 0.07\,M_\odot$, with orbital separation $A = 3.30 \pm 0.08\,R_\odot$. A $2.60\,M_\odot$ star of Hyades metallicity with a $0.84\,M_\odot$ core lies on the thermally-pulsing asymptotic giant branch, with radius (maximum in the thermal pulse cycle) which we estimate at $R_i = 680\,R_\odot = R_{1,L}$, making $A_i = 1450\,R_\odot$. With this combination of physical parameters, we derive an estimate of $\alpha_{CE}\lambda = 0.057$ for V471 Taui. Equation (3) then implies $\alpha_{CE} = 0.054$. This estimate of course ignores any mass loss prior to common envelope evolution (which would drive α_{CE} to lower values), or orbital evolution since common envelope evolution (which would drive α_{CE} to higher values). In any event, the status of V471 Tau would appear to demand only a very small efficiency of envelope ejection.[1]

The fact that V471 Tau is a double-lined eclipsing member of a well-studied cluster provides an exceptionally complete set of constraints on its prior evolution. In all other cases of short-period binaries with degenerate or compact components, available data are inadequate to fix simultaneously both the initial mass of the compact component and the initial binary separation, for example. To validate the energetic arguments outlined above,

[1] The anomalously small value of α_{CE} deduced for V471 Tau may be connected to its puzzlingly high white dwarf mass and luminosity: O'Brien et al. [39] suggest that it began as a heirarchical triple star, in which a short-period inner binary evolved into contact, merged (as a blue straggler), and later engulfed its lower-mass companion in a common envelope. An overmassive donor at the onset of common envelope evolution would then have a more massive core than produced by its contemporaries among primordially single stars, and it would fill its Roche lobe with a more massive envelope at somewhat shorter orbital period, factors all consistent with a larger value of α_{CE} having led to V471 Tau as now observed.

one must resort to consistency tests, whether demonstrating the existence of physically-plausible initial conditions that could produce some individual system, or else following a plausible distribution of primordial binaries wholesale through the energetics of common envelope evolution and showing that, after application of appropriate observational selection effects, the post-common-envelope population is statistically consistent with the observed statistics of the selected binary type. In the cases of interacting binaries, such as CVs, one should allow further for post-common-envelope evolution. Nevertheless, within these limitations, binary population synthesis models show broad consistency between the outcomes of common envelope evolution and the statistical properties of CVs and pre-CVs [4, 17, 24, 44, 63], as well as with most super-soft X-ray sources [5], for assumed common envelope ejection efficiencies typically of order $\alpha_{\rm CE} \approx 0.3$–$0.5$.

A useful tool in reconstructing the evolutionary history of a binary, used implicitly above in analyzing V471 Tau, is the mass-radius diagram spanned by single stars of the same composition as the binary. Figure 3 illustrates such a diagram for solar-composition stars from 0.08 $M_\odot$ to 50 $M_\odot$. In it are plotted various critical radii marking, as a function of mass, the transition from one evolutionary phase to the next.[2] Because the Roche lobe of a binary component represents a dynamical limit to its size, its orbital period fixes the mean density at which that star fills its Roche lobe,

$$\log P_{\rm orb}({\rm d}) \approx \frac{3}{2}\log(R_{\rm L}/R_\odot) - \frac{1}{2}\log(M/M_\odot) - 0.455 \ , \qquad (10)$$

to within a very weak function of the binary mass ratio. The mass and radius of any point in Fig. 3 therefore fixes the orbital period at which such a star would

[2] Not all evolutionary phases are represented here. In a binary, a donor initiates mass transfer when it first fills its Roche lobe; if it would have done so at a prior stage of evolution, then its present evolutionary state is 'shadowed', in the sense that it only occurs by virtue of the binary *not* having filled its lobe previously. Thus, for example, low- and intermediate-mass stars cannot in general initiate mass transfer during core helium burning, because they would have filled their Roche lobes on the initial ascent of the giant branch.

Fig. 1. The mass–radius diagram for stars of solar metallicity, constructed from the parametric models of stellar evolution by Hurley, Pols, and Tout [19] and models of thermally-pulsing asymptotic giant branch stars by Wagenhuber and Weiss [58]. Also plotted in the locus of asymptotic giant branch stars at the onset of the superwind, after G.H. Bowen (see Willson [64]) beyond this radius, systemic mass loss drives orbital expansion faster than nuclear evolution drives stellar expansion, and a binary will no longer be able to initiate tidal mass transfer. The unlabeled dotted line terminating at the junction between lines labeled 'helium core flash' and 'core helium ignition' marks the division between those helium cores (at lower masses) which evolve to degeneracy if stripped of their envelope, and those (at higher masses) which ignite helium non-degenerately and become helium stars.

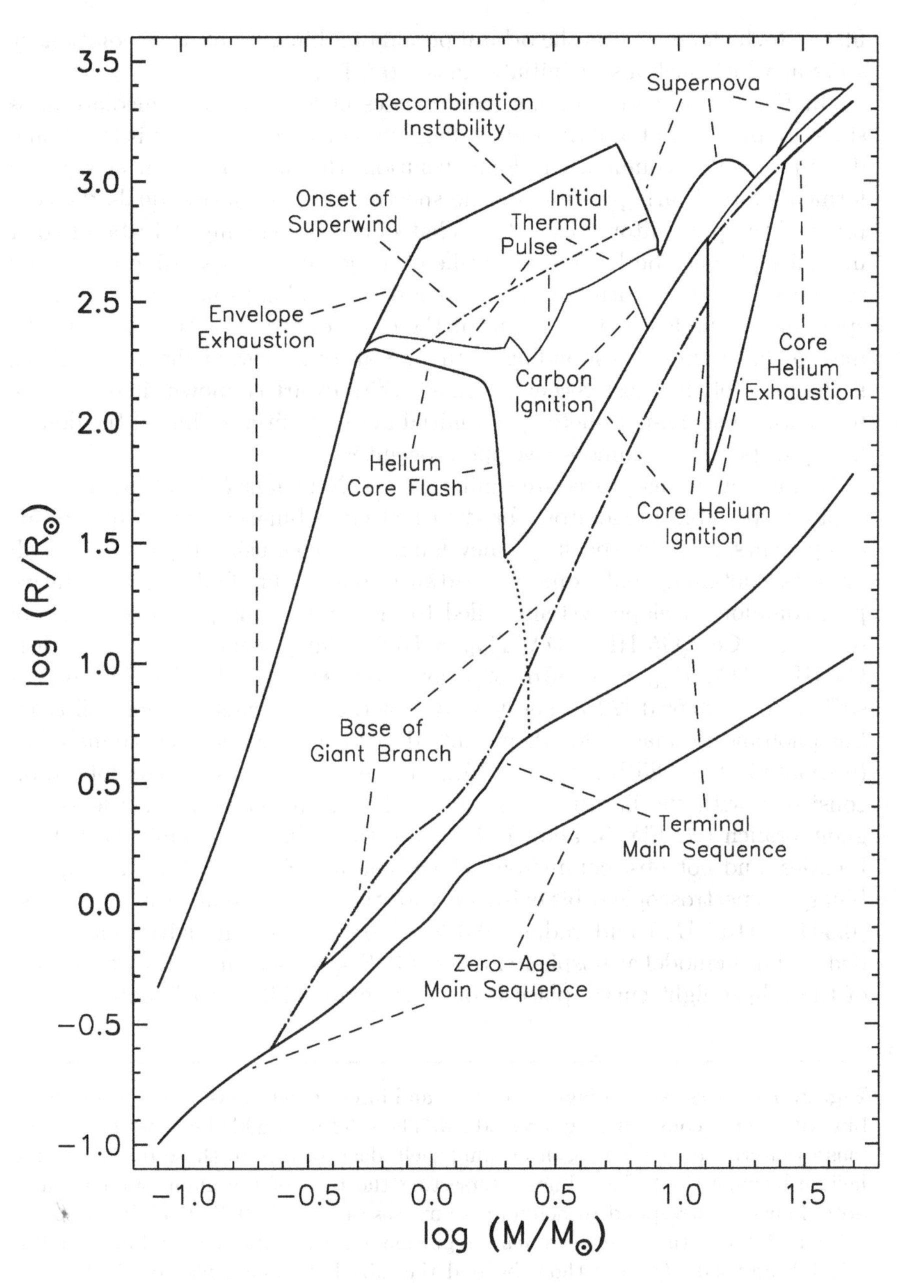
3.5
3.0
2.5
2.0
1.5
1.0
0.5
0.0
−0.5
−1.0
log (R/R⊙)
−1.0
−0.5
0.0
0.5
1.0
1.5
log (M/M⊙)
Supernova
Recombination
Instability
Onset of
Superwind
Initial
Thermal
Pulse
Envelope
Exhaustion
Core
Helium
Exhaustion
Carbon
Ignition
Helium
Core Flash
Core Helium
Ignition
Base of
Giant Branch
Terminal
Main Sequence
Zero−Age
Main Sequence

fill its Roche lobe, just as the orbital period of a binary fixes the evolutionary state at which such a star initiates mass transfer.

In Fig. 3, the corresponding core masses of low- and intermediate-mass stars are plotted in the mass-radius diagram. For a binary which is the immediate product of common envelope evolution, the mass of the most recently formed white dwarf (presumably the spectroscopic primary) equals the core mass of the progenitor donor star. That donor (presuming it to be of solar metallicity) must be located somewhere along the corresponding core mass sequence in Fig. 3, with the radius at any point along that sequence corresponding to the Roche lobe radius at the onset of the mass transfer, and the mass a that point corresponding to the initial total mass of the donor. Thus, if the mass of the most recently-formed white dwarf is known, it is possible to identify a single-parameter (e.g., initial mass or initial radius of the donor) family of possible common-envelope progenitors.

Using a mapping procedure similar to this, Nelemans & Tout [35] recently explored possible progenitors for detached close binaries with white dwarf components. Broadly speaking, they found solutions using (6) for almost all systems containing only one white dwarf component. Only three putative post-common-envelope systems failed to yield physically-plausible values of $\alpha_{\rm CE}\lambda$: AY Cet (G5 III + DA, $P_{\rm orb} = 56.80\,{\rm d}$ [49]), Sanders 1040 (in M67: G4 III + DA, $P_{\rm orb} = 42.83\,{\rm d}$ [56]), and HD 185510 (=V1379 Aql: gK0 + sdB, $P_{\rm orb} = 20.66\,{\rm d}$ [22]). The first two of these systems are non-eclipsing, but photometric masses for their white dwarf components are extremely low (estimated at $\sim 0.25\,M_\odot$ and $0.22\,M_\odot$, respectively), with Roche lobe radii consistent with the limiting radii of very low-mass giants as they leave the giant branch (cf. Fig. 3, above). They are thus almost certainly post-Algol binaries, and not post-common-envelope binaries. HD 185510 is an eclipsing binary; a spectroscopic orbit exists only for the gK0 component [9]. The mass ($0.304 \pm 0.015\,M_\odot$) and radius ($0.052 \pm 0.010\,R_\odot$) of the sdB component, deduced from model atmosphere fitting of IUE spectra combined with solution of the eclipse light curve, place it on a low-mass white dwarf cooling curve,

---

Fig. 2. The mass-radius diagram for low- and intermediate-mass stars, as in Fig. 1, but with loci of constant core mass added. The solid lines added correspond to core masses interior to the hydrogen-burning shell, dashed lines to those interior to the helium-burning shell. Solid lines intersecting the base of the giant branch (dash-dotted curve) correspond to helium core masses of to 0.15, 0.25, 0.35, 0.5, 0.7, 1.0, 1.4, and 2.0 $M_\odot$; those between helium ignition and the initial thermal pulse to 0.7, 1.0, 1.4, and 2.0 $M_\odot$, and those beyond the initial thermal pulse to 0.7, 1.0, and 1.4 $M_\odot$. Dashed lines between helium ignition and initial thermal pulse correspond to carbon-oxygen core masses of 0.35, 0.5, 0.7, 1.0, and 1.4 $M_\odot$. Beyond the initial thermal pulse, helium and carbon-oxygen core masses converge, with the second dredge-up phase reducing helium core masses above $\sim 0.8\ M_\odot$ to the carbon-oxygen core.

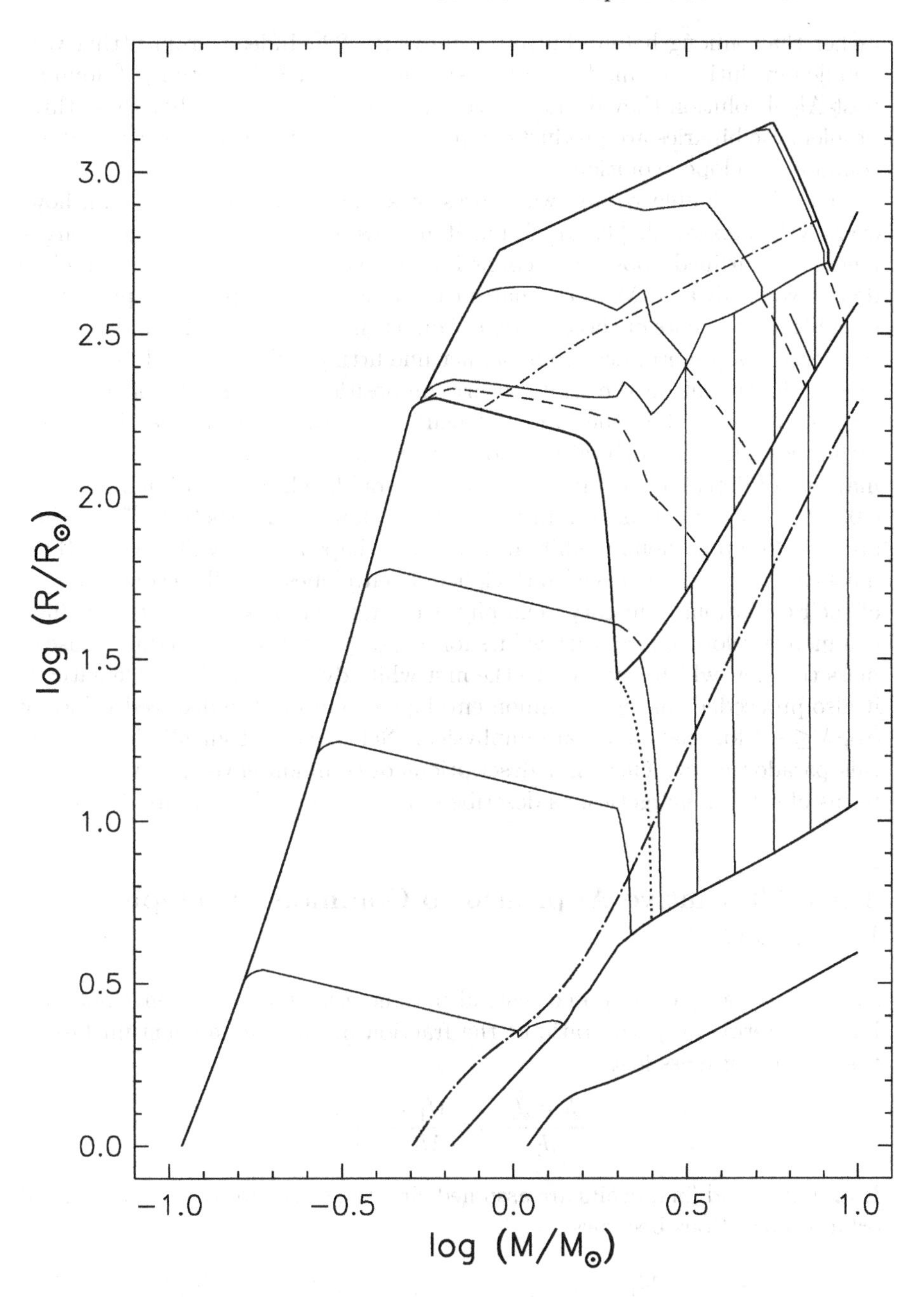
log (R/R$_\odot$)
log (M/M$_\odot$)
3.0
2.5
2.0
1.5
1.0
0.5
0.0
−1.0
−0.5
0.0
0.5
1.0

rather than among helium-burning subdwarfs [22]. Indeed, from fitting very detailed evolutionary models to this system, Nelson & Eggleton [38] found a post-Algol solution they deemed acceptable. It thus appears that these three problematic binaries are products of quasi-conservative mass transfer, and not common envelope evolution.

The close double white dwarfs present a more difficult conundrum, however. Nelemans et al. [35–37] found it impossible using the energetic arguments (6) outlined above to account for the existence of a most known close double white dwarfs. Mass estimates can be derived for spectroscopically detectable components of these systems from their surface gravities and effective temperatures (determined from Balmer line fitting). The deduced masses are weakly dependent on the white dwarf composition, and may be of relatively modest accuracy, but they are independent of the uncertainties in orbital inclination afflicting orbital solutions. These mass estimates place the great majority of detectable components in close double white dwarf binaries below $\sim 0.46\,M_{\odot}$, the upper mass limit for pure helium white dwarfs (e.g., [51]). They are therefore pure helium white dwarfs, or perhaps hybrid white dwarfs (low-mass carbon-oxygen cores with thick helium envelopes). While reconstructions of their evolutionary history yield physically-reasonable solutions for the final common envelope phase, with values for $0 < \alpha_{\rm CE}\lambda < 1$, the preceding phase of mass transfer, which gave rise to the first white dwarf, is more problematic. If it also proceeded through common envelope evolution, the deduced values of $\alpha_{\rm CE}\lambda \leq -4$ for that phase are unphysical. Nelemans & Tout [35] interpreted this paradox as evidence that descriptions of common envelope evolution in terms of orbital energetics, as described above, are fundamentally flawed.

4 An Alternative Approach to Common Envelope Evolution?

Nelemans et al. [36] proposed instead parameterizing common envelope evolution in terms of γ, the ratio of the fraction of angular momentum lost to the fraction of mass lost:

$$\frac{J_{\rm i} - J_{\rm f}}{J_{\rm i}} = \gamma \frac{M_1 - M_{1,\rm c}}{M_1 + M_2} \,. \tag{11}$$

Both initial and final orbits are assumed circular, so the ratio of final to initial orbital separations becomes

$$\frac{A_{\rm f}}{A_{\rm i}} = \left(\frac{M_1}{M_{\rm 1c}}\right)^2 \left(\frac{M_{\rm 1c} + M_2}{M_1 + M_2}\right) \left[1 - \gamma \left(\frac{M_1 - M_{\rm 1c}}{M_1 + M_2}\right)\right]^2 \,. \tag{12}$$

Among possible solutions leading to known close double white dwarfs, Nelemans & Tout [35] find values $1 < \gamma \lesssim 4$ required for the second (final) common envelope phase, and $0.5 \lesssim \gamma < 3$ for the first (putative) common envelope phase. They note that values in the range $1.5 < \gamma < 1.7$ can be found among possible solutions for all common envelope phases in their sample, not only those leading to known double white dwarfs, but those leading to known pre-CV and sdB binaries as well.

The significance of this finding is itself open to debate. At one extreme, it would seem implausible for any mechanism to remove less angular momentum per unit mass than the orbital angular momentum per unit mass of either component in its orbit (so-called Jeans-mode mass loss). At the other extreme, a firm upper limit to γ is set by vanishing final orbital angular momentum, J_f. If $M_{1\mathrm{c}}$ and M_2 can be regarded as fixed, the corresponding limits on γ are

$$\left(\frac{M_1 + M_2}{M_1 - M_{1\mathrm{c}}}\right) > \gamma > \left(\frac{M_1 + M_2}{M_1 - M_{1\mathrm{c}}}\right)\left[1 - \left(\frac{M_{1\mathrm{c}}}{M_1}\right)\left(\frac{M_1 + M_2}{M_{1\mathrm{c}} + M_2}\right)\right] . \qquad (13)$$

In a fairly typical example, $M_{1\mathrm{c}} = M_2 = \frac{1}{4}M_1$, γ is inevitably tightly constrainted for any conceivable outcome: $\frac{5}{3} > \gamma > \frac{5}{8}$. The ratio of final to initial orbital separation, $A_\mathrm{f}/A_\mathrm{i}$, is extremely sensitive to γ near the upper limit of its range. It is therefore not surprising to find empirical estimates of γ clustering as they do – their values merely affirm the fact that A_f must typically be much smaller than A_i.

The unphysically large or, more commonly, *negative* values of $\alpha_{\mathrm{CE}}\lambda$ noted above for the first mass transfer phase in the production of close white dwarf binaries [35] implies that the orbital energies of these binaries have *increased* through this phase (or, at any rate, decreased by significantly less than the nominal binding energies of their common envelopes). Such an increase in orbital energy is a hallmark of slow, quasi-conservative mass transfer, on a thermal or nuclear time scale. Thermal time scale mass transfer is driven by relaxation of the donor star toward thermal evolution; the re-expansion of the donor following mass ratio reversal is powered by the (nuclear) energy outflow from the core of the star. Likewise, the bulk expansion of the donor star in nuclear time scale mass transfer draws energy from nuclear sources in that star. It appears, therefore, that the first phase of mass transfer among known close double white dwarfs cannot have been a common envelope phase, but must instead have been a quasi-conservative phase, notwithstanding the difficulties that conclusion presents, as we shall now see.

The dilemma that the close double white dwarfs present is illustrated in Figs. 3 and 4. Figure 3 shows the distribution of immediate remnants of mass transfer among solar-metallicity binaries of low and intermediate mass, for a relatively moderate initial mass ratio. Conservation of total mass and orbital angular momentum have been assumed. The remnants of the intial primary include both degenerate helium white dwarfs, and nondegenerate helium stars which have lost nearly all of their hydrogen envelopes. The helium white dwarfs lie almost entirely along the left-hand boundary, the line labeled 'envelope exhaustion' in Fig. 3. (The extent of this sequence is more apparent in the distribution of remnant secondaries.) Their progenitors have enough angular momentum to accommodate core growth in the terminal phases of

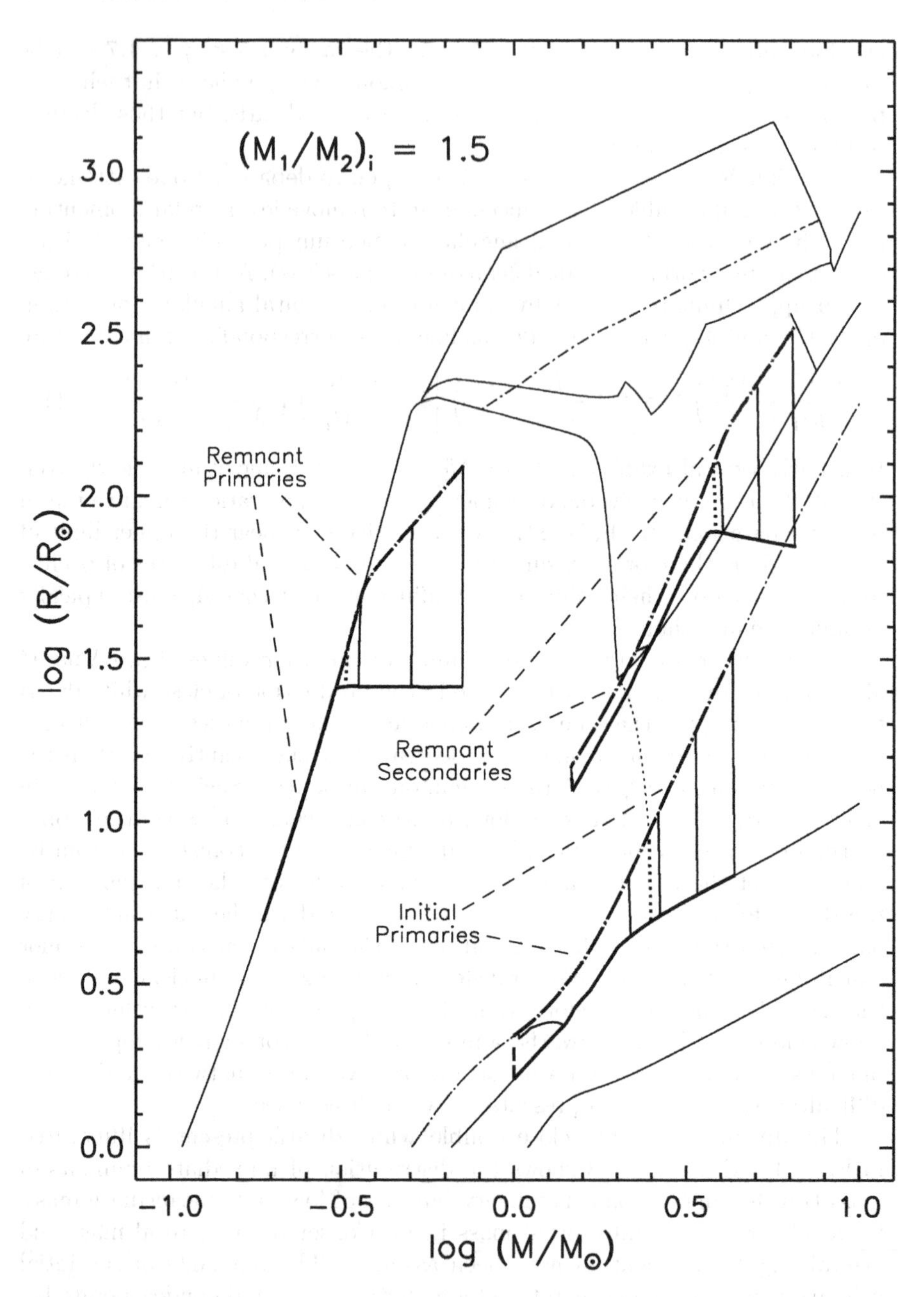
$(M_1/M_2)_i = 1.5$
Remnant Primaries
Remnant Secondaries
Initial Primaries
log (R/R⊙)
log (M/M⊙)
3.0
2.5
2.0
1.5
1.0
0.5
0.0
−1.0
−0.5
0.0
0.5
1.0

mass transfer. In the calculation shown, the least massive cores grow from $0.11\,M_\odot$ to $\sim 0.18\,M_\odot$ by the completion of mass transfer. In contrast, virtually all binaries leaving nondegenerate helium star remnants have too little angular momentum to recover thermal equilibrium before they have lost their hydrogen envelopes; for them, there is no slow nuclear time scale phase of mass transfer, and core growth during mass transfer is negligible. The lowest-mass helium star remnants have nuclear burning lifetimes comparable to their hydrogen-rich binary companions, now grown through mass accretion. Those more massive than $\sim 0.8\,M_\odot$ develop very extended envelopes during shell helium burning, and will undergo a second phase of mass transfer from primary to secondary, not reflected here; such massive white dwarfs are absent in the Nelemans & Tout [35] sample, and so are omitted here.

In Fig. 4, the remnants of the first phase of conservative mass transfer illustrated in Fig. 3 are followed through the second phase of mass transfer, using (6). Because the remnants of the first phase have second-phase donors much more massive their companions, and nearly all have deep convective envelopes, they are unstable to dynamical time scale mass transfer, and undergo common envelope evolution. The systems labeled 'Without Wind Mass Loss' have been calculated assuming that no orbital evolution or mass loss occurs between the end of the first phase of mass transfer and common envelope evolution. It is assumed furthermore that $\alpha_{\rm CE} = 1$, in principle marking the most efficient envelope ejection energetically possible. Binary orbital periods of 0.1, 1.0, and 10 days (assuming equal component masses) are indicated for reference.

Observed close double white dwarfs, as summarized by Nelemans & Tout [35], have a median orbital period of 1.4 d, and mass (spectroscopic primary) $0.39\,M_\odot$. Among double-lines systems, nearly equal white dwarf masses are strongly favored, with the median $q = 1.00$. Clearly, most observed double white dwarfs are too long in orbital period (have too much total energy and angular momentum) to have evolved in the manner assumed here. Furthermore, the computed binary mass ratios are typically more extreme than observed, with the second-formed core typically 1.3-2.5 times as massive as

---

Fig. 3. Products of mass- and angular momentum-conservative mass transfer for a typical initial mass ratio. The radii indicated refer to *Roche lobe radii* at the onset or termination of mass transfer, as appropriate. To avoid common envelope evolution, the donor stars (the region outlined in bold toward the lower right in the diagram) must have radiative envelopes, and so arise between the terminal main sequence and base of the giant branch. Their mass transfer remnants are outlined in bold at the center-left of the diagram, with the remnant accretors at upper right. The regions mapped are truncated in each case at a lower initial donor mass of $1.0\,M_\odot$ and upper initial donor core mass of $0.7\,M_\odot$. Lines of constant initial core mass (with values as in Fig. 2 are indicated for the initial and remnant primaries. Lines of constant remnant primary mass are indicated for the remnant secondaries.

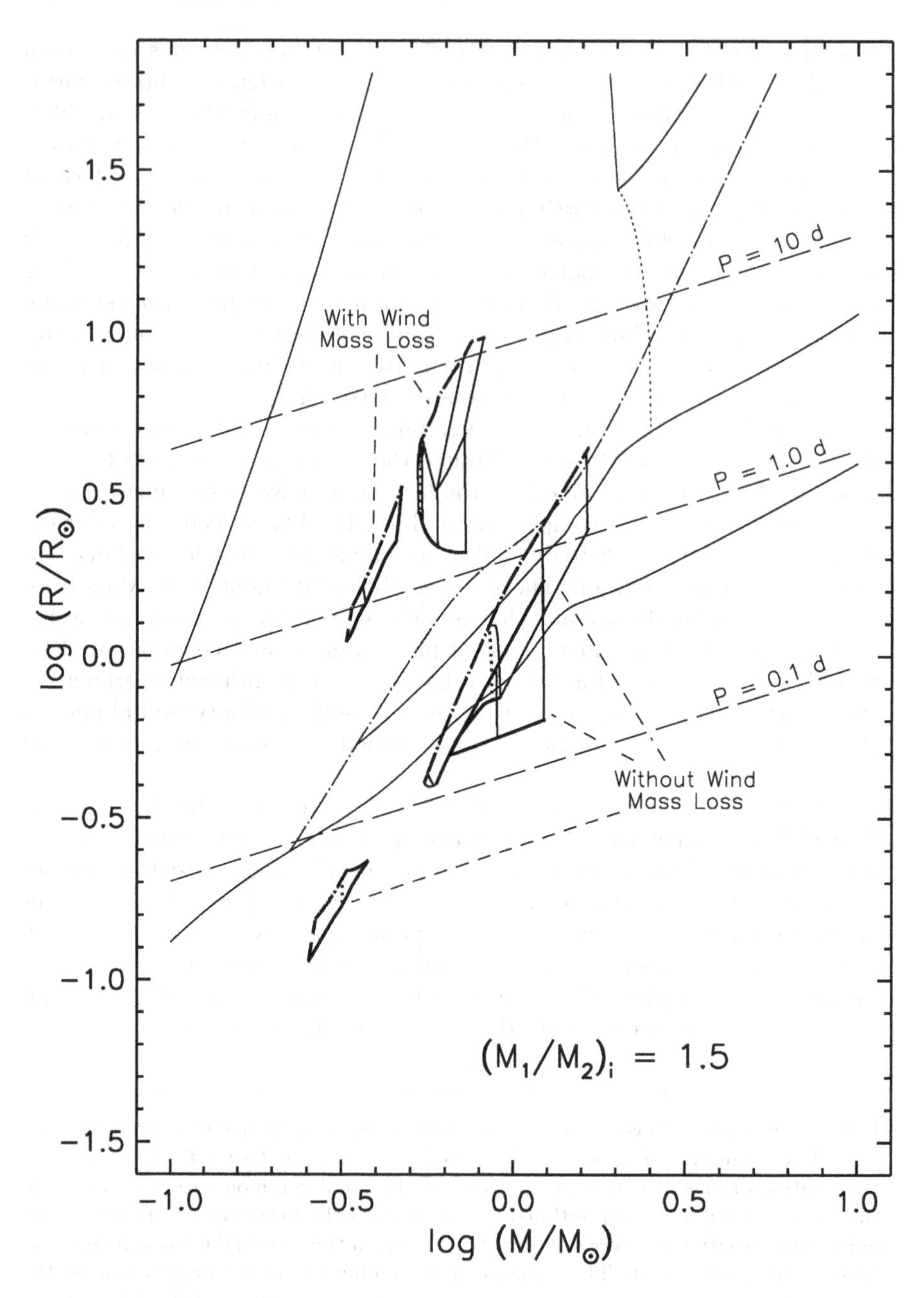

1.5
1.0
0.5
0.0
−0.5
−1.0
−1.5
log (R/R⊙)
−1.0
−0.5
0.0
0.5
1.0
log (M/M⊙)
P = 10 d
P = 1.0 d
P = 0.1 d
With Wind
Mass Loss
Without Wind
Mass Loss
(M1/M2)i = 1.5

the first. The problem is that, while remnant white dwarfs or low-mass helium stars with suitable masses can be produced in the first, conservative mass transfer phase, the remnant companions have envelope masses too large, and too tightly bound, to survive the second (common envelope) phase of interaction at orbital separations and periods as large as observed. Evidently, the progenitors of these double white dwarfs have lost a significant fraction of their initial mass, while gaining in orbital energy, prior to the final common envelope phase. These requirements can be fulfilled by a stellar wind, provided that the process is slow enough that energy losses in the wind can be continuously replenished from nuclear energy sources.

The requisite mass loss and energy gain are possible with stellar wind mass loss during the non-interactive phase between conservative and common envelope evolution, or with stellar winds in nuclear time-scale mass transfer or the terminal (recovery) phase of thermal time-scale mass transfer. Systemic mass loss during or following conservative mass transfer will (in the absence of angular momentum losses) shift the remnant regions to the left and upward in Fig. 3 (subject to the limit posed by envelope exhaustion), while systemic angular momentum losses shift them downwards. More extreme initial mass ratios shift them downwards to the left.

The net effect of wind mass loss is illustrated by the regions labeled 'With Wind Mass Loss' in Fig. 4. For simplicity, it is assumed here that half the remnant mass of the original secondary was lost in a stellar wind prior to the common envelope phase. Mass loss on this scale not only significantly reduces the mass of the second-formed white dwarf relative to the first, but the concomitant orbital expansion produces wider remnant double white dwarfs, bringing this snapshot model into good accord with the general properties of real systems. Losses of this magnitude might be unprecedented among single stars prior to their terminal superwind phase, but they have been a persis-

---

Fig. 4. Remnants of the second, common envelope, phase of mass transfer of the systems shown in Fig. 3. Masses refer to the final remnants of the original secondaries, and radii to their Roche lobe radii. Two groups of remnants are shown. Those at lower right labeled 'Without Wind Mass Loss' follow directly from the distributions of remnant primaries and secondaries shown in Fig. 3. Because the remnant secondaries straddle the helium ignition line in Fig. 3, across which core masses are discontinuous (see Fig. 2), the distribution of their post-common-envelopes remnants is fragmented, some appearing as degenerate helium white dwarf remnants (lower left), some as helium main sequence star remnants (lower center), and the remainder as shell-burning helium star remnants (upper center). These latter two groups overlap in the mass-radius diagram. The remnant distributions labeled 'With Wind Mass Loss' assume that the remnant secondaries of conservative mass transfer lose half their mass in a stellar wind prior to common envelope evolution. They too are fragmented, into degenerate helium white dwarf remnants (lower left) and shell-burning helium stars (upper right). Within each group of remnants, lines of constant remnant primary mass are shown, as in Fig. 3.

tent feature of evolutionary studies of Algol-type binaries [10, and references therein] and, indeed, of earlier studies of close double white dwarf formation [11]. In the present context, their existence appears inescapable, if not understood.

5 Long-Period Post-Common-Envelope Binaries and the Missing Energy Problem

If the properties of short-period binaries with compact components can be reconciled with the outcomes of common envelope evolution as expected from simple energetics arguments, a challenge to this picture still comes from the survival of symbiotic stars and recurrent novae at orbital separations too large to have escaped tidal mass transfer earlier in their evolution. Notwithstanding this author's earlier hypothesis that the outbursting component in the recurrent nova T CrB (and its sister system RS Oph) might be a nondegenerate star undergoing rapid accretion [29,59,61], it is now clear that the hot components in both of these systems must indeed be hot, degenerate dwarfs [1, 6, 18, 48]. Furthermore, the short outburst recurrence times of these two binaries demand that the degenerate dwarfs in each must have masses very close to the Chandrasekhar limit.

The complexion of the problem posed by these systems can be illustrated by a closer examination of T CrB itself. Its orbital period (P = 227.53 d) and spectroscopic mass function ($f(m) = 0.299 M_{\odot}$) are well-established from the orbit of the donor M3 III star [23]. The emission-line orbit for the white dwarf [25] now appears very doubtful [18], but the system shows very strong ellipsoidal variation (e.g., [1]), suggesting that the system is near a grazing eclipse. Following Hric, et al. [18], I adopt $M_{\rm WD} = 1.38\ M_{\odot}$ amd $M_{\rm M3} = 1.2\ M_{\odot}$. The Roche lobe radius of the white dwarf is then $R_{\rm L,WD} = 84\ R_{\odot}$, nearly an order of magnitude larger than can be accommodated from the energetics arguments presented above, even for $\alpha_{\rm CE} = 1$, assuming solar metallicity for the system (see Fig. 5). A similar discrepancy occurs for RS Oph.

It is evident that these long-period binaries are able to tap some energy source not reflected in the energy budget in (6). One possibility, discussed repeatedly in studies of planetary nebula ejection ([30, 42], more recently

---

Fig. 5. Post-common-envelope masses and Roche lobe radii for binaries consisting of a white dwarf or helium star plus a $1.2\,M_{\odot}$ companion, computed with $\alpha_{\rm CE} = 1$. Remnant systems inhabit the regions outlined in bold, and spanned vertically by lines of constant white dwarf/helium star mass of 0.25, 0.35, 0.5, 0.7, 1.0, 1.4, and $2.0\,M_{\odot}$. Other initial sequences, encoded as in Fig. 2, have been mapped through common envelope evolution. The location in this diagram of the white dwarf in the recurrent nova T CrB is also indicated.

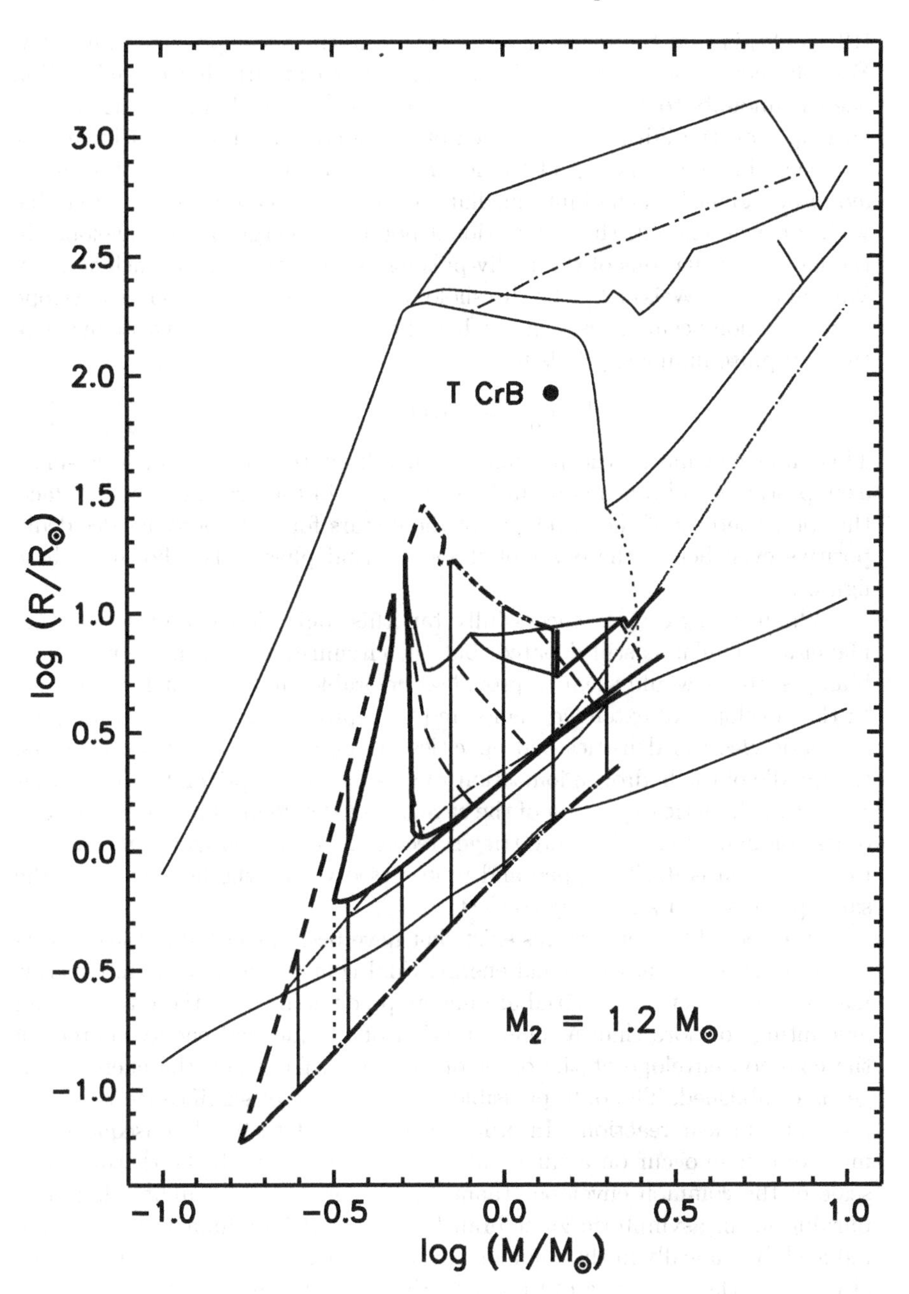
T CrB
M_2 = 1.2 $M_\odot$
log ($R/R_\odot$)
log ($M/M_\odot$)
3.0
2.5
2.0
1.5
1.0
0.5
0.0
−0.5
−1.0
−1.0
−0.5
0.0
0.5
1.0

[12,13,58]) is that the recombination energy of the envelope comes into play. For solar composition material (and complete ionization), that recombination energy amounts to $15.4\,\mathrm{eV\,amu^{-1}}$, or $1.49 \times 10^{13}\,\mathrm{erg\,g^{-1}}$. For tightly-bound envelopes on the initial giant branch of the donor, this term is of little consequence; but near the tip of the low-mass giant branch, and on the upper aymptotic giant branch of intermediate-mass stars, it can become comparable with, or even exceed, the gravitational potential energy of the envelope. In the model calculations of thermally-pulsing asymptotic giant branch stars by Wagenhuber & Weiss [58], the threshold for spontaneous ejection by envelope recombination occurs consistently when the stellar surface gravity at the peak thermal pulse luminosity falls to

$$\log g_{\mathrm{HRI}} = -1.118 \pm 0.042 \; . \tag{14}$$

This threshold marks the presumed upper limit to the radii of lower-mass asymptotic giant branch stars in Figs. 3 et seq. in the present paper. In fact, the total energies of the envelopes of these stars formally becomes decidedly positive even before the onset of the superwind phase, also shown in these figures.

Whether single stars successfully tap this ionization energy in ejecting planetary nebulae is still debated, but the circumstances of mass transfer in binary systems would seem to provide a favorable environment for doing so. In the envelopes of extended giants and asymptotic giant branch stars, photospheric electron densities and opacities are dominated by heavy elements; the middle of the hydrogen ionization zone is buried at optical depths of order $\tau \sim 10^5$. Adiabatic expansion of the envelope of the donor into the Roche lobe of its companion can therefore trigger recombination even as the recombination radiation is itself trapped and reprocessed within the flow, much as the same process occurs in rising convective cells.

Other possible energy terms exist that have been neglected in the energetics arguments above: rotational energy, tidal contributions, coulomb energy, magnetic fields, etc. But Virial arguments preclude most of these terms from amounting to more than a minor fraction of the internal energy content of the common envelope at the onset of mass transfer, when the energy budget is established. The only plausible energy source of significance is the input from nuclear reactions. In order for that input to be of consequence, it must of course occur on a time scale short compared with the thermal time scale of the common envelope. Taam [53] explored the possibility that shell burning in an asymptotic giant branch core could be stimulated by mixing induced dynamically in the common envelope (see also [52,54]). Nothing came of this hypothesis: mixing of fresh material into a burning shell required taking low-density, high-entropy material from the common envelope and mixing it downward many pressure scale heights through a strongly stable entropy gradient to the high-density, low-entropy burning region. In the face of strong buoyancy forces, dynamical penetration is limited to scales of order a pressure scale height.

6 Common Envelope Evolution with Recombination

The notion that recombination energy may be of importance specifically to common envelope evolution is not new. It has been included, at least parametrically in earlier studies, for example, by Han et al. [13], who introduced a second α-parameter, $\alpha_{\rm th}$, characterizing the fraction of the initial thermal energy content of the common envelope available for its ejection. The initial energy kinetic/thermal content of the envelope is constrained by the Virial Theorem, however, and it is not clear that there is a compelling reason for treating it differently from, say, the orbital energy input from the inspiraling cores. We choose below to formulate common envelope evolution in terms of a single efficiency parameter, labeled here $\beta_{\rm CE}$ to avoid confusion with $\alpha_{\rm CE}$ as defined above.

By combining the standard stellar structure equations for hydrostatic equilibrium and mass conservation, we can obtain an expression for the gravitational potential energy, $\Omega_{\rm e}$, of the common envelope:

$$\Omega_{\rm e} \equiv -\int_{M_{\rm c}}^{M_*} \frac{GM}{r}\,{\rm d}M = 3\,PV\Big|_{R_{\rm c}}^{R_*} - 3\int_{V_{\rm c}}^{V_*} P\,{\rm d}V\ , \tag{15}$$

where subscripts c refer to the core-envelope boundary, and $*$ to the stellar surface. This is, of course, the familiar Virial Theorem applied to a stellar interior.

It is convenient to split the pressure in this integral into non-relativistic (particle), $P_{\rm g}$, and relativistic (photon), $P_{\rm r}$, parts. The envelopes of giants undergoing common envelope evolution are sufficiently cool and non-degenerate to make the classical ideal gas approximation an excellent one for the particle gas. One can then write

$$P = P_{\rm g} + P_{\rm r} = \frac{2}{3}u_{\rm g} + \frac{1}{3}u_{\rm r}\ , \tag{16}$$

where $u_{\rm g}$ and $u_{\rm r}$ are kinetic energy densities of particle and radiation gases, respectively. The *total* internal energy density of the gas is

$$u = u_{\rm g} + u_{\rm r} + u_{\rm int}\ , \tag{17}$$

where the term $u_{\rm int}$ now appearing represents non-kinetic contributions to the total energy density of the gas, principally the dissociation and ionization energies plus internal excitation energies of bound atoms and molecules. The overwhelmingly dominant terms in $u_{\rm int}$ are the ionization energies: $u_{\rm int} \approx \rho\chi_{\rm eff}$.

Integrating over the stellar envelope, we obtain for the total energy $E_{\rm e}$ of the envelope:

$$\begin{aligned} E_{\rm e} &= \Omega_{\rm e} + U_{\rm e} \\ &= \left(3\,PU|_{R_{\rm c}}^{R_*} - 2U_{\rm g} - U_{\rm r}\right) + (U_{\rm g} + U_{\rm r} + U_{\rm int}) \\ &= -4\pi R_{\rm c}^3 P_{\rm c} - U_{\rm g} + U_{\rm int}\ , \end{aligned} \tag{18}$$

where we explicitly take $P_* \to 0$. In fact, experience shows that, for red-giant like structures, R_c is so small that the first right-hand term in the last equality can generally be neglected. In that case, we get the familiar Virial result, but with the addition of a term involving the ionization/excitation/dissociation energy available in the gas, $U_{\rm int} \approx M_e \chi_{\rm eff}$, which becomes important for diffuse, loosely-bound envelopes.

In the context of common envelope evolution, it is of course the dissipated orbital energy, $E_{\rm orb}^{(i)} - E_{\rm orb}^{(f)}$, that must unbind the envelope. However, the inclusion of $U_{\rm int}$ in E_e now opens the possibility that the common envelope began with *positive* total energy; that is, in the usual $\alpha_{\rm CE}$-prescription, it is possible for λ^{-1} to be zero or even negative, which has the undesirable consequence that $\alpha_{\rm CE}$ need not lie in the interval $0 \leq \alpha_{\rm CE} \leq 1$ for all physically-possible outcomes. However, the gravitational potential energy of the envelope, Ω_e, is negative-definite, and by comparing it with all available energy sources (orbital energy released plus internal energy of the envelope), we can define an ejection efficiency $\beta_{\rm CE}$ that has the desired property, $0 \leq \beta_{\rm CE} \leq 1$:

$$\beta_{\rm CE} \equiv \frac{\Omega_e}{(E_{\rm orb}^{(f)} - E_{\rm orb}^{(i)}) - U_e} = \frac{4\pi R_c^3 P_c + 2U_g + U_r}{E_{\rm orb}^{(i)} - E_{\rm orb}^{(f)} + U_g + U_r + U_{\rm int}} \ . \tag{19}$$

By analogy to the form factor λ in the conventional $\alpha_{\rm CE}$ formalism above, we can define separate form factors λ_Ω for the gravitational potential energy and λ_P for the gas plus radiation contributions to the (kinetic) internal energy of the envelope:

$$\Omega_e = \frac{GMM_e}{\lambda_\Omega R} \quad \text{and} \quad U_g + U_r = \frac{GMM_e}{\lambda_P R} \ , \tag{20}$$

In contrast, the recombination energy available can be written simply in terms of an average ionization energy per unit mass,

$$U_{\rm int} = M_e \chi_{\rm eff} \ . \tag{21}$$

The ratio of final to initial orbital separation then becomes

$$\frac{A_f}{A_i} = \frac{M_{1c}}{M_1}\left[1 + 2\left(\frac{1}{\beta_{\rm CE}\lambda_\Omega r_{1,L}} - \frac{1}{\lambda_P r_{1,L}} - \frac{\chi_{\rm eff} A_i}{GM_1}\right)\left(\frac{M_1 - M_{1c}}{M_2}\right)\right]^{-1} \ . \tag{22}$$

In the limit that radiation pressure P_r, ionization energy ($U_{\rm int}$), and the boundary term ($4\pi R_c^3 P_c$) are all negligible, then $2\lambda_\Omega \to \lambda_P \to \lambda$ and $\beta_{\rm CE} \to 2\alpha_{\rm CE}/(1 + \alpha_{\rm CE})$.

Fig. 6. Post-common-envelope masses and Roche lobe radii as in Fig. 5, but with recombination energy included, computed from (22) with $\beta_{\rm CE} = 1$, with the approximation $2\lambda_\Omega = \lambda_P = \lambda$ from (3). At small separations, the differences are inconsequential, but substantially larger final separations are allowed when $A_f \gtrsim 10\,R_\odot$ ($R_L \gtrsim 3\,R_\odot$).

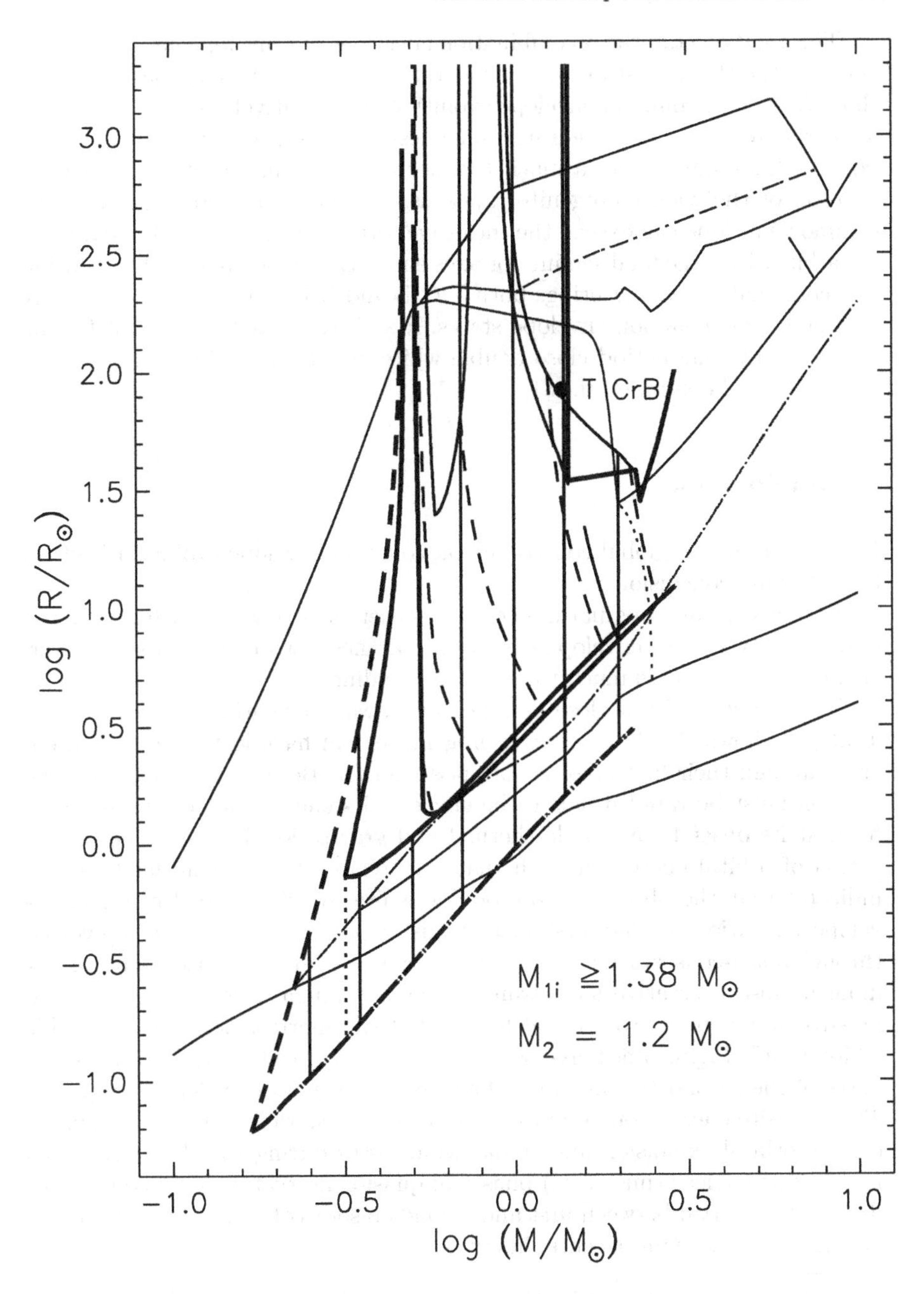
T CrB
$M_{1i} \geqq 1.38\ M_\odot$
$M_2 = 1.2\ M_\odot$
log (R/R⊙)
log (M/M⊙)
3.0
2.5
2.0
1.5
1.0
0.5
0.0
−0.5
−1.0
−1.0
−0.5
0.0
0.5
1.0

The ability to tap the recombination energy of the envelope has a profound effect on the the final states of the longest-period intermediate-mass binaries, those that enter common envelope evolution with relatively massive, degenerate carbon-oxygen (or oxygen-neon-magnesium) cores. As is evident in Fig. 6, possible final states span a much broader range of final orbital separations. Indeed, for the widest progenitor systems, the (positive) total energy of the common envelope can exceed the (negative) orbital energy of the binary, making arbitrarily large final semimajor axes energetically possible.[3] The inclusion of recombination energy brings both T CrB and RS Oph within energetically accessible post-common-envelope states. It suffices as well to account for the exceptionally long-period close double white dwarf binary PG 1115+166, as suggested by Maxted, et al. [31].

7 Conclusions

Re-examination of global constraints on common envelope evolution leads to the following conclusions:

Both energy and angular momentum conservation pose strict limits on the outcome of common envelope evolution. Of these two constraints, however, energy conservation is much the more demanding.

The recent study of close double white dwarf formation by Nelemans & Tout [35] shows clearly that their progenitors can have lost little orbital energy through their first episodes of mass transfer. Because common envelope ejection must be rapid if it is to be efficient, its energy budget is essentially fixed at its onset by available thermal and gravitational terms. The preservation of orbital energy through that first phase of mass transfer therefore indicates that the observed close double white dwarfs escaped common envelope formation in that first mass transfer phase. They evidently evolved through quasi-conservative mass transfer. However, strictly mass- and angular momentum-conservative mass transfer leaves remnant accretors that are too massive and compact to account for any but the shortest-period close double white dwarfs. Significant mass loss and the input of orbital energy prior to the onset of the second (common envelope) phase of mass transfer are required. The requisite energy source must be of nuclear evolution, which is capable of driving orbital expansion and stellar wind losses during the slower (thermal recovery or nuclear time scale) phases of quasi-conservative mass transfer, or during the interval between first and second episodes of mass transfer. Details of this process remain obscure, however.

[3] The final orbit remains constrained by the finite initial orbital angular momentum of the binary. Final semimajor axes much in excess of the initial semimajor axis may be energetically allowed, but the finite angular momentum available means that they cannot be circular – see (9) – an effect which has been neglected in Fig. 6.

Long-period cataclysmic variables such as T CrB and RS Oph pose a more extreme test of common envelope energetics. With their massive white dwarfs, the evident remnants of much more massive initial primaries, they are nevertheless too low in total systemic mass to be plausible products of quasi-conservative mass transfer, but too short in orbital period to have escaped tidal mass transfer altogether. They must be products of common envelope evolution, but to have survived at their large separations, they demand the existence of a latent energy reservoir in addition to orbital energy to assist in envelope ejection. It appears that these binaries efficiently tap ionization/recombination energy in ejecting their common envelopes. That reservoir is demonstrably adequate to account for the survival of these binaries. Its inclusion requires only a simple revision to the parameterization of common envelope ejection efficiency.

8 Acknowledgments

This work owes its existence to both the encouragement and the patience of Gene Milone, to whom I am most grateful. Thanks go as well to Ron Taam and to Gijs Nelemans for useful discussions of possible loopholes in common envelope theory, and to Jarrod Hurley for providing the source code described in Hurley, et al. (2000). This work was supported in part by grant AST 0406726 to the University of Illinois, Urbana-Champaign, from the US National Science Foundation.

References

1. K. Belczyński, J. Mikołajewska: MNRAS **296**, 77 (1998)
2. H.E. Bond: Binarity of Central Stars of Planetary Nebulae. In *Asymmetrical Planetary Nebulae II: From Origins to Microstructures*, ed by J.H. Kastner, N. Soker, S. Rappaport (ASP Conf. Ser. Vol. 199, San Francisco 2000), pp. 115-123
3. H.E. Bond, W. Liller, E.J. Mannery: ApJ **223**, 252 (1978)
4. M. de Kool: A&A **261**, 188 (1992)
5. R. Di Stefano, S. Rappaport: ApJ **437**, 733 (1994)
6. D. Dobrzycka, S.J. Kenyon, D. Proga, J. Mikołajewska, R.A. Wade: AJ **111**, 2090 (1996)
7. P.P. Eggleton: ApJ **268**, 368 (1983)
8. P. Eggleton: *Evolutionary Processes in Binary and Multiple Stars* (CUP, Cambridge)
9. F.C. Fekel, G.W. Henry, M.R. Busby, J.J. Eitter: AJ **106**, 2170 (1993)
10. G. Giuricin, F. Mardirossian: ApJS **46**, 1 (1981)
11. Z. Han: MNRAS **296**, 1019 (1998)
12. Z. Han, P. Podsiadlowski, P.P. Eggleton: MNRAS **270**, 12 (1994)
13. Z. Han, P. Podsiadlowski, P.P. Eggleton: MNRAS **272**, 800 (1995)

14. R. Härm, M. Schwarzschild: ApJS **1**, 319 (1955)
15. M.S. Hjellming, R.E. Taam: ApJ **370**, 709 (1991)
16. M.S. Hjellming, R.F. Webbink: ApJ **318**, 794 (1987)
17. S.B. Howell, L.A. Nelson, S. Rappaport: ApJ **550**, 897 (2001)
18. L. Hric, K. Petrik, Z. Urban, P. Niarchos, G.C. Anupama: A&A **339**, 449 (1998)
19. J.R. Hurley, O.R. Pols, C.A. Tout: MNRAS **315**, 543 (2000)
20. I. Iben, Jr., M. Livio: PASP **105**, 1373 (1993)
21. I. Iben, Jr., A.V. Tutukov: ApJ **284**, 719 (1984)
22. C.S. Jeffery, T. Simon: MNRAS **286**, 487 (1997)
23. S.J. Kenyon, M.R. Garcia: AJ **91**, 125 (1986)
24. U. Kolb: A&A **271**, 149 (1993)
25. R.P. Kraft: ApJ **127**, 625 (1958)
26. R.P. Kraft: ApJ **135**, 408 (1962)
27. R.P. Kraft: ApJ **139**, 457 (1964)
28. Y. Lebreton, J. Fernandes, T. Lejeune: A&A **374**, 540 (2001)
29. M. Livio, J.W. Truran, R.F. Webbink: ApJ **308**, 736 (1996)
30. L.B. Lucy: AJ **72**, 813 (1967)
31. P.F.L. Maxted, M.R. Burleigh, T.R. Marsh, N.P. Bannister: MNRAS **334**, 833 (2002)
32. M. Morris: ApJ **249**, 572 (1981)
33. M. Morris: PASP **99**, 1115 (1987)
34. D.C. Morton: ApJ **132**, 146 (1960)
35. G. Nelemans, C.A. Tout: MNRAS **356**, 753 (2005)
36. G. Nelemans, F. Verbunt, L.R. Yungelson, S.F. Portegies Zwart: A&A **360**, 1011 (2000)
37. G. Nelemans, L.R. Yungelson, S.F. Portegies Zwart, F. Verbunt: A&A **365**, 491 (2001)
38. C.A. Nelson, P.P. Eggleton: ApJ **552**, 664 (2001)
39. M.S. O'Brien, H.E. Bond, E.M. Sion: ApJ **563**, 971 (2001)
40. D.E. Osterbrock: ApJ **118**, 529 (1953)
41. B. Paczynski: B. Common Envelope Binaries. In *Structure and Evolution of Close Binary Systems*, ed by P.P. Eggleton, S. Mitton, J.A.J. Whelan (Reidel, Dordrecht 1976), pp 75-80
42. B. Paczyński, J. Ziółkowski: AcA **18**, 255 (1968)
43. L. Pastetter, H. Ritter: A&A **214**, 186 (1989)
44. M. Politano: ApJ **465**, 338 (1996)
45. F.A. Rasio, M. Livio: ApJ **471**, 366 (1996)
46. S. Refsdal, M.L. Roth, M.L., A. Weigert: A&A **36**, 113 (1974)
47. E.L. Sandquist, R.E. Taam, A. Burkert: ApJ **533**, 984 (2000)
48. P.L. Selvelli, A. Cassatella, R. Gilmozzi: ApJ **393**, 289 (1992)
49. T. Simon, F.C. Fekel, D.M. Gibson, Jr.: ApJ **295**, 153 (1985)
50. N. Soker: ApJ **496**, 833 (1998)
51. A.V. Sweigart, L. Greggio, A. Renzini: ApJ **364**, 527 (1990)
52. R.E. Taam: The Common Envelope Phase of Binary Evolution. In *Interacting Binary Stars*, ed by A.W. Shafter (ASP Conf. Ser., Vol. 56, San Francisco 1994), pp 208-217
53. R.E. Taam: private communication (2007)
54. R.E. Taam, P. Bodenheimer: ApJ **337**, 849 (1989)
55. R.E. Taam, E.L. Sandquist: ARA&A **38**, 113 (2000)

56. M. van den Berg, F. Verbunt, R.D. Mathieu: A&A **347**, 866 (1999)
57. G. Vauclair: A&A **17**, 437 (1972)
58. J. Wagenhuber, A. Weiss: A&A **290**, 807 (1994)
59. R.F. Webbink: Nature **262**, 271 (1976)
60. R.F. Webbink: The Evolutionary Significance of Recurrent Novae. In *Changing Trends in Variable Star Research, IAU Colloq. No. 46*, ed by F.M. Bateson, J. Smak, I.H. Urch (U. Waikato, Hamilton, NZ 1979), pp 102-118
61. R.F. Webbink, M. Livio, J.W. Truran, M. Orio: ApJ **314**, 653 (1987)
62. R.F. Webbink: Late Stages of Close Binary SystemsClues to Common Envelope Evolution. In *Critical Observations Versus Physical Models for Close Binary Systems*, ed by K.-C. Leung (Gordon & Breach, New York 1988), pp 403-446
63. B. Willems, U. Kolb: A&A **419**, 1057 (2004)
64. L.A. Willson: ARA&A **38**, 573 (2000)
65. H.W. Yorke, P. Bodenheimer, R.E. Taam: ApJ **451**, 308 (1995)

Index

absorption dips, 111
acceptance probability, 199
accreted–matter nuclear–burning timescale, 98, 245, 247
accreting black holes, 4, 10, 13, 15, 18, 20
accreting neutron stars, xii, 11, 87, 88, 90
accreting white dwarfs, 142
accretion
  flow(s), xii, 16–18, 87*ff*
    models, 14, 126
  from secondary star wind, 126, 144
  low rates, 137, 139, 140
  of common envelope, 236
  rates in *LARPS*, 144
  spin equilibrium, 95, 97
  spin–up, 91, 95–97
  states, 139, 143, 151
  torques, 94, 96
accretion column(s), 111, 112, 116*ff*, 123, 137, 138, 144, 148, 150
  model, 123
accretion disk(s), xii, xiii, 6, 7, 9, 10, 16, 20, 87*ff*, 102, 112, 137*ff*, 147*ff*, 161*ff*, 205
  and Reynolds number, 171, 172
  annulus line spectrum synthesis, 163
  clumpiness, 101–103
  continuum, 7–9, 15*ff*, 138*ff*, 161, 166*ff*
  database, 162, 164
  effective temperature, 162
  emission, 8, 16, 18, 19, 120, 205
  external irradiation, 16, 164, 165, 167, 174
  geometry, xii, 113
  heating, 162, 174
  in cataclysmic variables, 137, 162*ff*
    as dominant *CV* source, 150
  luminosity, 18, 101, 105
  Lyα emission, 140, 141
  models, 9, 18, 143, 161, 162, 165, 169
    annulus model calculation, 163
    "standard model", 162, 169, 170
  optical depth, 118, 164–167, 171, 172
  polarization, 16, 210
  properties table, 165
  radius, 16, 114, 116, 119, 165
  shadowing effects, 120, 121, 167
  synthetic spectrum, 162*ff*
  thermal structure, 9, 19, 94, 161, 162, 170, 172
  viscosity parameter, 163, 166, 172
accretion–powered systems, 87*ff*, 111*ff*
  *MSP*s (millisecond pulsars), 88*ff*
  *QPO*s (quasi-periodic oscillations), 7*ff*, 93, 94, 99, 103
  brightness oscillations, 90
  X-ray *MSP*s, 91
Active Galactic Nuclei (*AGN*), 16
Adaptive Random Search algorithm, 198
adjustable 3–body parameters, 194

Algol(s), 154
cool, 154
mass loss, 231, 232, 244, 246, 247
model, 208, 219
paradox, xi, 154, 208, 231, 233
systems, 153, 154, 208, 219, 234, 244
as progenitors of non–magnetic *CV*s, 154, 158
temperature, 154
All Sky Monitor on *RXTE*, 184
angular momentum, x, xi, 4, 87, 128, 129, 235, 245, 247
accreted, 87
black hole, 4, 65
constraint, 87
coupling radius, 94, 95
gravitational lenses, 64, 65
neutron star, 93, 97
losses, 234–237
from magnetic braking, 139
orbital, 55
total, 56
transfer, 93
ATLAS, 209
ATM2000, 195

Balmer jumps
in dwarf novae, 161
in *CV*s, 161
Binary Maker, 193, 196, 197
BINSYN, 163, 165, 167–170, 174, 209
bipolar structure, 235
black hole(s), xi, 3, 7, 87
accreting, 4, 9, 10, 13, 15, 18
angular momentum, 4, 65
binaries, xi, 3, 4, 6, 7, 10, 11, 13–17, 26, 33, 65, 74, 75, 80, 100, 154, 191
immersed in magnetic fields, 83
in globular clusters, 83
in the Milky Way, 6
list of, 5
scale drawings, 6
vs. neutron star binaries, 14
candidates, 3, 6, 11
event horizon, 4
Kerr black holes, 4, 15, 16
spin, 4, 14, 15, 18, 20, 21
minimum mass, 5
Penrose process, 15
scale length, 65
Schwarzschild black holes, 4, 16, 17
spin, 14*ff*, 64
stellar sized, 3, 5, 13, 20, 21
X–ray spectrum, 6, 7
bloating factor of secondary stars, 152
blue straggler as merger product, 237
boundary layer emission, 14, 139, 141, 144, 174
broken–contact states, 215–217
brown dwarfs, 207
as secondary stars, 137, 144, 157
Burgess Shale, 206
burning front thickness, 97
Burst and Transient Source Experiment (*BATSE*), 121, 126, 184
burst tail oscillations, 98

cataclysmic variable(s) (*CV's*), xii, xiii, 135*ff*, 147*ff*, 161*ff*, 231, 233, 236, 242
accretion disks, xiii, 162*ff*
circumbinary disk, 157, 158
components, 137, 150, 234
disk models, 139, 162, 168, 172
emission lines, 170, 171
calculated from irradiated secondaries, 154
cyclotron emission, 137, 144
evolution, 139, 147, 149, 154
dynamically stable, 162
IR spectroscopy, 151–154
K band diagnostic features, 151
long-period, 255
low transfer rates, 139
model visibility key, 168
orbital periods, 147, 148
minimum periods, 140, 148
secondary star, 137, 147
age at contact onset, 149
core mass-radius relation, 149
internal energy generation, 148
secondary Roche Lobe, 152
variability, 137, 147
magnetic systems, 137, 144, 147, 148, 153, 156
field induced currents, 156

spectral energy distributions, 138, 139, 142, 144, 151
synthetic spectra, 167, 168
TOAD class, 157
X-ray emission, 137
Chandra X–ray Observatory, 14
Chandrasekhar limit, 25, 248
channeled polar flow, 93
circumbinary disk(s), 157, 158
in *CV*s, 157
in interacting binaries, 158
in short–period magnetic *CV*s, 157
clusters (stellar)
47 Tuc, 193, 211
Hyades, 234, 237
M67, 240
NGC 6440, 89
CNO processing, 141, 149, 153, 154
onset of burning, 149, 150
CO emission in *CV* systems, 157
coalescing compact systems, 23
black holes, 15, 21
neutron stars, 24, 27, 38, 44, 45, 48
coherent X-ray oscillations, 10, 98
colliding black holes, 33
common envelope(s),
ejection, 234*ff*
efficiencies, 234, 236, 238, 248
energy content, 236, 250, 251
energy density, 251
gravitational potential energy, 250–252
evolution, 231*ff*
alternative, 242
constraints, 237, 254
energetics, 235–237
entropy differences, 236, 250
final orbital energy, 234*ff*
flaw, 242
recombination energy, 233, 250*ff*
second phase, 245
time scales, 234
initial conditions
angular momentum, 235, 236, 254
envelope energy, 235
orbital eccentricity, 237
orbital energy, 235
particle *vs.* photon pressures, 251, 252
phases, 238, 243, 254
duration, 4
Compton Gamma Ray Observatory, 184
condensed polytrope models, 235
conservative mass transfer, 242, 243, 245, 247
contact vs. over–contact designations, x
continuum–fitting method, 18
cool Algols, 154
Coriolis effects on neutron star, 97, 98
coronae in *CV*s, 139, 144, 174
critical surface gravity of asymptotic giants, 250
cyclotron emission in CV's, 137, 144

detached binary star system(s), x
evolution phase, 233
differential corrections, 181, 182, 192, 193, 195, 197
direct distance estimation method, 181, 183, 194
direction set method(s), 191, 192
disk instability model (*DIM*), 166, 169–173
parameters, 114, 163
precession, xii, 114, 115, 117, 118
Doppler modeling, xi, 168, 207, 208, 210
double pulsar system, xii, 24, 53*ff*, 63, 65, 74, 78–81
mass-mass diagram, 60
double white dwarf systems, 233, 242*ff*
first transfer phase, 233, 242, 243, 245, 254
median orbital period, 243
system formation, 254
dwarf novae, 142, 148, 161, 162, 170
outbursts, 154, 162, 170, 173
dynamical time scale mass exchange, 233
dynamically stable *CV*s, 161*ff*

eccentric orbit(s)
circularization of, 34
in binary pulsar systems, 58
in X-ray binary GX301-2, 126, 129
modeling of, 179, 182, 185, 194, 225
of PSR J0737-3039, 54, 63, 79
usefulness of, 54
eclipse mapping, 162, 205

Eddington approximation, 164
Eddington critical rate, 93
Eddington flux units, 167
Eddington limit, 10
ellipsoidal variation
  in GP Vel, 183, 184
  in T CrB, 248
envelope ejection, 250
equatorial rings, 235
equilibrium spin
  frequency, 94
  period, 95
ESPaDOnS, 210
event horizon, 4, 13, 14
EXOSAT, 16, 112, 126
expanding shell model, 207
extra–solar planets, 191
extreme ultraviolet, 120
Extreme Ultraviolet Explorer (EUVE), 112

fan beam, 115, 123, 125, 126
Fe K fluorescence line, 7, 8, 15, 16, 20, 21
fluid ocean on neutron stars, 97
flux calibrations, 181

gamma–ray bursts, 3, 47
gas stream, 127–129, 148
Gemini Telescopes, 157
general relativity (*GR*), 4–7
  predictions, 53, 54, 105
  polarization, effects on, 16
  ray–tracing methods, 17
  resonances, 11, 17
  Shapiro delay, 59, 75–77
  strong–field limit, 6, 87
    probes of, 10, 87
  test of, 54, 58–60
genetic algorithms (GA), 191, 205, 206
globular cluster(s), x, 83, 89, 97, 193, 211
Gould, Stephen J., 206
graphical user interface, 197
gravitational lensing, 63
  attenuation, 80, 81
  by black holes, 64, 74
  geometry, 66
  intensity enhancement, 77
  light bending, 123
  microlensing, 64
  of pulsars, xii, 64, 65, 70, 73, 82
  weak, 64
gravitational radiation, xii, 33, 96, 139, 147
  excitation of r-waves, 96
  gravitational waves, 13, 21, 23*ff*
  quadrupole, 59
  search parameters, 37–39, 42
  spin-breaking effects, 87, 96, 139
  torques, 96

Hall effect in *CV*s, 174
helium enriched objects
  degenerate objects, 147, 154, 243, 247
  sub–dwarfs, 242
  white dwarfs, 242, 243
  white dwarf remnants, 247
helium ignition, 238, 240
High Mass X-ray Binaries (*HMXBs*), 126, 179, 186, 188
high state (high mass transfer rate), 10, 112, 120, 143, 148, 162, 170, 171, 173
high–frequency timing noise, 14
hot spot
  on secondary stars, 224, 225
  retention in white dwarfs, 137–139, 144, 168, 224
  visibility, 139
hybrid white dwarfs, 242
hydrogen ionization zone, 250
hypernovae, 3

ignition on neutron star surface, 97
impact parameter, 64, 65
incident flux at secondary *CV* components, 155
inner accretion disks, xii, 7, 9, 17
inner disk edge, xii, 114–119
innermost stable orbit(s), 104
  circular orbit(s), 4, 18, 87, 95, 105
interacting binary stars, ix, 147, 154, 233
intermediate polars (*IP*s), 138, 148
inverse compton emission, 10
IRWG infrared passbands, 209, 210

Jeans–mode mass loss, 243

Keck Telescope, 157
Kerr solution
  black hole, 4, 16, 83
  metric tensor, 13, 15
Kerr spacetime, 64
KERRBB, 20, 21

ℓ_1 norm, 192
ℓ_2 norm, 192
Lagrangian point(s), 196
  L_1, 147, 148
Lens-Thirring metric, 64
Levenberg-Marquardt (damped least-squares) program, 192, 200, 201, 205
  convergence engine, 192
  method, 192
light curve(s), x, xii, 7, 79, 180, 181, 183, 184, 191, 206, 217, 222
  35-day duration, 113
  analysis, x, xi, 184, 191, 194, 207, 209, 222
  broken contact, 216
  CV, xiii, 137–139, 144, 162, 167
  disk precession, xii
  distorted, 217, 218
  double pulsar, 74, 79–83
  ellipsoidal, 183
  infrared, 209
  LMXB, 112–119, 122, 128, 129
  observables, 192, 193, 208, 209
  parameters, 5, 180*ff*, 191*ff*, 218*ff*
  solutions, 205, 217–219, 223, 226, 240
  synthetic, 167, 168, 196*ff*, 222
  W UMa type, 216
  X-ray, 7, 183, 184, 188
Light Synthesis Program, 192
light-time effect, 182
LIGO, 13, 30*ff*
limb–darkening coefficients, 195
  applications, 167
limit cycle instability model, 162
line–broadening effects, 163
line profile(s), 15, 16, 21, 155, 171, 198
line profile asymmetries, 16
LISA, 13, 30, 50
long–period
  cataclysmic variables (*CV*s), 255
  double white dwarf binaries, 254
  interacting binaries, 233
  post–common envelope binaries, 248
Low Accretion Rate Polars (*LARP*s), 144
low mass
  giants, 234, 240, 250
  secondaries, 148, 154, 156, 158
  transfer rates, 138
low mass systems
  cataclysmic variables, 148, 156, 158, 234
  helium star remnants, 245, 247
  X-ray binaries (*LMXB*s), xii, 87*ff*, 111, 154
low state(s) (low mass transfer rate), 139, 152
  polar depiction, 156, 157
  in nova–like variables, 173
Lyα line in cataclysmic variables, 140

MACHOS, 69
magnetic accretion annulus, 147
magnetic and non–magnetic *CV* secondary star differences, 153
magnetic braking, 147
  by dipole field, 95
  in cataclysmic variables, 139
  of accretion torque, 87
magnetic field(s), 91, 94, 103, 144, 150
  around black holes, 83
  decrease timescales, 94
  disk, 123, 173
  in cataclysmic variables, 137, 144*ff*, 156, 173, 174
  of neutron stars, 54, 87*ff*
  of pulsars, 54, 80, 95–97
  of secondary star, 174
magnetically funnelled accretion streams, 148
magnetically-induced stellar activity in *CV* secondaries, 158
magneto-rotational instability (*MRI*), 162, 173, 174
magnification factor, 63*ff*
mass constraints, 130, 131
mass function, 4, 60, 68, 75, 126, 248
mass loss
  in Algol systems, 232, 244
  in CVs, 149

in double white dwarfs, 233, 254
mechanisms, xi
via stellar winds, 128, 137, 247
mass ratio reversal, 241
mass transfer
conservative, 217, 245, 247
core helium burning, 236
dynamical time-scale, 245
in Algols, 231
in CVs, 137*ff*, 172–174
in double white dwarfs, 243
in dwarf novae, 162
in neutron star-white dwarf binaries, 94
in polars, 137, 144
in W UMa systems, 225
nuclear timescale, 143, 247
onset, 139, 235*ff*
quasi-conservative, 242, 243, 254, 255
rates, large, 142, 144, 148, 162, 170, 171
rates in nova–like systems, 140, 162, 173
second-phase, 245, 254
slow episode, 233
stream direction, 137
terminal phase, 243*ff*
thermal timescale, 243
tidal, 248, 255
mass–radius diagram, 238, 241, 245
mass–radius relation
calculated, 125
for *CV* secondaries, 148, 149
for non-rotating stars, 104
massive white dwarfs, 233, 234, 245, 255
matched filtering, 36, 38, 40
maximum entropy techniques, 209
maximum likelihood estimator, 192
mean density at RLO constrained by orbital period, 238
Metropolis criterion, 199
microlensing, 63*ff*, 77, 83
Milky Way black hole binaries, 6
millisecond pulsar(s) (*MSP*s), 88*ff*
magnetic fields, 88, 91*ff*
production, 54, 92*ff*
progenitor spin rates, 87, 91, 92, 97
spin frequency, 88*ff*, 184
spin–relaxation timescale, 94
missions
Compton Gamma Ray Observatory, 184
EUVE (Extreme Ultraviolet Explorer), 122
EXOSAT, 16, 112
FUSE, 137, 139, 140, 143–145, 164, 170, 173
GALEX, 137, 144
GINGA, 123
HST, 137, 139–142, 144, 164, 170
IUE (International Ultraviolet Explorer), 240
ROTSE-1 sky patrol, 216
*R*XTE (Rossi X-ray Timing Explorer), xii, 7, 10, 112*ff*, 123, 127–129, 184, 185
TENMA, 127
modeling of disk systems, 143
molecular features in *CV*s, 151, 157, 158, 251
Monte Carlo
search methods, 33, 46, 206
simulations, 43, 59

natural selection, 205, 206
near–contact binaries, 217
Nelder & Mead simplex algorithm (*NMS*), 198
neural networks, 193, 206
neutron star(s), xi, xii, 5, 14, 87*ff*, 96–100, 103, 105
accretion luminosity, 87, 101, 104, 141
binaries, xi, xiii, 14, 23*ff*, 46, 54, 65, 120
burning front, 97
colliding, 33, 35, 46, 48
double, 54
emission, 14, 114*ff*, 123
equation of state, 38, 47, 60, 104
flares, 126
gamma ray burst association, 47
gravitational lensing, 63*ff*, 123
gravitational waves, 23*ff*
in common envelope systems, 54
light curve models, 111*ff*, 183, 184, 209
magnetic fields, 54, 88, 91*ff*
mass, 5, 53, 59, 83

mass-radius constraint, 47, 103, 104, 125, 126, 128
pulsars, 54
quadrupole configuration, 34
radius, 53, 64, 103, 104, 116, 125
recycled, 54
side–view, 93
size effects, 47
spin rates, 54, 87, 88, 90, 96
spin–up, 94
surface
fluid ocean, 97
phenomena, 14
r-waves, 98
variable-frequency QPOs, 11, 13
vertical oscillations, 101
neutron-star–white-dwarf binary systems, 94
nova–like (*NL*) systems, 142, 143, 161, 162, 173
nova(e), ix, xi
dwarf, 139, 142, 148, 150, 154, 161, 162, 170, 173
recurrent, 233, 248
X-ray, 3, 4,7, 10, 14
nuclear evolution energy source, 247, 250, 252, 254
nuclear time–scale, 243, 245, 247, 254
nuclear–powered X-ray sources, 88, 89, 90
emissions, 93
oscillations, 88*ff*, 97*ff*
pulsars, 89, 90, 94

occultation(s), 115*ff*, 211
oldest white dwarfs in *CV*s, 145
optically thick accretion disks, 161, 164, 166, 171–174
optically thin accretion disks, 151, 164, 170–173, 208
orbital
elements/parameters, 58, 64, 65, 70, 79, 126, 148, 205
inclination, 5, 60, 65, 83, 130, 168, 169, 242
orbital angular momentum, 55, 234, 236, 243, 254
over–contact (or 'contact') binaries, vi, x, 180, 205, 207, 209, 215
as merging stars, x
designation, x
in hierarchical systems, x

P Cygni features (absence), 173
pencil beam, 115, 123, 125, 126
Penrose Process, 15
period evolution time–scale, 96, 140, 255
period–radius relation for *CV* secondaries, 152
phased–locked oscillations, 93
`PHOEBE`, 192, 193, 197
`PHOENIX`, 208, 209
photometry calibration, 181
photon index, 7, 8
Pikaia gracilens, 206
planetary nebula(e), 234, 235, 248, 250
bipolar structure, 235
eclipsing systems, 234
ejection, 234, 248, 250
planetary systems
eclipses, 191
transits, 210, 211
polar jets, 235
polarimetry, 15–18, 20
Polar(s), 137, 144, 145, 148, 150, 151, 156
artist's view, 156
emitting components, 150, 205
Intermediate, 138, 148
low-accretion rate (LARP), 144
models, 114*ff*, 144, 145
prototype (AM Her), 144
short–period systems, 157
synthetic spectrum, 169
post–Algol systems
binary examples, 240, 242
evolution, 153
post-common-envelope evolution, 238
post-Keplerian parameters, 58, 59, 60
power density spectrum, 7
pre-cataclysmic variables, 238, 243
precession
neutron star model, 123
of disk, 114*ff*
of jets; model, 207, 208
of pulsar spin axis, 56

prograde orbits, 4
pulsar(s) (*see also* neutron star(s))
  in globular clusters, 89, 97
  pulse profiles, 64*ff*, 112, 120, 123, 125, 184
  spin frequencies, 55, 57, 58, 87*ff*
pulsating white dwarfs, 137, 142
  instability strip, 137, 142
pulse shape (X–ray sources)
  changes, 122, 123
  evolution, 115
pulse-arrival times, 65, 182

quasi–conservative mass transfer, 240, 241, 252
quasi–periodic gravitational waves, 23, 33
quasi–periodic oscillations (*QPO*s), xii, 7, 10*ff*, 89, 90–93, 99–105
quiescent
  accretion disks, 139, 171, 174
  CV systems, 172, 173
  dwarf novae, 170

radio–emitting pulsars, 87, 96
recombination
  energy, 233, 250*ff*
  in common envelope, 233, 250, 251, 254, 255
  trigger, 250
red shift parameter, 58
reflection effect, xi, 180
relativistic jets, 13, 15, 19, 20
Reynolds number, 163, 166, 171, 172, 175
  in accretion disks, 163, 166, 172
  relation to viscosity parameter α, 175
Roche lobe, xi, 130, 153, 230, 258
  constraints on
    density of secondary, 238
    evolution, 238*ff*
    inclination, 131
    period, 238–240
  expansion into, by
    companion, 250
  filling by primary, 235–237
  filling by secondary, 121, 122, 131, 147, 152, 168, 207, 208, 235–238
  limits on secondary, 152, 153, 238
  overflow, xi, 235
  radius, 130, 152, 155, 234*ff*
Roche model used for *CV*s, 168
Roche surface, 130
Rossi X-ray Timing Explorer (*RXTE*), xii, 87, 88, 184
rotation–powered pulsars, 91, 92, 95, 97
  emission spin–down, 91, 92
  MSP production, 92*ff*
    radio sources, 91

Schwarzschild
  black hole, 4, 16, 17
  horizon, 72
  lens, 66, 67, 71
  spacetime, 64, 66, 73
second dredge-up phase, 238
secondary star(s)
  adiabatic expansion, 247
  as mass transfer, 137, 147, 207
  as seen from a *CV*'s white dwarf, 155
  CNO processing, 153
  degeneracy, 139
  envelope expulsion, 54
  H_α emission, 155
  in *CV*s, 147*ff*
  IR spectroscopy, 151, 153
  irradiation by accretion disk, 167
  irradiation effects, 154–156
  luminosities, 150
  magnetic field effects, 174
  main sequence properties, 149, 150
  mass function provider, 5
  mass-period relation, 148, 149
  mass-radius relation, 148
  model, 209
  nuclear evolution, 54
  radial velocities, 228
  radius-period relations, 152
  spectral features, 138
  spectral types, 151–153, 158, 162
  starspot activity, 148, 156, 225
  synthetic spectrum, 166, 168
  wind, 144
semi–detached binary star system, xi, 180
shadowing by accretion disk, 167
Shapiro delay, 58, 59, 69*ff*, 83

shell–burning
helium star remnants, 247
hypothesis in asymptotic giants, 250
SHELLSPEC, 207–209
short-period binary creation, 63, 233
signal template, 38
simplex
algorithm, 192*ff*
method, 191*ff*
program, 200*ff*
simulated annealing, 191*ff*
Sloan Digital Sky Survey (*SDSS*), 137, 138, 142, 170
sonic–point model
beat frequency model, 101, 103, 105
modifications, 101–103
radius, 101
with spin–resonance, 101
sources (*see also* stars/individual)
4U 0614+09, 90, 104
4U 1223−62, 126
4U 1543–475, 5, 6, 20, 89, 104
4U 1608−52, 89, 96, 101
4U 1630–12, 47
4U 1636−53, 89, 90, 100
4U 1702–429, 89
4U 1728–34, 89
4U 1916–05, 89
A 0620–003, 5, 6
A 1744–316, 89
Abell 63, 234
Aql X-1, 89
B1820−30A, 97
B1821−24, 97
Cir X–1, 90, 91, 100
Cyg X–1, 5, 6, 11, 16
EXO 0748–676, 89, 96
Fin 381, 225
GRO J0422+32, 5, 6
GRO J1655–40, 5*ff*, 19, 20
GRS 1009–45, 5, 6
GRS 1915+105, 5, 6, 11, 12, 17, 19, 20
GS 1354–64, 5
GS 1826–238, 89
GS 2000+251, 5, 6
GS 2023+25 (V404 Cyg), 5, 6
GX 301–2, 111, 126*ff*
GX 339–4, 5–7, 10, 17
GX 5−1, 91
H 1705–250 (Nova Oph 77), 5, 6
H 1743–322, 11, 12
Hercules X-1, 111*ff*
HETE J1900.1–2455, 89
HS 2331+3905, 142
IGR J00291+5934, 89
IGR J17191–2821, 89
KS 1731-260, 89
LMC X–1, 5, 6
LMC X–3, 4–6, 19, 20
M33 X–7, 5
MXB 1659–298, 89
PG 1115+166, 254
PSR (B) 1913+16, xii, 24, 54
PSR J0737-3039AB, xii, 53*ff*, 63, 65, 74, 78*ff*
RX J0806.3+1527, xiii
RX J1914.4+2456, xiii
Sanders 1040 (in M67), 240
SAX J1748.9−2021, 89
SAX J1750.8−2900, 89
SAX J1808.4−3658, 88–93, 99, 100
SAX J1819.3−2525, 5, 6
Sco X–1, 100, 101
SDSS J0131, 138, 142
SDSS J0748, 138
SDSS J0809, 169, 170, 173
SDSS J1553, 138, 144
SWIFT J1756.9–2508, 89
Vela X-1/GP Vel, 179, 183–185, 187
pulse arrival variation, 185
X 1743–29, 89
XTE J0929−314, 89, 103
XTE J1118+480, 5, 6, 9
XTE J1550–564, 5, 6, 11, 12, 17
XTE J1650–500, 5, 17
XTE J1655–40, 5, 6, 8, 9, 17, 19, 20
XTE J1739−285, 89, 96
XTE J1751–305, 89, 103
XTE J1807.4−294, 89–92, 96, 99, 100
XTE J1814−338, 88, 89, 93, 103
XTE J1859+226, 5, 6
Southeastern Association for Research Astronomy (SARA), 219, 222
spectral resolution of *IR* spectrographs, 151
spectropolarimetry, 209, 210

spin–evolution
  diagram, 95
  spin parameter, 17–21, 58, 65
  time scales, 96
Spitzer Space Telescope, 157
star(s)
  classes
    δ Scuti, 205
    Algol systems, 153, 154, 219, 234
    AM Her stars (*polars*), 137, 144, 145, 148*ff*
    cataclysmic variables (*q.v.*)
    dwarf nova(e) (*q.v.*)
    FK Comae stars, 210
    hypernova(e), 3
    intermediate polars, 138, 148
    nova(e), ix, xi
    nova-like systems, 142, 161, 162, 173
    polars (AM Her stars) (*q.v.*)
    recurrent nova(e), 233, 248
    RS CVn–type systems, 155
    supernova(e), xi, xiii, 24
    SU UMa–type systems, 140, 141, 172
    SW Sex systems, 170
    symbiotic stars, 248
    W UMa systems, ix, 215*ff*
    X-ray nova(e), 3, 4, 7, 10, 14
  individual
    HD 163151, 225
    HD 185510, 240
    HD 199178, 210
    HD 209458b, 211
    Procyon, 153
    Sirius, 153
    FF Aqr, 154
    V1379 Aql (HD 185510), 240
    BH Cas, 205
    HT Cas, 172
    AY Cet, 154, 240
    T CrB, 81, 233, 248, 254, 255
    W Crv, 215*ff*
    SS Cyg, 139, 151
    V404 Cyg (H2023+25), 5
    AB Dor, 210
    AS Eri, 234
    EF Eri, 144
    AM Her, 144
    HZ Her (Her X–1), 111*ff*
    VW Hyi, 162
    AR Lac, 209
    RT Lac, 209
    SS Lac, 205
    GW Lib, 141, 142
    MV Lyr, 143, 170
    V651 Mon, 154
    Nova Mus 1991, 5, 6, 10
    RS Oph, 233, 248, 254, 255
    V2388 Oph, 215, 218, 225–228
    Nova Oph 1977 (H1705+250), 5
    IP Peg, 205
    WZ Sge, 140, 157, 171
    V4641 Sgr (SAX J1819.3–2525), 5, 17
    RW Sex, 161
    SW Sex, 170
    θ^2 Tau, 205
    V471 Tau, 155, 234, 237, 238
    V781 Tau, 199
    DW UMa, 143
    SW UMa, 140, 141
    GP Vel (Vela X–1), 179, 183–185, 187, 188
    IX Vel, 161
starspot–activity cycles, xi
  in *CV's*, 148, 207
stellar activity in *CV* secondaries, 154–156, 158
stellar atmosphere model(s), 18, 98, 161, 163, 167, 179, 180, 183, 186, 194, 219
stellar envelope total energy, 251
stellar wind, xi, 111, 126, 127, 129
  accretion, 126
  burst signal source, 24
  mass loss, 237, 247, 254
    in double white dwarf progenitors, 246, 247
  reprocessing, 127
  structure, 111
stochastic
  background signals, 24, 25, 33
  effects of dynamic tides, 183
  methods, 205, 206
  noise, 74
Stokes parameters, 16

supernova(e)
core–collapse models, 33
explosions, xi, xiii, 53, 54, 56
superoutburst, 157, 162
symbiotic stars, 248
SYNSPEC, 139, 163, 166–169, 209
systemic losses
angular momentum loss, 234, 236, 247
mass loss, 233, 234, 236, 238, 247, 255

Tabu search method, 206, 207
thermal catastrophe, 164, 174
thermal evaporative instability, 174
thermal relaxation oscillations (*TRO's*), 216, 141, 162
model, 215
thermal timescale
mass transfer, 243, 247
vs. common envelope evolution, 235, 236
thermally–pulsing asymptotic giants, 237–239, 250
thermonuclear bursts
as X-ray sources, 92, 93
in *MSP*s, 88
of accelerating neutron stars, 88
type I, 14
TLUSTY, 139, 163*ff*
TOAD class of *CV*s, 157
tomography, 207
tuning parameters, 199, 200
type I bursts, 14, 97, 98
X–ray bursts, 89, 93, 97
Type Ia supernova(e), xiii

vertical energy transport in accretion disks, 174
vertical viscosity profile, 174
Very Large Telescope (*VLT*), 157
virial theorem
constraint, 251
for stellar interiors, 251
viscosity, 162*ff*
dissipation source in accretion disks, 163
due to magnetorotational instability, 173
effect on accretion disk temperature, 164
kinematic, 163, 164
model options, 166
parameterization, 162
shear in disks, vertical variation effect, 174
visibility key, 68
visibility of low accretion rate systems, 137, 139, 140

W UMa systems, x, xi, 215*ff*
A– and W–types, 216
ages, x
binaries, ix
frequency of occurrence, 216
in multiple systems, x, 215, 218, 226
mass–luminosity relation, 216
white dwarf(s), xi, 147, 231
accreting, xii, 137, 140, 147
binaries, xiii, 138, 233, 234, 242
Chandrasekhar mass limit, 94, 140
composition, 141, 242
cool disk accretion, 140, 141
cooling age, 234, 237
cooling curve, 145, 240
double, 243, 245, 247
heating with high accretion rates, 143
helium, 234, 242*ff*, 254
hot spot models, 144
hybrid, 242
instability, strip(s), 137, 142
magnetic fields, 137, 144, 147, 156–158
magnetic poles, 148
mass, 25, 147, 153, 237, 240, 245, 255
mass-ratio relation, 139, 240
modeling, 137*ff*
primary star, 147
progenitors, 149, 234, 237, 241, 254
pulsating model, 142
Roche lobe radius, 248
rotation, 141
spectra, 138*ff*, 156
SED models, 140, 141, 162
temperatures, 137*ff*, 154, 156, 162
in Polars, 144
wind loss, 247

Wilson–Devinney model, 184, 191
  program, 192–197, 200, 208, 210, 211
  WD2007, 192*ff*, 204
  wd98k93, 194, 195, 211

X–ray(s)
  absorption by winds, 127
  binary systems, 3, 87*ff*, 111*ff*
    individual systems (*see* sources/stars)
    orbital elements, 182
    distances, 179
  burst(s), 89, 93, 97, 98
  continuum fitting, 15
  eclipse, 55, 131, 179*ff*
  heating of accretion disk, 174
  heating of secondary, 120
  irradiation of accretion disks, 174
  jets, 15
  line profiles, 16, 88
  luminosities, 16
  neutron star emission, 126, 127
  novae, 3, 4, 7, 10, 14
  occultation by accretion disk, 123
  oscillations, 87, 88, 90, 93, 97*ff*
  polarimetry, 15, 16
  pulsars, 90, 91, 123, 183–185
    distance estimation, 179*ff*
    modeling, xii
  pulse(s), 111, 182, 183, 184
    arrival times, 184
    profile(s), 112, 120
    timing, 111
  reflection, 116, 120, 121, 126
  spectrum, 16, 18, 19, 127
  states, 8
    hard state, 9, 10, 16, 116, 119, 120
    high and low states, 111, 112, 119, 120
    steep power law, 10
XSPEC, 16, 17, 20

Yonsei–Yale isochrones, 193

Zeeman Doppler imaging, 210
  modeling, 207
zero–potential surface, xi